26th Summer School and International Symposium on the Physics of Ionized Gases

(SPIG 2012)

AA002627

Journal of Physics: Conference Series Volume 399

Zrenjanin, Serbia
27-31 August 2012

ISBN: 978-1-62276-865-3
ISSN: 1742-6588

Printed from e-media with permission by:

Curran Associates, Inc.
57 Morehouse Lane
Red Hook, NY 12571

Some format issues inherent in the e-media version may also appear in this print version.

Printed by Curran Associates, Inc. (2013)

For permission requests, please contact the Institute of Physics at the address below.

Institute of Physics
Dirac House, Temple Back
Bristol BS1 6BE UK

Phone: 44 1 17 929 7481
Fax: 44 1 17 920 0979

techtracking@iop.org

Additional copies of this publication are available from:

Curran Associates, Inc.
57 Morehouse Lane
Red Hook, NY 12571 USA
Phone: 845-758-0400
Fax: 845-758-2634
Email: curran@proceedings.com
Web: www.proceedings.com

TABLE OF CONTENTS

ATOMIC COLLISION PROCESSES

012001 Plasma-Wall Interactions in the Cesiated SNS H⁻Ion Source 1
Martin P. Stockli
012002 Photoionization of Atomic and Molecular Positively Charged Ions 11
J. M. Bizau, D. Cubaynes, M. M. Al Shorman, S. Guilbaud, C. Blancard, J. Lemaire, R. Thissen, A. Giuliani, C. Nicolas, A. R. Milosavljevic
012003 Vibrationally Inelastic Collisions of Slow Electrons with Polyatomic Molecules 17
R. Curik, P. Carsky
012004 Measurements of Differential Cross Sections for Elastic Electron Scattering and Electronic Excitation of Silver and Lead Atoms 24
S. D. Tosic

PARTICLE AND LASER BEAM INTERACTIONS WITH SOLIDS

012005 Preliminary Results from Recent Experiments and Future Roadmap to Shock Ignition of Fusion Targets 32
D. Batani, G. Malka, G. Schurtz, X. Ribeyre, E. Lebel, L. Giuffrida, V. Tikhonchuk, L. Volpe, A. Patria, P. Koester, L. Labate, L. A. Gizzi, L. Antonelli, M. Richetta, J. Nejdl, M. Sawicka, D. Margarone, M. Krus, E. Krousky, J. Skala, R. Dudzak
012006 Properties of Plasmas Produced by Short Double Pulse Laser Ablation of Metals 40
J. Hermann, L. Mercadier, E. Axente, S. Noel
012007 Application of keV and MeV Ion Microbeams Through Tapered Glass Capillaries 52
T. Ikeda, T. M. Kojima, T. Kobayashi, W. Meissl, V. Mackel, Y. Kanai, Y. Yamazaki
012008 The Non-Resonant Neutralization Dynamics of the Multiply Charged Rydberg Ions Escaping Solid Surfaces 61
S. M. D. Galijas, N. N. Nedeljkovic, M. D. Majkic
012009 Intermediate Stages of the Neutralization of Multiply Charged Ions Interacting with Solid Surfaces 68
M. D. Majkic, N. N. Nedeljkovic, S. M. D. Galijas
012010 Electron Emission from a Metal Nano-Tip by Ultrashort Laser Pulses 75
G. Wachter, C. Lemell, J. Burgdorfer

LOW TEMPERATURE PLASMAS

012011 Modeling of Plasma and Plasma-Surface Interactions for Medical, Environmental and Nano Applications 82
A. Bogaerts, R. Aerts, R. Snoeckx, W. Somers, W. Van Gaens, M. Yusupov, E. Neyts
012012 Experimental Transition Probability Measurements in Pulsed Lamps: Critical Points 94
Juan A. Aparicio, Maria T. Belmonte, Ramon J. Pelaez, Stevica Djurovic, Santiago Mar
012013 Cathode Sheath and Hydrogen Balmer Lines Modelling in a Micro-Hollow Gas Discharge 102
Dj Spasojevic
012014 Physico-Chemistry of Planetary Atmospheric Entry Plasmas 112
Arnaud Bultel, Julien Annaloro, Vincent Morel
012015 Diagnostics and Biomedical Applications of Radiofrequency Plasmas 121
Sasa Lazovic
012016 Near-threshold Photoelectron Angular Distributions from Two-Photon Resonant Ionisation of He and Ne Atoms 128
A. Mihelic, M. Zitnik, P. O'Keeffe, P. Bolognesi, A. Moise, R. Richter, L. Avaldi
012017 Breakdown and Discharge Regimes in Standard and Micrometer Size DC Discharges 135
N. Skoro

GENERAL PLASMAS

012018 On the Role of Electron-Driven Processes in Planetary and Cometary Atmospheres 143
Laurence Campbell, Oddur Ingolfsson, Michael J. Brunger

012019 Plasma Spectroscopy in the Conditions of the ITER Tokamak 153
J. Rosato, Y. Marandet, V. Kotov, D. Reiter, H. Capes, L. Godbert-Mouret, M. Koubiti, R. Hammami, R. Stamm

012020 Gyrokinetic Simulation Studies for Non-Axisymmetric Plasma Confinement: Turbulent Transport and Entropy Transfer 161
T. H. Watanabe, H. Sugama, M. Nunami, M. Nakata

012021 The Structure of the Si IV Region in Be Stars: A Study of Si IV Spectral Lines in Be 68 Stars 171
A. Antoniou, E. Danezis, E. Lyratzi, D. Stathopoulos

012022 Stellar Population in the Sample of Type 2 Active Galactic Nuclei 178
Natasa Gavrilovic Bon, Edi Bon, Luka C. Popovic

012023 The Properties of Emission Lines and Their Correlations in Spectra of Active Galactic Nuclei 184
Jelena Kovacevic, Luka C. Popovic

012024 Laboratory Studies of Charging Properiteis of Dust Grains in Astrophysical/Plenetary Environments 190
D. Tankosic, M.M. Abbas

Author Index

PREFACE

This volume of *Journal of Physics: Conference Series* contains the general invited lectures, topical invited lectures and progress reports presented at the 26th Summer School and International Symposium on the Physics of Ionized Gases – SPIG 2012. The conference was held in Zrenjanin, Serbia, from 27–31 August. The SPIG conference has a 52 year long tradition.

The structure of the papers in this volume cover the following sections: atomic collision processes, particle and laser beam interactions with solids, low temperature plasmas and general plasmas. As these four topics often overlap and merge in numerous fundamental studies and, more importantly applications, SPIG in general serves as a venue for exchanging ideas in the related fields. We hope that this volume will be an important source of information about progress in plasma physics and will be useful, first of all, for students, but also for plasma physics scientists.

The Editors would like to thank the invited speakers for their participation at SPIG 2012 and for their efforts writing contributions for this volume. We also express our gratitude to the members of Scientific and Organizing committees for their efforts in organizing this SPIG. Especially we would like to thank the Ministry of Education, Science and Technological Development of Republic of Serbia, Provincial Secretariat for Science and Techonological Development, Province of Vojvodina, Institute Français de Serbie and Biser Zrenjanin for financial support as well as the European Physical Society (EPS) for supporting the award for the best poster of a young scientist and American Elements, USA.

Milorad Kuraica
Zoran Mijatovic
October 2012
Editors

Peer review statement

All papers published in this volume of *Journal of Physics: Conference Series* have been peer reviewed through processes administered by the proceedings Editors. Reviews were conducted by expert referees to the professional and scientific standards expected of a proceedings journal published by IOP Publishing.

Plasma-Wall Interactions in the Cesiated SNS H$^-$ Ion Source

Martin P. Stockli

Spallation Neutron Source, Oak Ridge National Laboratory, Oak Ridge, TN 37830, USA

E-mail: stockli@ornl.gov

Abstract. High-current H$^-$ ion beams are needed to drive high-power accelerators as well as to heat future fusion reactors with neutral beams. The most productive H$^-$ sources enhance the production of H$^-$ ions with caesium. Cs lowers the work functions of the metal walls, which increases the probability of ions to capture a second electron when bouncing back from the metal plasma walls. However, caesium also causes voltage break downs that can be severe and cause significant downtime. SNS has developed a frugal caesium management, which uses a single injection of ~4 mg of caesium to produce ~50 mA of H$^-$ beam without decay for up to 6 weeks. This paper presents calculations and experimental data, which suggest the persistence of the caesium enhanced H$^-$ beams are due to 1) thermally limiting the Cs emission, and 2) conditioning for high purity hydrogen plasma, which eliminates the sputtering of Cs by non-hydrogen ions, and 3) hydrogen ions being too light to sputter Cs.

1. Introduction

High current beams of negative hydrogen ions are needed for producing neutral beams to heat the plasma in fusion reactors and in high-power proton accelerators with accumulator rings [1]. For example, SNS produces a 1 MW proton beam that impacts 60 times per second on a mercury target. Every pulse produces copious amounts of neutrons, which are used in neutron scattering experiments at the end of up to 16 neutron beam lines [2].

Briefly, to produce these neutrons, the ion source shown in figure 1 has to deliver ~50 mA of H$^-$ ions for 1 ms at 60 Hz. The ion source voltage accelerates the beam to 65 kV. The two-lens electrostatic beam transport shown in figure 1 refocuses the beam into a radio frequency quadrupole, which accelerates the beam to 2.5 MeV. The medium energy beam transport refocuses the beam into the linear accelerator, which boosts the energy close to 1 GeV. A gentle bend filters the beam energy and brings the beam to the accumulator ring. As the H$^-$ beam enters a bending magnet of the accumulator ring, a thin foil strips the 2 electrons and the stripped protons can join the trajectories of the already stored proton beam. Accumulating close to 1000 beamlets allows the ring to accumulate tens of amps of 1 GeV protons.

A chopper integrated into the second lens shown in figure 1 removes the beam for ~300 ns about every μs. These gaps accumulate in the accumulator ring and allow for a clean extraction of the beam. After accumulating for ~1 ms, the extraction kickers direct the accumulated beam onto the liquid mercury target.

This paper discusses the interactions between the walls and the plasma, which play crucial roles in the production of the H$^-$ beam with the SNS ion source, its persistence and its lifetime.

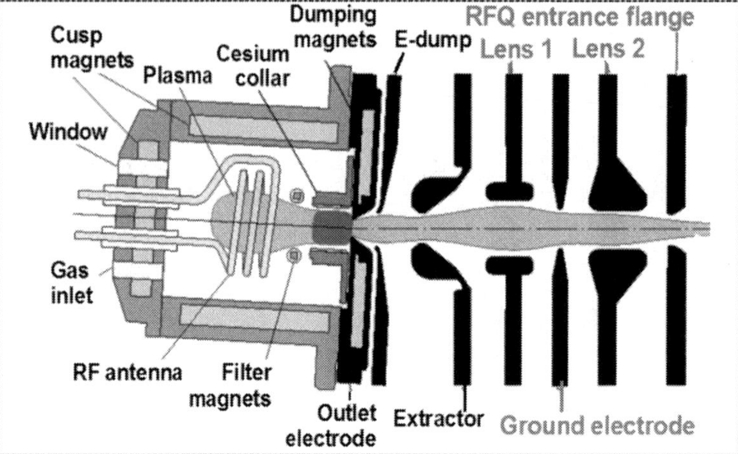

Figure 1. SNS H^- ion source and low-energy beam transport.

2. The Production and Extraction of Negative Hydrogen Ions from Pure Hydrogen Plasma

The production of negative hydrogen ions is challenged by the mismatch between the ions' small electron affinity of 0.75 eV and the ~15 eV binding energies found in hydrogen atoms and molecules. The hydrogen's limited atomic structure makes the capture of free electrons difficult, resulting in a cross section of $\sim 5 \cdot E^{-22}$ cm^2 for the radiative capture. The increased complexity of molecules allows for negative hydrogen ions to be produced with a dissociative attachment in collisions with low-energy electrons (E_e <10 eV) with a cross section of $\sim 10^{-20}$ cm^2 [3].

Accordingly, in 1977 it was a surprise when Bacal found large signals from negative ions in hydrogen plasma [4]. Extensive research has shown that negative hydrogen ions can be efficiently produced with a cross section of $\sim 3 \cdot 10^{-20}$ cm^2 when slow electrons (E_e <1 eV) break up highly rovibrationally excited hydrogen molecules ($4 \leq v \leq 9$). However the efficient production of highly rovibrationally excited molecules with a cross section of $\sim 5 \cdot 10^{-18}$ cm^2 requires fast electrons with energies >15 eV. Unfortunately, such fast electrons destroy the negative hydrogen ions with a cross section of $\sim 3 \cdot 10^{-15}$ cm^2, severely limiting the lifetime and usability of the produced negative ions [4].

This dilemma was reduced with the introduction of "tandem type sources" like the SNS source shown in figure 1 [5]. It shows on the left the multicusp magnet lined plasma chamber containing a 2.5-turn porcelain-coated copper antenna. About 350 W of 13 MHz radio frequency generate continuous plasma. This dim plasma accelerates the breakdown produced by the 50-70 kW, 1 ms long pulses from a 2 MHz amplifier at 60 Hz. After the first 5 μs of each pulse, the frequency is up-shifted by ~1% to adjust the match for the increased plasma inductance [6]. The high-power RF produces a broad distribution of electron energies, which in turn excite numerous molecules. Near the open end of the antenna are filter magnets that generate a transverse dipole field of ~250 Gauss. This field returns energetic electrons towards the antenna, while it allows low-energy electrons to diffuse towards the outlet of the source. The exited molecules drift unimpeded through the filter field and produce negative hydrogen ions when colliding with low-energy electrons. The lower plasma temperature near the outlet improves the lifetime of the negative ions, which have more time to drift into the outlet where they are accelerated by the -65 kV applied to the ion source.

3. The Plasma assisted Production of Negative Ions on Metal Surfaces

Metals suitable for plasma chambers have work functions in the range of 4 to 5.6 eV, and those more loosely bound electrons are better matched for being captured by a hydrogen atom to form a negative ion. This fact motivated the successful development of compact surface plasma H^- sources (magnetron and penning sources) [7] and surface converter H^- sources [8]. The SNS source was conceived as a hybrid source, where the large outlet boosts the output from the volume-produced H^- ions, and a converter supplements surface-produced H^- ions [5].

The 7 mm inner diameter source outlet is shown in detail in figure 2. On the plasma side of the outlet is the Mo converter, which is thermally attached to the Cs collar, which will be discussed later. During operation, positive hydrogen ions drifting into the plasma sheath are accelerated by the plasma potential. When the ions impact on the surface, they rapidly neutralize through auger transitions. Being energetic, most atoms will bounce back into the sheath, and there is a chance that a surface electron gets captured in the field of the fast atom. However, the chance is small because the 4.6 eV work function of Mo is much larger than the 0.75 eV electron affinity of the hydrogen atom. The electron capture probability increases with the velocity of the escaping atom [9]. The so produced negative ions initially tend to move away from the outlet, but diffusion, resonant charge transfer with hydrogen atoms, and the filter field allows for a fraction of the ions to reach the outlet. A fraction of the negative ions is lost due to associative detachment of the electron in collisions with energetic hydrogen atoms or due mutual neutralization in collisions with protons [3]. These losses are minimized by placing the converter surface as close as possible to the outlet and by using a cone with a 45° taper.

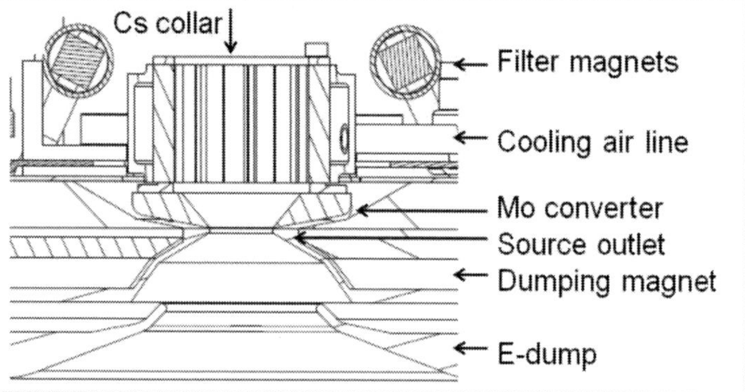

Figure 2. Details of the outlet of the SNS H^- ion source.

4. Increasing the Negative Ion Yield with Caesium

Around 1970 Novosibirsk was able to produce unprecedented currents of negative hydrogen ions by adding Cs to their magnetron source [7]. This started a large interest in the surface production of negative ions and the efficiency of various surfaces.

Briefly, alkali metals have lower work functions in the range of 2 to 3 eV due to their single, loosely bound electron in their outer shell. When being adsorbed on a substrate metal, the work function can be even lower than that of the bulk alkali metal. Figure 3 shows the work function of a

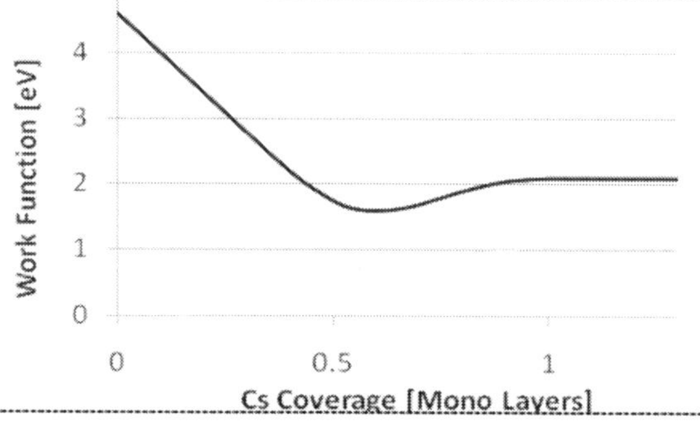

Figure 3. The work function of a Mo surface with adsorbed Cs.

Mo surface as a function of the amount of adsorbed Cs in units of mono layers, meaning one 1-atom thick, dense layer. Without any Cs the work function starts out with 4.6 eV, the work function of the clean substrate Mo surface. Gradually adsorbing Cs slowly lowers the work function until reaching a minimum of 1.6 eV near 0.6 mono layers. Adding more Cs gradually increases the work function up to 2.1 eV, the work function of bulk Cs [10].

To demonstrate the effect of the changing work function on the production of negative hydrogen ions, the SNS ion source was equipped with an external Cs reservoir. After outgassing the Cs supply system and plasma-conditioning the plasma chamber, the Cs ampoule was cracked and a small Cs flow was started by heating the reservoir to ~110°C. As the Cs gradually adsorbed on the 55°C converter surface, the unanalyzed H^- beam current gradually grew from ~10 mA to over 40 mA as shown in figure 4. However, after about an hour the current started to gradually decay to 36 mA. This is consistent with the work function initially decreasing, passing through a minimum and then increasing after exceeding the optimal fraction of a monolayer [11].

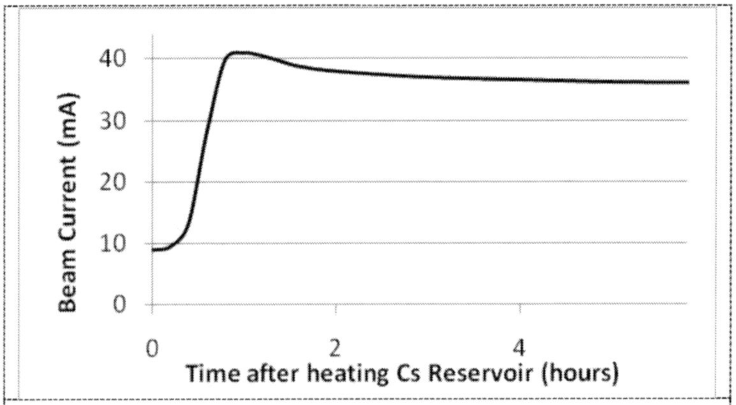

Figure 4. Beam current versus time when gradually applying Cs to the converter.

This suggests that the minimum in the work function is relevant for the production of H^- with an ion source, even so it employs imperfect, technical surfaces rather than the perfectly structured surfaces that were used to measure such work function data.

Clearly the H^- production on surfaces benefits greatly from the addition of Cs. Lowering the work function also leads to larger electron emission from negatively charged surfaces such as the negative ion source itself, associated high-voltage electrodes like the E-dump, which collects the co-extracted electrons, and electrostatic lenses needed to focus the expanding beam as shown in figure 1. For example, the load current of the first electrostatic lens normally increases when Cs is released, which is a useful Cs indicator. However, excess Cs has led to many high voltage breakdowns and sometimes required extensive conditioning before the operational high voltages could be restored. For this reason some ion sources avoid Cs and settle for lower yields of H^- ions. The highly frugal Cs management developed with the SNS H^- source is a successful compromise [12].

5. Managing a Fractional Monolayer of Caesium

Creating a ~0.6 monolayer of Cs on a Mo substrate and maintaining it under plasma condition is a challenging task and took many optimizations to find a reliable process. This is complicated by the fact that Cs likes to stick to metal surfaces and therefore much Cs needs to be deployed before the partial monolayer is formed on the desired surface. For example, the SNS source requires ~4 mg of Cs to reliably and consistently apply a partial monolayer on the Mo converter, which accounts for only ~0.1 μg of Cs. This initial 4 mg dose produces a persistent H^- beam current for up to 6 weeks, apparently consuming less than 0.12 mg per day or less than 2 mg per plasma-day when normalized to the plasma duty factor.

Other cesiated H^- sources use significantly more Cs, like a converter H^- source, which consumes ~1 g per day [13] or ~8 g per plasma-day. Low duty factor H^- sources consume much less Cs, such as 0.7 mg per day achieved with a frugally managed magnetron source [14], although that adds up to 37 g per plasma-day. Apparently, SNS operates its H^- source without a significant consumption of Cs.

Cs can be lost through sputtering by ions from the plasma, which will be discussed later. Thermal emission is the other loss mechanism, which normally can be described by with an exponential decay with an exponent $-(E_{Cs}/k \cdot T)$, where E_{Cs} is the bond energy of the adsorbed Cs atom, k the Boltzmann constant, and T the absolute temperature in °K [15]. Such an exponential decay could be used to depopulate the surface until reaching the desired fractional monolayer and then drastically lowering the temperature to prevent further losses.

However, Cs behaves quite differently because its binding energy decreases with the amount of Cs that is adsorbed on the surface as seen figure 5 for a polycrystalline Mo surface (Kaminsky) [16] and a 110 Mo surface (Hansen) [17]. This unique behaviour appears to be caused by the ionic radius of Cs, which is significantly larger than the metallic radius of most metals except for some alkali and alkaline earth metals [18].

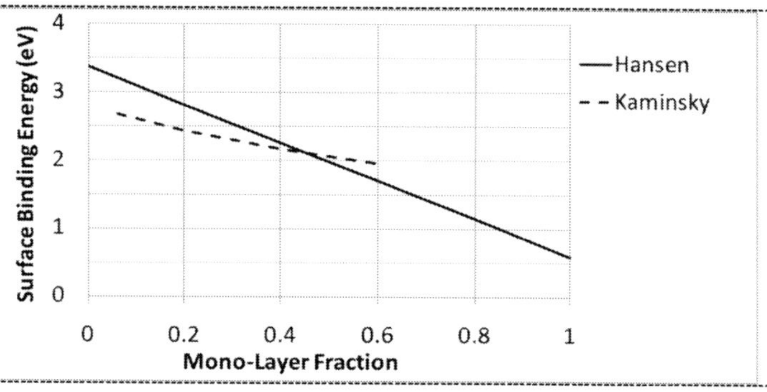

Figure 5. Cs binding energy as a function of adsorbed Cs on Mo.

As thermal emission depopulates the Cs, the increasing binding energy drastically reduces the thermal emission. The resulting loss rate and remaining layer can be calculated by integrating the time it takes to lose a certain fraction [11]. Using the Hansen approximation, and starting with one monolayer, figure 6 shows the results for nine different temperatures of the Mo substrate [11].

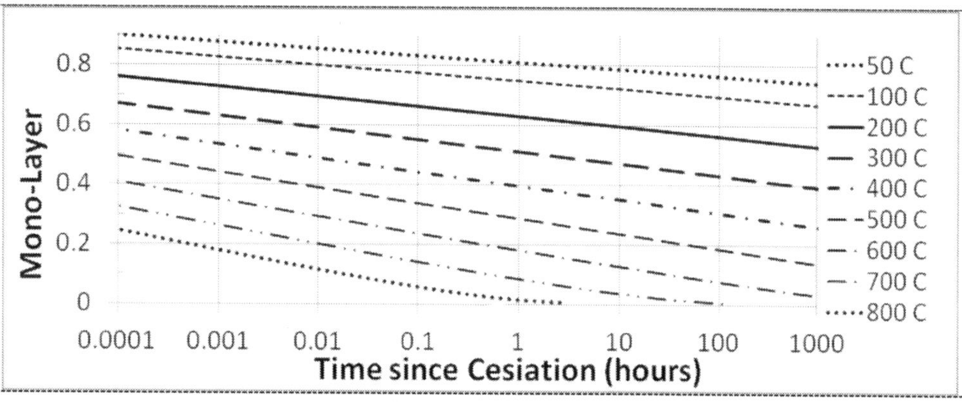

Figure 6. Residual Cs monolayer versus time on a Mo surface for different temperatures.

Because logarithmic time scales are unusual, figure 7a shows the same data for three different temperatures for the first 5 hours and figure 7b for the first five week [12]. Figure 7a shows all drastic changes to happen in the first few minutes, and some noticeable changes over the first two days before the system stabilizes. Figure 7b shows that small changes could be noted over the first 1-2 weeks before the system stabilizes. These calculations shows how the temperature of the substrate can be used to rapidly obtain a desired monolayer, and that no adjustment is needed to maintain this monolayer within a narrow range.

The SNS Cs collar is normally set to ~170°C without further adjustments. The H^- beam normally grows for a few days before it becomes persistent and remains at this value for the remainder of the source service cycle.

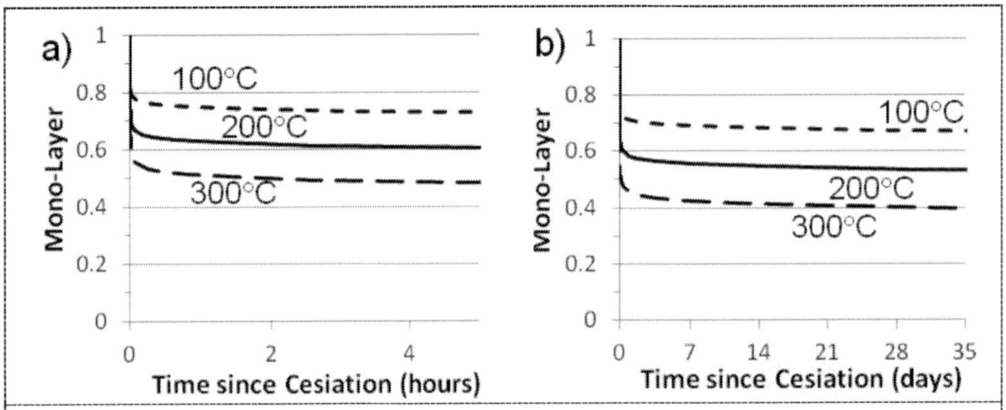

Figure 7. As figure 6 but with linear time scales for the first a) 5 hours and b) 5 weeks.

6. Achieving and Loosing the Persistence of the H^- Beams

Mono layers are fragile and therefore it was no surprise when initially the H^- beam kept decaying every time after applying a dose of Cs [19]. The surprise came about a year later when it was noticed that decay slowed down with every cesiation as shown in figure 5 [20]. The figure also shows that the H^- beam became persistent after applying a dose of Cs the following morning.

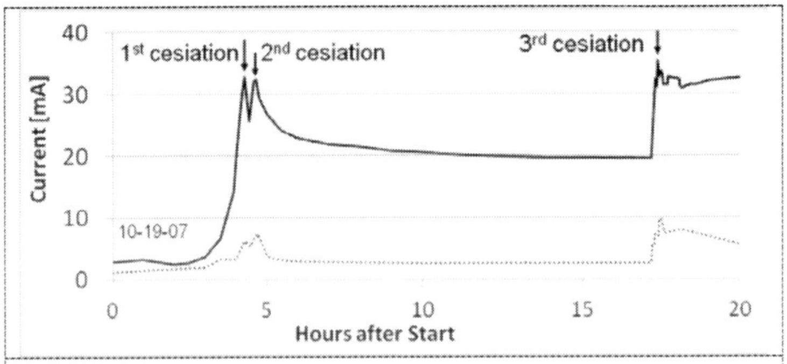

Figure 8. The beam current and lens-1 load current during a 2007 source start up.

In 2008, the full-power plasma conditioning was extended from 30 to 150 minutes, which started to produce persistent beams after the first cesiation. Apparently, the plasma sputter-cleaned the surface and the Cs strongly bonded to the Mo surface and resisted sputtering. Without thorough sputter cleaning, the Cs is adsorbed on water and other adsorbates forming much weaker bonds.

Conditioning for about three hours yielded persistent beams until a ~1% beam loss per hour was encountered in June of 2011 [20]. RGA data showed about 1 ppm NH_3, which disappeared when the plasma was switched off. A post-service leak check revealed a E-10 scc/s leak in a widow. Apparently, the heat from the plasma enlarged the air leak to E-6 scc/s, which contaminated the hydrogen plasma with about 1 ppm of products from air.

Applying another dose of Cs would restore the H^- beam current, but the decay continued at the same rate. The unchanging decay rate means that the Cs keeps bonding to the surface with the same bond energy, which one would expect if the Cs is being sputtered while the surface remains unchanged. This suggests that ions from ammonia sputter Cs, while ions from hydrogen do not.

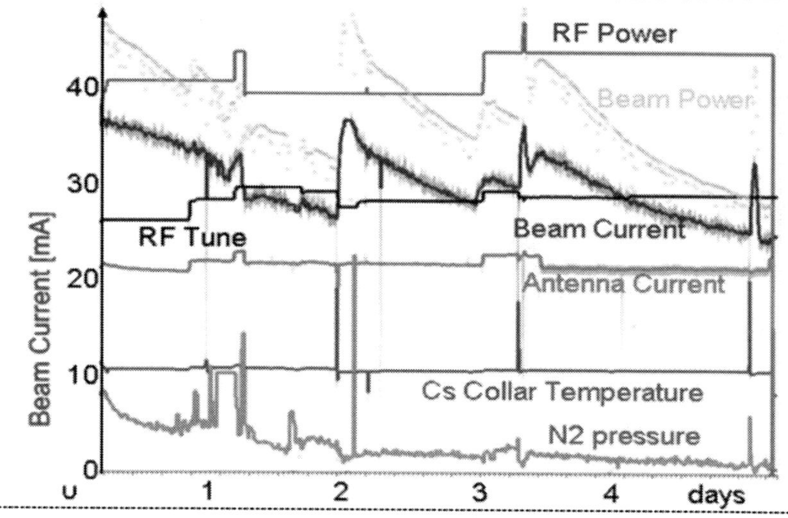

Figure 9. The beam current and lens-1 load current during a 2007 source start up.

7. The Sputtering of Caesium

Sputtering plays an important role in ion sources. Its thresholds and yields are governed by the adsorbate mass m_a and surface bond energy E_a, and the ion mass m_i, and ion energy E_i, which is normally dominated by the plasma potential. The threshold E_{th} was approximated as [21]:

$$E_{th} \approx 8 \cdot E_a \cdot (m_i/m_a)^{2/5} \qquad \text{for } m_i > 0.3 \cdot m_a \tag{1a}$$

$$E_{th} \approx E_a/(\gamma \cdot (1-\gamma)) \quad \text{with } \gamma = 4 \cdot m_i \cdot m_a/(m_i+m_a)^2 \qquad \text{for } m_i \leq 0.3 \cdot m_a \tag{1b}$$

The ratio of these approximate thresholds divided by the bond energy for Cs is plotted in figure 10 versus the square root of the ion mass m_i. It clearly shows the hydrogen ions to have a much higher threshold than nitrogen or any heavier ions. Using the 2 eV bond energy shown in figure 2 suggests that the plasma potential exceeds 14 V to provide nitrogen ions with enough energy to sputter Cs. On the other hand, the normal persistence of the beam suggests the plasma potential to be less than 24 V to inhibit the hydrogen ions from sputtering the Cs.

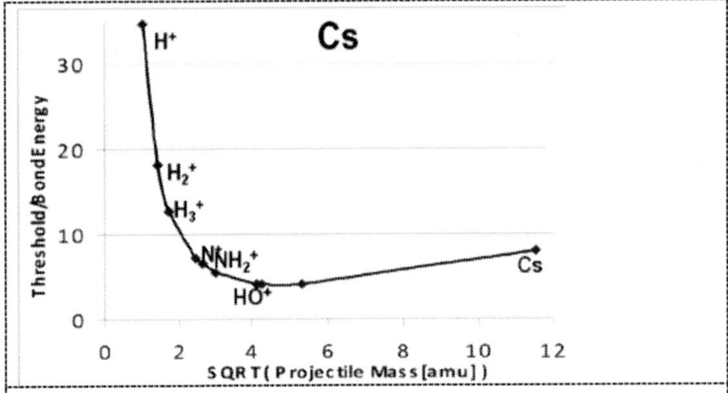

Figure 10. Cs sputtering threshold divided by the surface bond energy for different ion masses

8. The Sputtering of the RF Antenna

There were significant concerns that plasma ions bombarding the RF antenna limit the lifetime of RF sources with antennas immersed in the plasma [22]. Optical spectroscopy was implemented on the SNS H^- source to monitor the plasma for impurities. The spectrometer features two gratings, one optimized for short wavelengths and the other for long wavelengths [23]. Figure 11 shows the spectra emitted from a SNS H^- source shortly after deploying Cs. It is dominated by the Balmer lines of hydrogen and two bands from hydrogen molecules. There are no significant traces of oxygen or OH, while the Cs I line is clearly visible at 852 nm. The intense line at 589 nm is a superposition of the Na I line and a line from the molecular hydrogen band. The porcelain coating of the RF antenna is the only significant source of Na.

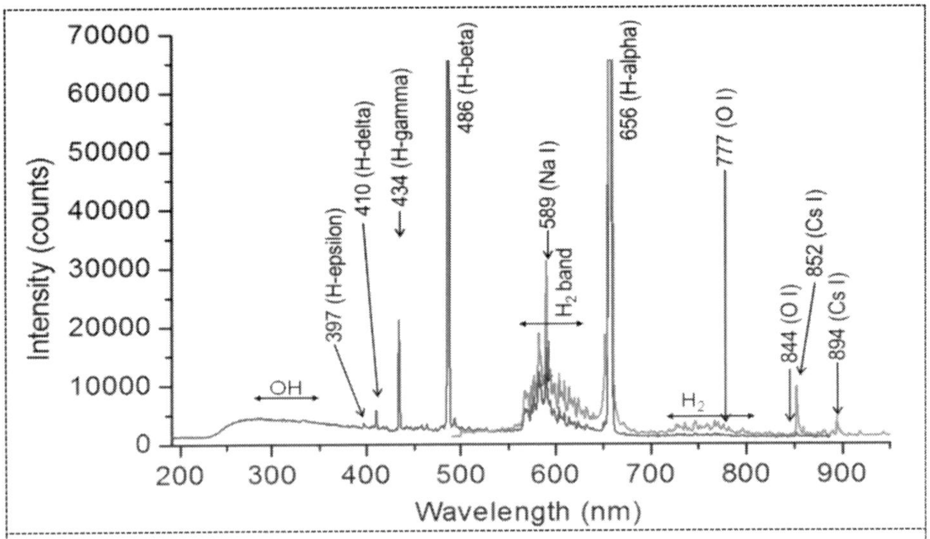

Figure 11. Optical emission from the plasma shortly after deploying caesium.

Figure 12 shows the time evolution of selected lines starting about 30 minutes after starting up the source. The start up procedure uses 50 kW RF power, but after deploying Cs the power had to be reduced to avoid plasma outages, most likely caused by discharges. The shown hydrogen delta line at 410 nm is a good indicator of the injected RF power. The bands from molecular hydrogen grow with time during the first ten hours or so as it can be seen by comparing the line at 582 nm with the delta line. Oxygen at 777 nm, a fragment of water, is rapidly conditioned away and disappears when Cs is introduced. The Cs at 852 nm spikes during the 12-minute cesiation period and falls to a more moderate level. It then gradually falls to the background level over the next few hours. The data between 17 and 23 hours show the hydrogen molecular lines at 582 nm and 589 nm to be of equal strength. Accordingly, the difference between the 589 nm line and the 582 nm line is due to Na. This suggests that Na is present in the plasma during conditioning and for about 15 hours after cesiation. Raising the RF power and/or retuning the source briefly returns Na into the plasma. This coincides with the time periods when heavy impurities are in the plasma, namely water during conditioning and Cs after cesiation and retuning.

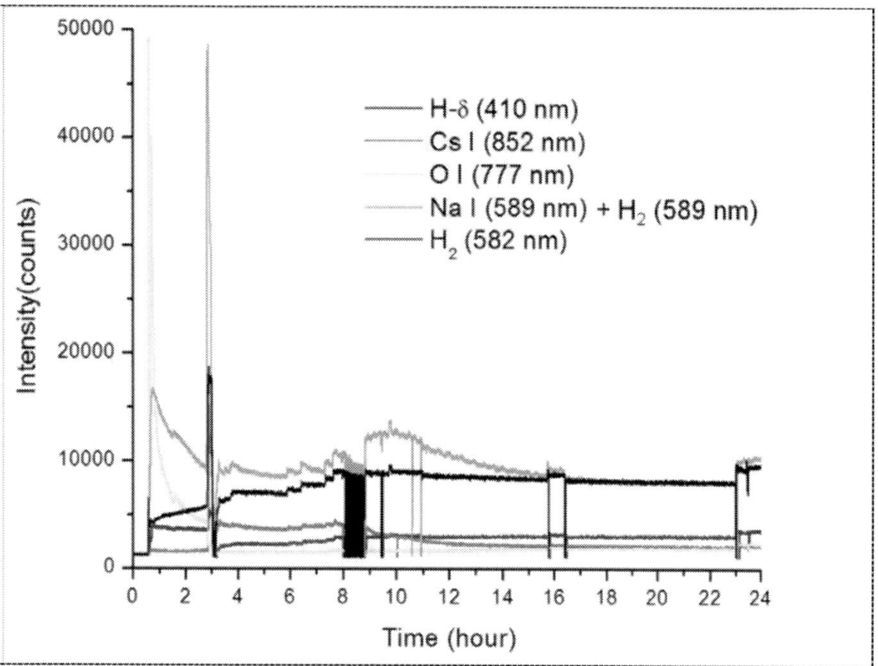

Figure 12. H-δ, Cs I, O I, Na I and H_2 line emissions during the first day of a source.

The sputter approximation of formula 1 predicts Na to be more likely sputtered by hydrogen ions than heavy ions such as H_2O and Cs. Apparently, Na sputters not as single atoms but as a cluster of several atoms, making its effective mass much heavier.

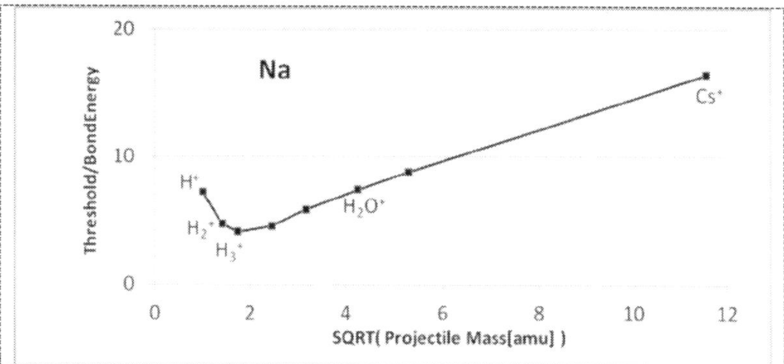

Figure 13. Na sputtering threshold divided by the surface bond energy for different ion masses

9. Summary

The interaction between the hydrogen plasma and the caesiated Mo converter contributes a large fraction of the produced H^- ion beam and therefore is critical to the success of SNS. Spectroscopic data show significant impurities in the plasma during the start up of the source and those impurities appear to sputter the caesium as well as the antenna. However, these impurities disappear before or about at the same time as the Cs disappear from the plasma. Because all impurities become negligible within a day and hydrogen is to light to sputter the caesium and the antenna, sputtering does not appear to limit the lifetime of the SNS H- sources.

10. References

[1] Takeiri Y and Tsumori K 2011 *Second International Symposium on Negative Ions, Beams and Sources* (AIP Conference Proceeding 1390)
[2] http://neutrons.ornl.gov/facilities/SNS/
[3] Zhang H 1999 *Ion Sources* (Berlin: Springer)
[4] Bacal M, Hatayama A and Peters J 2005 *IEEE Trans. On Plasma Science* **33** 1845
[5] Keller R, Thomae R, Stockli M and Welton R 2002 *AIP Conference Proceedings* **639** 47
[6] Han B, Hardek T, Kang Y, Murray S N, Pennisi T R, Piller C, Santana M, Welton R F and Stockli M P 2011 *AIP Conference Proceedings* **1390** 216
[7] Dudnikov V G 1992 *Rev Sci Instrum* **63** 2660
[8] Sherman J, Chacon-Golcher E, Geros E, Jacobson E, Lara P, Meyer B, Naffziger P, Rouleau G, Schaller S, Stevens R and Zaugg T 2005 *AIP Conference Proceedings* **763** 254
[9] Rasser M, Wunnik JNM and Los J 1982 *Surf. Sci.* **118** 697
[10] Graham W G 1980 *Proceedings on the 2nd International Symposium on the Production and Neutralization of Negative Ions and Beams* 126 (New York: Brookhaven National Laboratory)
[11] Stockli M P, Han B, Murray S N, Pennisi T R, Santana M, Welton R F 2011 *AIP Conference Proceedings* **1390** 123
[12] Stockli M P, Han B, Murray S N, Pennisi T R, Santana M and Welton R F 2010 *Rev Sci Instrum* **81** 02A729
[13] Keller R, Tarvainen O, Chacon-Golcher E, Geros E G, Johnson K F, Rouleau G, Stelzer JE and Zaugg T J 2009 *AIP Conference Proceedings* **1097** 161
[14] Peters J 2009 *AIP Conference Proceedings* **1097** 236
[15] Lee T J and Stickney R E 1972 *Surf. Sci.* **32** 100
[16] Kaminsky M. 1965 *Atomic and Ionic Impact Phenomena on Metal Surfaces* (New York: Springer)
[17] Hansen L K 1978 *Thermionic Converters and Low temperature Plasma* (Information Centre: US-DOE-tr-1)
[18] See http://www.webelements.com/periodicity/ for a collection of atomic data
[19] Stockli M P, Han B X, Murray S N, Newland D, Pennisi TR, Santana M and Welton RF 2009 *AIP Conference Proceedings* **1097** 223
[20] Stockli M P, Han B X, Hardek T W, Kang Y W, Murray S N, Pennisi TR, Piller C, Santana M and Welton R 2012 *Rev Sci Instrum* **83** 02A732
[21] Bohdansky J and Roth J 1980 *J. Appl. Phys.* **51** 2861
[22] Peters J 2004 *Proceedings from EPAC'04* 1727
[23] Jaz from Ocean Optics, Dunedin, FL 34697, USA

26th Summer School and International Symposium on the Physics of Ionized Gases (SPIG 2012) IOP Publishing
Journal of Physics: Conference Series **399** (2012) 012002 doi:10.1088/1742-6596/399/1/012002

Photoionization of atomic and molecular positively charged ions

J M Bizau[1], D Cubaynes[1], M M Al Shorman[1], S Guilbaud[1], C Blancard[2], J Lemaire[3], R Thissen[4], A Giuliani[5,6], C Nicolas[6] and A R Milosavljević[7]

[1] Institut des Sciences Moléculaires d'Orsay (ISMO), CNRS UMR 8214, Université Paris-Sud, Bât. 350, F-91405 Orsay cedex, France
[2] CEA-DAM-DIF, Bruyères-le-Châtel, F-91297 Arpajon Cedex, France
[3] Laboratoire de Chimie Physique, CNRS UMR 8000, Université Paris Sud, Bât 350, F-91405 Orsay cedex, France
[4] UJF-Grenoble 1 / CNRS-INSU, Institut de Planétologie et d'Astrophysique de Grenoble (IPAG) UMR 5274, Grenoble, F-38041, France
[5] UAR1008, Cepia, Institut National de la Recherche Agronomique (INRA), BP 71627, F-44316 Nantes Cedex 3, France
[6] Synchrotron SOLEIL, L'Orme des Merisiers, Saint Aubin, BP 48, F-91192 Gif-sur-Yvette cedex, France
[7] Institute of Physics, University of Belgrade, Pregrevica118, 11080 Belgrade, Serbia

E-mail: jean-marc.bizau@u-psud.fr

Abstract. We compare advantages and drawbacks of the two techniques mainly used for the experimental study of gas phase photoionization processes on ionic targets in combination with synchrotron radiation, the merged beam and ion trap. The comparison will be illustrated by recent results obtained on atomic and small molecular positively charged ions using both techniques at SOLEIL, the French synchrotron radiation facility.

1. Introduction

Photoionization (PI) of ionic species is a key process for the modeling of many plasmas, including laboratory plasmas, planetary ionospheres and astrophysical plasmas. The knowledge of PI cross sections is required over extended ionic charge stages and photon energies, both parameters of increasing magnitude with increasing plasma temperature. Until recently, relatively few experiments were performed on PI processes, facing the difficulty to combine a high density ionic target with a high flux of X ray radiation. Therefore, most of the data on these processes are known from models using more or less sophisticated approximations for the description of electron correlation effects [1-3]. Only the advent of high photon flux available at synchrotron radiation facilities has allowed the experiments to develop. First, merged beam setups have been used, allowing the determination in absolute values of PI cross sections for ions with charge stage up to 9 [4,5]. One difficulty of this technique consists in an ionic target often composed of a mixture of ions in the ground and metastable states. During the last five years, a new approach showed rapid expansion: the use of ion traps. It has allowed already the measurements of PI cross sections on relaxed ions [6], on more highly charged ions [7] and large molecular ions and clusters [8,9].

Journal of Physics: Conference Series **399** (2012) 012002 doi:10.1088/1742-6596/399/1/012002

We present here our recent results obtained at SOLEIL, the French synchrotron radiation facility, on the PI of atomic and small molecular positively charged ions, using both a merged beam setup and an ion trap. To conclude, we will summarize the advantages and drawbacks of the two techniques.

2. Merged beam technique

The merged beam technique was first developed by Peart *et al* [10] for the study of electron impact processes on atomic ions. It was adapted later by Lyon *et al* [11] for measuring PI cross sections at the Daresbury synchrotron radiation source. The principle is to merge a beam of ions, produced in an ion source and charge over mass selected using a magnetic device, with a monochromatic synchrotron radiation beam. After the interaction, the charge of the ions is analyzed, the parent ions being collected in a Faraday cup and the ions which have increased their charge by one (or several) unit(s) during the interaction with the photons (the so-called photoproducts) are counted. Compared to similar studies on neutral species, the main advantage is a target density easy to measure, allowing determination of PI cross sections in absolute values. In return, the target density is five or six orders of magnitude lower and a high background is often produced by collisional ionization processes between the fast ion beam and the residual gas in the interaction chamber. To improve the signal to noise ratio, high photon flux is mandatory, requiring the use of undulators as synchrotron radiation source. A long interaction region (20-50 cm) is also welcome, and ultra-high vacuum is requested. Once these conditions are achieved, 10 to 20% accuracy is typically obtained on the cross section measurements, the main contribution to the uncertainty coming from photon flux determination and statistics when the signal/noise ratio is low.

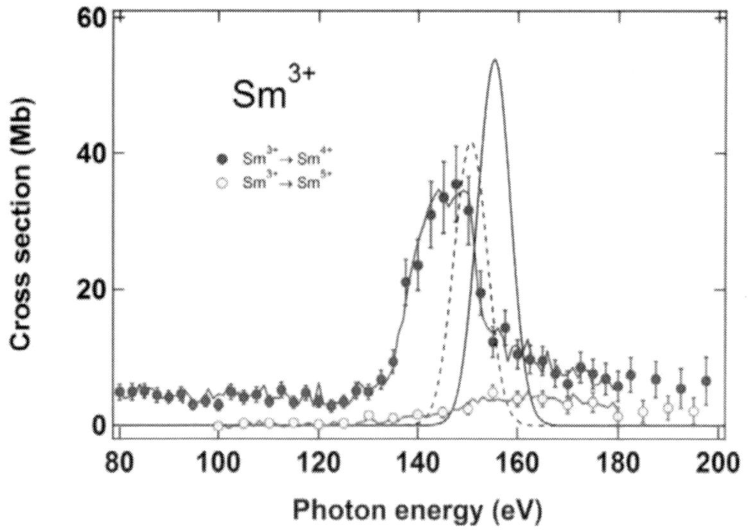

Figure 1: Single (blue dots) and double (red dots) measured absolute PI cross sections for Sm^{3+} ion, compared to the results of a UTA code (black curves).

Figure 1 shows an example of PI cross section that we have measured recently for Sm^{3+} ion between 80 and 200 eV photon energy using the merged beam setup permanently installed on the PLEIADES beam line at SOLEIL. The ions were produced using an Electron Cyclotron Resonance Ion Source (ECRIS). Single PI (corresponding to the detection of Sm^{4+} ions, blue curve and filled points) and double PI (corresponding to detection of Sm^{5+} ions, red curve and open points) have been obtained. Electronic configuration of Sm^{3+} ion in its ground state is [Kr] $4d^{10}$ $5s^2$ $5p^6$ $4f^5$. The large structure centered around 145 eV in the single PI cross section corresponds mainly to photoexcitation of a 4d inner-shell electron to the partially filled 4f subshell, leaving a Sm^{3+} ion in highly excited states [Kr] $4d^9$ $5s^2$ $5p^6$ $4f^6$ which will further decay via autoionization to the energetically open continua. Because of the high temperature of the plasma in the ECRIS, all the states with the same configuration as the Sm^{3+} ground state, extending on an excitation energy range larger than 6 eV, are populated in the source and have lifetime long enough to contribute to our spectra. Combined with the presence of open 4d and 4f subshells in the Sm^{3+} $4d^9$ $5s^2$ $5p^6$ $4f^6$ intermediate states, all this leads to a huge number of allowed 4d → 4f photoexcitations. Due to the large number of transitions and their large natural width resulting from the very fast decay via super Coster-Kroning transition to the Sm^{4+} [Kr] $4d^{10}$ $5s^2$ $5p^6$ $4f^4$ ground configuration, the line spacing is statistically lower than their natural width and only very broad structures, called Unresolved Transition Arrays (UTA), can be observed. We are trying to reproduce our observations using the statistical model developed by J. Bauche [12], assuming a statistical population of the initial states. Very preliminary results, not including yet the effect of electron correlations, are shown on Figure 1 (black continuous line). The agreement of theory with experiment is rather satisfactory. However, the experimental UTA profile is broader and asymmetric compared to the calculated profile. One explanation might be that the Sm^{3+} [Kr] $4d^{10}$ $5s^2$ $5p^5$ $4f^6$ states are metastable and contribute also to the measured spectra, broadening the structure on the low energy side. Calculated 4d → 4f UTA starting from the metastable configuration is also shown as the dashed line on Figure 1.

Double PI channel in mainly populated via cascade Auger decay. It is weak at the energy of the 4d → 4f UTA because few Sm^{4+} configurations, involving one or two holes in the 5s subshell, are open above the Sm^{5+} thresholds. It is more intense at higher energy, where more excited 4d → nf, n>4 excitations are involved.

3. Ion trap

Contribution of ions in various metastable state is a quite general problem in the merged beam experiments, especially when using high temperature plasma ion source like ECRIS. It makes more difficult the interpretation of the experimental spectra and their comparison with theoretical predictions, in particular because the relative population of the different states is most of the time unknown. A commonly used procedure to minimize the contribution of metastable states is to store the ions before interaction, for a time long enough allowing them to radiatively decay into the ground state. The first coupling of an ion trap with synchrotron radiation was performed by Kravis *et al* in the 90's at NSLS facility in Brookhaven, using a Penning ion trap to measure broadband PI of Ar^{2+} ion [13,14]. To demonstrate the feasibility that a PI experiment on cooled ions can be performed in association with synchrotron radiation, we brought a small movable Fourier Transform Ion Cyclotron Resonance (FT-ICR) ion trap first at ELETTRA [6], then on the DESIRS beam line at SOLEIL [15]. In this device, 1.2 T magnetic fields produced by permanent magnet is combined with an electrostatic potential to trap the ions inside a 2 cm^3 cubic cell [16]. FT-ICR traps are widely used for mass spectrometry studies on large molecules, often using superconducting magnets to produce very high magnetic fields. Nevertheless, even our small trap is able to achieve 70,000 mass resolving power.

Figure 2 shows PI cross sections measured for Xe^+ ion between 19.5 and 23 eV photon energy. The upper spectrum was obtained with the ASTRID merged beam setup [5]. The sharp lines correspond to photoexcitations of a 5p electron to empty nd or ns orbitals, leaving a Xe^+ ion in $5p^4$ nd,ns highly excited levels which will decay via autoionization to the ground state of Xe^{2+} ion [6,17]. The vertical bars above the spectrum indicate the position of the Xe^+ $5p^{-1}$ ionization thresholds [18]. A strong signal is visible at lower photon energy than the first ionization threshold at 20.98 eV, clearly demonstrating the contribution of Xe^+ ions in metastable states to our spectrum. Actually, only the Xe^+ $5p^5$ $^2P_{1/2}$ term is metastable, with a lifetime of 48.7 ms [19], and lines observed below 20.98 eV can be attributed to photoexcitation of a 5p electron starting from this $^2P_{1/2}$ excited term.

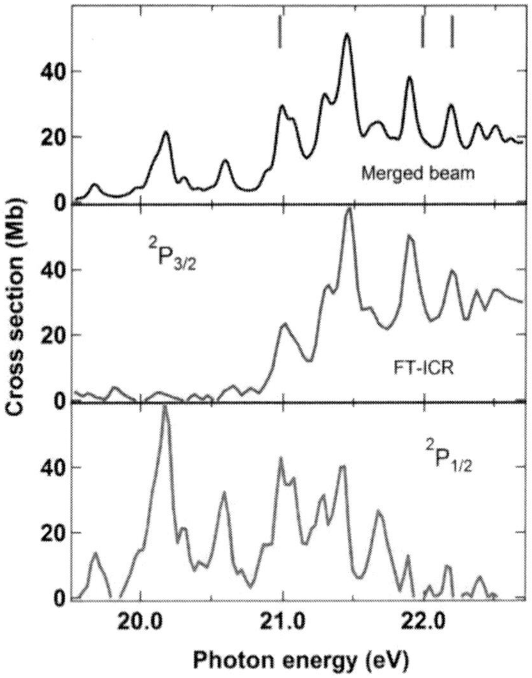

Figure 2: Comparison of the PI cross sections measured on Xe^+ ion with a merged beam setup (upper panel) and a FT-ICR trap (middle panel). Combination of the two spectra allows estimating the cross section for ions in the pure metastable state (lower panel).

The spectrum recorded with the FT-ICR trap is shown on the middle panel of Figure 2. The Xe^+ ions were produced directly inside the trap by electron impact, a pulse of electrons ionizing a pulse of xenon gas. The ions were trapped for 1 sec, allowing the metastable species to decay, and then irradiated by the monochromatized synchrotron radiation. The signal below 20.98 eV has clearly vanished, demonstrating that we measured the PI cross section of Xe^+ ion in the pure ground state. As it is not possible to determine the density of ions inside the trap, the cross section is only obtained in relative value. It was normalized on the measurements performed with the merged beam setup, assuming for the latter a statistical population of the ground and metastable terms.

From the difference between the two spectra, we can derive the cross section for the ions in the pure metastable state, assuming once again a statistical population of the two terms in the merged beam spectrum. It is shown on the bottom panel of Figure 2. The cross section in the high energy part is dominated by the direct PI process and is not expected to be of very different magnitude for the ions in the ground and metastable terms, contrary to what is observed. However, it should be noted that the two spectra were recorded on two different beam lines, with slightly different resolution, and there is also contribution of higher orders light diffracted by the monochromator on the ASTRID beam line.

As stated earlier, owing to very high *m/z* resolution, FT-ICR ion traps are particularly well suited for the study of molecular ions. Only studies on CO^+ ions [20,21] and some endohedral ions [22,23], have been performed up to now on molecular ions using merged beam setups. One difficulty is that often doubly-charged molecular ions (so-called dications) are generally unstable with respect to dissociation into two singly charged ions, due to the coulombic repulsion which dominates the long range interaction between the two fragments. Nevertheless, short range covalent bonding can overcome the repulsive coulombic forces and metastable or even stable molecular dications are observed. For example, in CO_2^{2+} lifetimes up to 4s have been measured [24]. Figure 3 gives the variation of the cross section we have measured for CO_2^+ ion with the FT-ICR ion trap. Detection of the dications was performed in the 10-1000 ms time window after their production by the photons, in such a way only dications in long lived states can be detected. This first result on molecular ion has been obtained in the very last hour of our last beam time on DESIRS beam line. Although very preliminary, it is

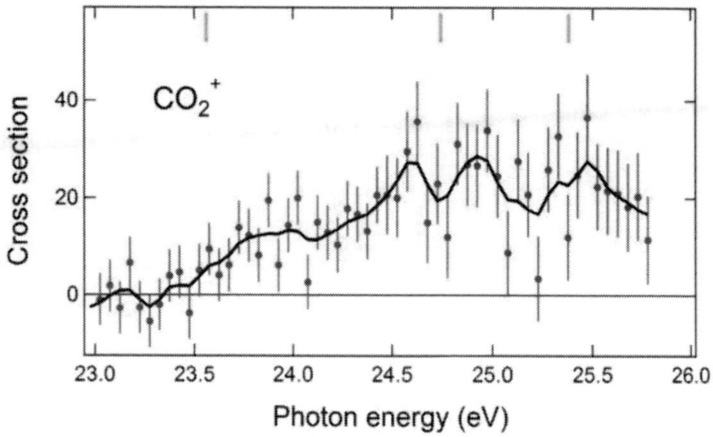

Figure 3: Relative PI cross section for CO_2^+ molecular ion measured with the FT-ICR ion trap.

consistent already with the electronic state energies of the dication measured by Slattery *et al* [25], shown as the vertical bars on Figure 3.

4. Conclusion

We have given examples of PI cross sections of ionic species measured at SOLEIL synchrotron radiation facility using two different techniques, the merged beam and ion trapping, and we have tried to show their complementarities. Merged beam setups are perfectly adapted for the determination of cross sections in absolute values, but interpretation of their results is often obscured by metastable ions contribution. In contrast, ion traps provide with relative cross sections on cold ions, which can be normalized using the absolute values. Merged beam technique is in use since the 80's and has been applied to the study of singly (both positively and negatively) and multiply-charged atomic ions, with charge stage up to 9 [26]. At present, such setups are in operation in the USA (ALS) [27], Denmark (ASTRID) [5] and in France (SOLEIL) [28,29], and one is under commissioning in Germany (PETRA). They are huge setups requiring a permanent installation on a synchrotron radiation beam line, and can support several detection techniques like ion and electron spectrometry [30,31]. In contrast, ion traps are light and easily movable, but limited to mass spectrometry, often with a very high mass resolution making them a perfect tool for the study of molecular ions. They have been in fast development during the last five years, mainly at SOLEIL in France and BESSY II in Germany. Different traps have been used: FT-ICR for the study of singly-charged atomic and small molecular ions as shown in this paper, linear traps for the study of singly-charged atomic ions [32], large molecular [9] and metallic clusters ions [8], RF trap for the study of large molecular ions of biological interest [33-35], electron beam ion trap (EBIT) for the study of highly-charged atomic ions [7,36]. To finish, let us note the construction of an electrostatic ring to store the ions at ASTRID II in Denmark, which will combine some of the advantages of both merged beam setup and ion trap.

References

[1] R. F. Reilmann and S. T. Manson 1979 *ApJS* **10**, 815
[2] D. A. Verner and D. G. Yakovlev 1995 *Astron. Astrophys. Suppl. Ser.* **109** 125
[3] N R Badnell, M A Bautista, K Butler, F Delahaye, C Mendoza, P Palmeri, C J Zeippen and M J Seaton 2005 *Month. Not. Royal Atron. Soc.* **360** 458
[4] J. B. West 2001 *J. Phys. B: At. Mol. Opt. Phys.* **34** R45
[5] H. Kjeldsen 2006 *J. Phys. B: At. Mol. Opt. Phys.* **39** R325

[6] R. Thissen , J M Bizau, C Blancard, M Coreno, C Dehon, P Franceschi, A Giuliani, J Lemaire and C Nicolas 2008 *Phys. Rev. Lett.* **100** 223001
[7] M. C. Simon *et al* 2010 *J. Phys. B: At. Mol. Opt. Phys.* **43** 065003.
[8] K. Hirsch *et al* 2009 *J. Phys. B: At. Mol. Opt. Phys.* **42** 154029
[9] A. R. Milosavljevi, C. Nicolas, J. Lemaire, C. Dehon, R. Thissen, J-M. Bizau, M. Réfrégiers, L. Nahon and A. Giuliani 2011 *Phys. Chem. Phys*. **13** 15432
[10] B. Peart, J.G. Stevenson and K. Dolder 1973 *J. Phys. B: At. Mol. Phys.* **6** 146
[11] I.C. Lyon, B. Peart, J.B. West and K. Dolder 1986 *J. Phys. B: At. Mol. Phys.* **19** 4137
[12] C. Bauche-Arnoult, J. Bauche and M. Klapisch 1978 *J. Opt. Soc. Am.* **68** 1136
[13] S.D. Kravis *et al* 1991 *Phys. Rev. Lett.* **66** 2956
[14] S.D. Kravis *et al* 1992 *Phys. Rev. A* **45** 6379
[15] L. Nahon, N. de Oliveira, G. Garcia, J. F. Gil, B. Pilette, O. Marcouille, B. Lagarde, F. Polack 2012 *J. Synchrotron Rad* **19** 508
[16] G. Mauclaire, J. Lemaire, P. Boissel, G. Bellec and M. Heninger 2004 *Eur. J. Mass Spectrom.* **10** 155
[18] J.M. Bizau *et al* 2011 *J. Phys. B: At. Opt. Mol. Phys*. **44** 055205
[19] Yu Ralchenko, A.E. Kramida, J. Reader and NIST ASD Team 2010. NIST Atomic Spectra Database (ver. 4.0.0), [Online]. Available: http://physics.nist.gov/asd [2010, September 30]. National Institute of Standards and Technology, Gaithersburg, MD.
[20] S. Jullien , J. Lemaire, S. Fenistein, M. Heninger, G. Mauclaire and R. Marx 1993 *Chem. Phys. Lett.* **212** 340
[21] T. Andersen, H. Kjeldsen, H. Knudsen and F. Folkmann 2001 *J. Phys. B: At. Opt. Mol. Phys.* **34** L327
[22] G. Hinojosa *et al* 2002 *Phys. Rev. A* **66** 032718
[23] S.W.J. Scully et al 2007 *Phys. Rev. Lett.* **98** 179602
[24] A.L.D. Kilcoyne *et al* 2010 *Phys. Rev. Lett.* **105** 213001
[25] D. Mathur, L. H. Andersen, P. Hvelplund, D. Kella and C. P. Safvan 1995 *J. Phys. B* **28**, 3415
[26] A.E. Slattery, T.A. Field, M. Ahmad, R.I. Hall, J. Lambourne, F. Penent, P. Lablanquie and J.H.D. Eland 2005 *J. Chem. Phys.* **122** 084317
[27] M. Habibi, D.A. Esteves, R.A. Phaneuf, A.L.D. Kilcoyne, A. Aguilar and C. Cisneros 2009 *Phys. Rev. A* **80** 033407
[28] A.M. Covington *et al* 2002 *Phys. Rev. A* **66** 062710
[29] M.F. Gharaibeh *et al* 2011 *J. Phys. B: At. Opt. Mol. Phys.* **44** 175208
[30] C. Blancard *et al* 2012 *Phys. Rev. A* **85** 043408
[31] J.M. Bizau *et al* 1991 *Phys. Rev. Lett.* **67** 576
[32] A. Gottwald, S. Anger, J.-M. Bizau, D. Rosenthal, and M. Richter 1997 *Phys. Rev. A* **55** 3941
[33] A. R. Milosavljević, C. Nicolas, J.-F. Gil, F. Canon, M. Réfrégiers, L. Nahon and A. Giuliani 2012 *J. Synchrotron Rad.* **19** 174
[34] A.R. Milosavljević, F. Canon, C. Nicolas, C. Miron, L. Nahon and A. Giuliani 2012 *J. Phys. Chem. Lett.* **3** 1191
[35] K. Hirsch *et al* 2012 *Phys. Rev. A* **85** 062501
[36] S. Bari, O. Gonzalez-Magaña, G. Reitsma, J. Werner, S. Schippers, R. Hoekstra and T. Schlathölter 2011 *J. Chem. Phys.* **134** 024314
[37] M.C. Simon *et al* 2010 *Phys. Rev. Lett.* **105** 183001

Vibrationally inelastic collisions of slow electrons with polyatomic molecules

R Čurík[1], **P Čársky**[1]

[1] J. Heyrovský Institute of Physical Chemistry, Academy of Sciences of the Czech Republic, v.v.i., Dolejškova 3, 182 23 Prague 8, Czech Republic

E-mail: `roman.curik@jh-inst.cas.cz`

Abstract. We present a computational method developed for calculations of vibrationally inelastic collisions of electrons with polyatomic molecules. The method is first applied to calculate vibrational excitations of cyclopropane molecule. In second step we demonstrate how the method can be altered for collisions at higher energies, namely 50–300 eV. In the case of furan molecule we compare a size of summed vibrationally inelastic cross sections with the elastic data. For the two molecules studied we also compare our calculated results with the available experimental data.

1. Introduction

During the last two decades the modeling of elastic collisions of electrons with polyatomic molecules became a well-established task. Many authors report fixed-nuclear elastic data obtained by use of R-matrix method [1, 2], complex Kohn variational method [3, 4], Schwinger multichannel method [5, 6, 7], or optical potential method with single-center expansion of electron wave function [8, 9].

However, the situation is quite different in computational modeling of the inelastic processes during the collision. In case of vibrational excitation there have been many rigorous and successful attempts for diatomic [10, 11, 12, 13] and triatomic [14, 15, 16] molecules but very few results are available for the polyatomic systems. We believe that at present state a treatment of electron-impact vibrational excitation of polyatomic molecules requires several numerically controlled approximations. In the presented method (Discrete Momentum Representation) we show that it is still possible to retain both the rigor of the theory and computational feasibility.

Some of the lighter hydrocarbons are known as sources of carbon atoms during chemical vapor decomposition reactions. The cyclopropane was detected in cooler edges of the fusion plasmas. The most marked feature observed by experiments [17] has to do with the excitation of ν_3 vibration, the C-C ring stretching. The molecule was previously studied computationally [18], however due to the model limitations only three full-symmetric vibrational modes were studied. Therefore, we employed Discrete Momentum Representation (DMR) method [16] to calculate cross sections for excitation of all the vibrational modes.

The motivation to study electron interactions with furan molecules (C_4H_4O) in recent years predominantly comes from radiation damage research, since its structure may be considered as a simple representative of the sugar component in the DNA backbone. Hence, the aim of the present study is to simplify the DMR method for higher collision energies (by use of unitarized

Born approximation) and attempt to estimate a probability of deposition of electron collision energy into vibrational motion of the target furan molecule. The collision energy range of our interest is 50–300 eV.

2. Theory

2.1. Overview of the DMR Method

The vibrationally inelastic DMR method was discussed in detail in Ref. [16] and only a brief summary is given here. We made use of the following approximations:

- We introduce one-electron optical potential $\hat{V}$ for the interaction between the scattered electron and the charge density of the molecule. Moreover, we retain only the first term of the optical potential expansion (see for example Ref. [19]), ending up with the static-exchange (SE) approximation.
- The SE approximation is corrected by a model DFT potential V_{cp} that accounts for orbital relaxation of the bound electrons and for the correlation between the scattered electron and the bound electrons. We used the interpolation formula suggested by Perdew and Zunger [20].
- Nuclear dynamics is described by the rotationally frozen and vibrationally harmonic approximations. Moreover, for the vibrational space of each normal mode we use only a two-state approximation.

These approximations allow us to describe the scattering problem *via* the two-channel Lippmann-Schwinger equation in the three-dimensional momentum space:

$$\langle\chi_1\boldsymbol{k}_1|\hat{T}|\chi_0\boldsymbol{k}_0\rangle = \langle\chi_1\boldsymbol{k}_1|\hat{U}|\chi_0\boldsymbol{k}_0\rangle + \sum_{i=0}^{1}\int \mathrm{d}\boldsymbol{k}\,\frac{\langle\chi_1\boldsymbol{k}_1|\hat{U}|\chi_i\boldsymbol{k}\rangle\langle\chi_i\boldsymbol{k}|\hat{T}|\chi_0\boldsymbol{k}_0\rangle}{k_0^2 - 2E_i - k^2 + \mathrm{i}\varepsilon}\,, \tag{1}$$

where $\hat{U}$ stands for twice the interaction potential $\hat{V}$, $E_0 = 0$ (for the elastic channel) is the energy of the vibrational ground state and E_1 is the energy of the first excited vibrational state. These two states are described by the vibrational functions χ_0 and χ_1, respectively. The vectors $\boldsymbol{k}_0$ and $\boldsymbol{k}_1$ represent the plane-wave functions for the incoming and outgoing electrons, respectively.

Numerical discretization (p and i run through abscissas of the radial and angular quadratures, respectively)

$$\int \mathrm{d}k \int \mathrm{d}\hat{\boldsymbol{k}}\, g(\boldsymbol{k}) \rightarrow \sum_{p}^{NRAD} w_p \sum_{i}^{NANG} w_i\, g(\boldsymbol{k}_{pi}) \tag{2}$$

of the integral on the r.h.s of the Eq. (1) leads to a set of two coupled matrix equations:

$$\begin{pmatrix} \mathrm{T}_{00} & \mathrm{T}_{01} \\ \mathrm{T}_{10} & \mathrm{T}_{11} \end{pmatrix} = \begin{pmatrix} \mathrm{U}_{00} & \mathrm{U}_{01} \\ \mathrm{U}_{10} & \mathrm{U}_{11} \end{pmatrix} + \begin{pmatrix} \mathrm{U}_{00} & \mathrm{U}_{01} \\ \mathrm{U}_{10} & \mathrm{U}_{11} \end{pmatrix} \cdot \begin{pmatrix} \mathrm{G}_0 & 0 \\ 0 & \mathrm{G}_1 \end{pmatrix} \cdot \begin{pmatrix} \mathrm{T}_{00} & \mathrm{T}_{01} \\ \mathrm{T}_{10} & \mathrm{T}_{11} \end{pmatrix}, \tag{3}$$

with the interaction matrix elements defined as follows:

$$\begin{aligned} [\mathrm{U}_{00}]_{pi,qj} &= \langle\chi_0\,\boldsymbol{k}_{pi}|\hat{U}|\chi_0\,\boldsymbol{k}_{qj}\rangle \\ [\mathrm{U}_{11}]_{pi,qj} &= \langle\chi_1\,\boldsymbol{k}_{pi}|\hat{U}|\chi_1\,\boldsymbol{k}_{qj}\rangle \\ [\mathrm{U}_{01}]_{pi,qj} &= \langle\chi_0\,\boldsymbol{k}_{pi}|\hat{U}|\chi_1\,\boldsymbol{k}_{qj}\rangle = [\mathrm{U}_{10}]_{pi,qj}\,. \end{aligned} \tag{4}$$

The body-fixed scattering amplitudes for the vibrational transitions are obtained by the matrix inversion in equation (3). More details about the evaluation of the matrix elements (4) and the quadrature scheme (2) of the singular kernels on r.h.s of the Eq. (1) can be found in Refs. [16, 21].

2.2. Unitarized First Born Approximation

In a case of weak interactions or higher collision energies the Eq. (1) may be simplified by neglecting the second term on the r.h.s. Such an approximation is called first Born approximation (FBA). It is clear that in the FBA elastic and inelastic channels are considered independent as their scattering amplitudes are given by simple potential matrix elements

$$f_{\rm ela}^{\rm FBA}(\nu_0\boldsymbol{k}_1 \leftarrow \nu_0\boldsymbol{k}_0) = -4\pi^2\langle\, \nu_0\boldsymbol{k}_1\,|\hat{V}|\,\nu_0\boldsymbol{k}_0\,\rangle\;, \tag{5}$$

$$f_{\rm inela}^{\rm FBA}(\nu_1\boldsymbol{k}_1 \leftarrow \nu_0\boldsymbol{k}_0) = -4\pi^2\langle\, \nu_1\boldsymbol{k}_1\,|\hat{V}|\,\nu_0\boldsymbol{k}_0\,\rangle\;. \tag{6}$$

One of drawbacks of the FBA model is that it breaks the unitarity of the scattering matrix which is connected to a flux conservation during the scattering process. Therefore we implemented unitarized first Born approximation (UFBA). The UFBA represents the FBA of a K-matrix (instead of T-matrix approximated above). In this way the approximated hermitian K-matrix leads to a unitary S-matrix and therefore conservation of the scattering flux.

The usefulness of this approach was first demonstrated by Itikawa [22], and it has been later noted that the UFBA often yields results for scattering cross sections that are significantly more accurate than those obtained by the FBA [23]. The two-channel FBA K-matrix can be written as (we use the plane-wave normalization instead of the more common energy normalization)

$$\langle\, \nu_i\boldsymbol{k}_1\,|K|\,\nu_j\boldsymbol{k}_0\,\rangle \approx -\pi\langle\, \nu_i\boldsymbol{k}_1\,|\hat{V}|\,\nu_j\boldsymbol{k}_0\,\rangle\,, \quad i,j=0,1\;. \tag{7}$$

The unitarized first Born T-matrix $T^{\rm U}$ is then obtained by means of Heitler's integral equation [24]:

$$\pi T^{\rm U} = -K\,(1-{\rm i}K)^{-1}\;. \tag{8}$$

The elastic and inelastic scattering amplitudes are directly proportional to the parts of $T^{\rm U}$ as

$$f_{\rm ela}^{\rm U}(\nu_0\boldsymbol{k}_1 \leftarrow \nu_0\boldsymbol{k}_0) = -4\pi^2\langle\, \nu_0\boldsymbol{k}_1\,|T^{\rm U}|\,\nu_0\boldsymbol{k}_0\,\rangle\;, \tag{9}$$

$$f_{\rm inela}^{\rm U}(\nu_1\boldsymbol{k}_1 \leftarrow \nu_0\boldsymbol{k}_0) = -4\pi^2\langle\, \nu_1\boldsymbol{k}_1\,|T^{\rm U}|\,\nu_0\boldsymbol{k}_0\,\rangle\;. \tag{10}$$

It should be pointed out that the UFBA is an approximation using full static-exchange potential and it should not be confused with the (first) dipole Born approximation used in the literature, as the latter employs only a dipole potential that predicts zero cross sections for infrared inactive modes.

3. Results

3.1. Cyclopropane

Electronic state of the target molecule was described at Hartree-Fock level. We used Gaussian type orbital basis set of DZP quality. Polarizability tensor components used in the present calculations were obtained as linear response functions in Kohn-Sham DFT [29] calculations with B3LYP hybrid functional and Sadlej's polarized VTZ basis sets [30] as implemented in program Dalton Release 2.0 (2005). Body frame of reference was chosen by principal axes, thus the calculated polarizability tensor is diagonal:

$$\boldsymbol{\alpha}(\mathrm{C_3H_6}) = \begin{pmatrix} 38.8 & & \\ & 38.8 & \\ & & 33.8 \end{pmatrix}\;.$$

The tensor compares favorably with the experimentally measured value of spherical polarizability $\alpha_0 = 38.2$ a.u. [31].

Figure 1 displays a computed integral cross sections with the present DMR method at static-exchange (DMR SE) and static-exchange + polarization (DMR SEP) levels. It can be seen

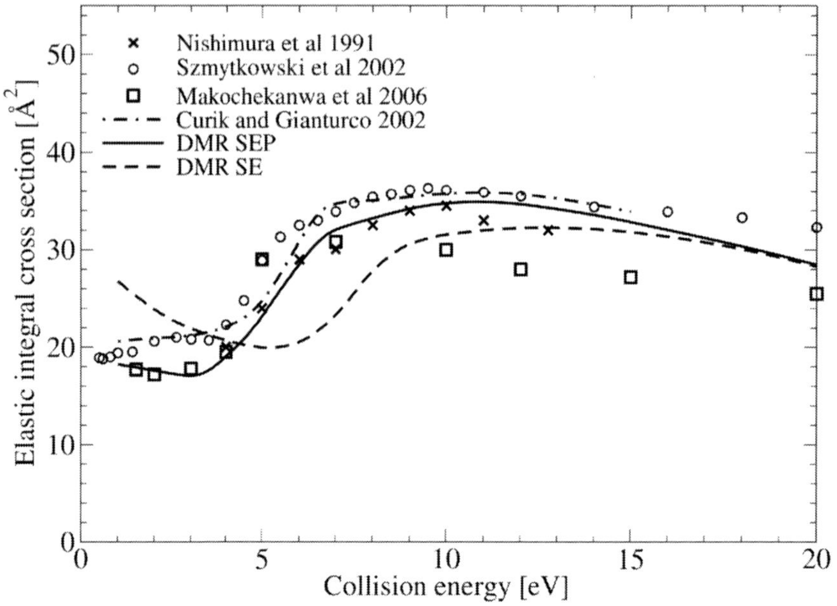

Figure 1. Elastic integral cross section of cyclopropane. The experimental data of Nishimura *et al* 1991 [25] are denoted by crosses, circles are taken from Szmytkowski *et al* 2002 [26] and the rectangles are from Makochekanwa *et al* 2006 [27]. Previous computations of Curik and Gianturco 2002 [28] are shown as dot-dashed curve. Present results in static-exchange approximation are displayed by dashed curve while the calculations with static-exchange + polarization model are shown as full curve.

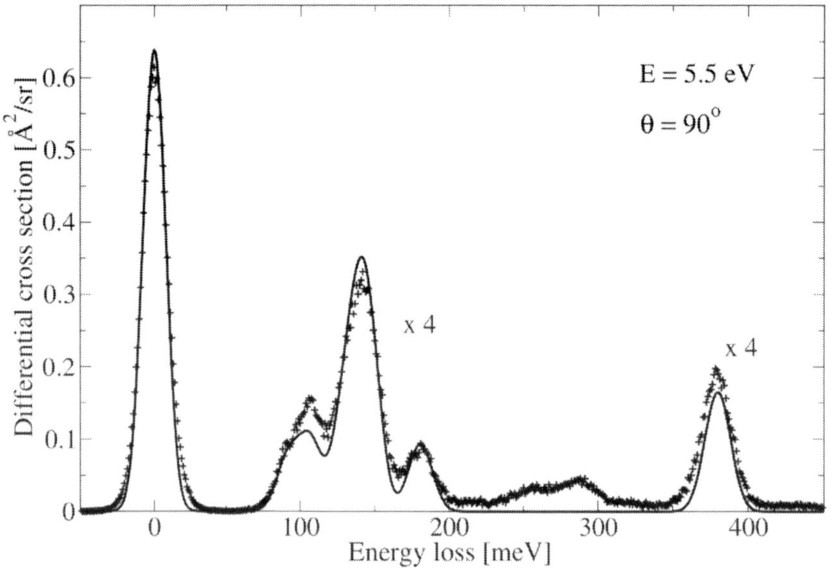

Figure 2. The measured spectrum shown as crosses was taken from Ref. [17]. Computed EELS spectrum (with half-width of 16 meV) is displayed as full line. Experimental data [17] were normalized at the elastic peak.

that correlation-polarization forces strongly affect even the elastic scattering process at collision energies under 15 eV. Our results are compared with previous experimental and theoretical data.

The cyclopropane molecule has 21 vibrational modes. Out of them only 14 are unique due to high symmetry of the molecule. We calculated vibrational excitation cross sections of all the modes. In order to compare the data with an experiment we have decided to reconstruct an experimentally measured electron energy-loss spectrum (EELS). The energy resolution half-width was set to 16 meV. As a shape for the resolution function we choose the Gaussian function. A comparison of such computed EELS and the experimental measurement is shown in Fig. 2. Because the experimental EELS data in [17] were not in absolute values we normalized them to our computed elastic peak. We conclude that the agreement is exceptionally good.

3.2. Furan

Electronic target state is once again described by Hartree-Fock theory. The Slater-type wave function then generates electrostatic and exchange interaction with continuum electron. Since the collision energies of our interest here start at 50 eV and reach 300 eV we can neglect the correlation-polarization interaction with impunity. Furan belongs to C_{2v} symmetry, therefore it has 21 non-degenerate vibrational modes and most of them are infrared active.

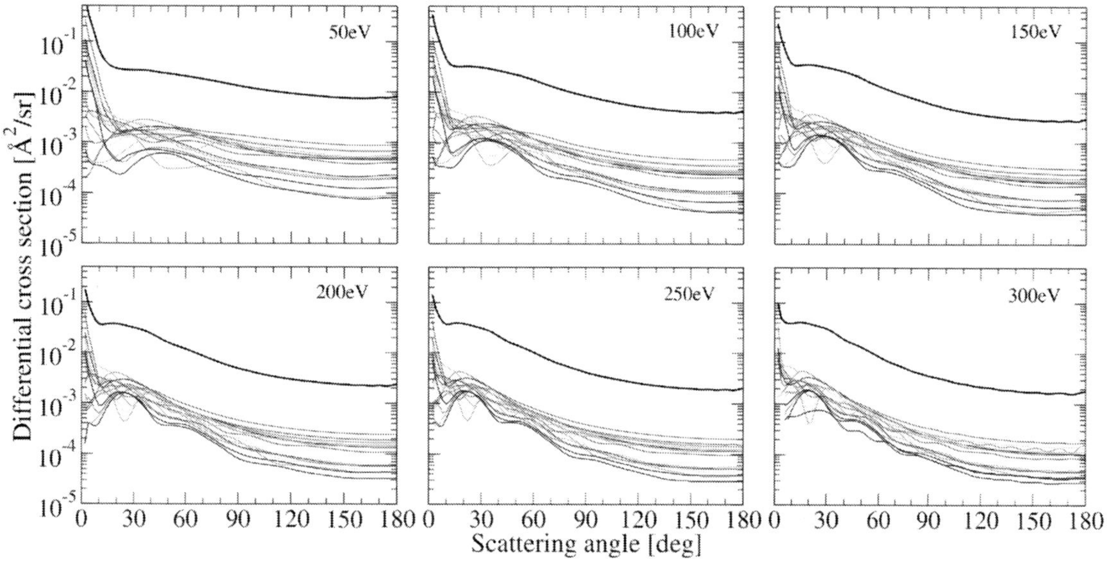

Figure 3. Rotationally summed $0 \rightarrow 1$ vibrationally inelastic absolute differential cross section for scattering of the electrons by furan. All 21 excitation modes are plotted. Sum of all the vibrational excitation cross sections is displayed as a thick curve.

It has been discussed previously [32] that vibrational excitations should not affect significantly elastic differential cross section (DCS) in the present high energy range. Still, according to out knowledge, an explicit estimation of the absolute DCSs for vibrational excitation of similar molecules (representing building blocks of large biopolymers) and for the present incident energy range has not been reported. The calculated UFBA rotationally summed $0 \rightarrow 1$ vibrationally inelastic differential cross sections are shown in Fig. 3. The selected set of collision energies is 50, 100, 150, 200, 250 and 300 eV. The cross sections were calculated according to formula (10).

The resulting data were used in Ref. [33] to estimate the ratio between a probability of the deposition of collision energy into vibrational motion of furan during the collision process and a probability that the electron undergoes elastic scattering. Authors concluded that the above mentioned ratio as a function of the scattering angle may reach about 10% in the examined collision energy range of 50–300 eV.

4. Summary

In present work we applied the DMR method for two different inelastic studies. We summarized the approximations used in the method and then we have shown that their use still leads to results of a very good quality.

In first we calculated electron impact vibrational excitation of cyclopropane. All the vibrational modes were computed. The elastic and inelastic data agree very well with the available experimental data. In second we attemped to estimate the size of cross section for deposition of collision energy into vibrational modes of furan. The basic DMR method for this problem was simplified by UFBA that can be well justified by a weak nature of the inelastic interaction together with higher collision energies (50–300 eV). We conclude that in this energy range the size of total vibrationally inelastic cross section reaches about 10% of the elastic probability.

Acknowledgments

The work was supported by the Ministry of Education of the Czech Republic (Grant No. OC10046), the Grant Agency of the Czech Republic (Grant No. P208/11/0452), and motivated by the COST Action CM0805.

REFERENCES

[1] Tennyson J 2010*Phys. Rep.* **491** 29
[2] Masin Z and Gorfinkiel J D 2011 *J. Chem. Phys.* **135** 144308
[3] Rescigno T N, McCurdy C W, Orel A E, Lengsfield III B H in *Computational Methods for Electron-Molecule Collisions* (W. M. Huo and F. A. Gianturco, Eds.), (Plenum Press, New York, 1995)
[4] Trevisan C S,Orel A E, Rescigno T N 2006 *Phys. Rev. A* **74**, 042716
[5] Takatsuka K and McKoy V 1980 *Phys. Rev. Lett.* **45** 1734
[6] Palihawadana P, Sullivan J, Brunger M, Winstead C, McKoy V, Garcia G, Blanco F, and Buckman S 2011 *Phys. Rev. A* **84** 062702
[7] Khakoo M A, Muse J, Ralphs K, da Costa R F, Bettega M H F, Lima M A P 2010 *Phys. Rev. A* **81** 062716
[8] Gianturco F A, Thompson D G and Jain A, in *Computational Methods for Electron-Molecule Collisions* (W. M. Huo and F. A. Gianturco, Eds.), (Plenum Press, New York, 1995)
[9] Baccarelli I, Bald I, Gianturco F A, Eillenberger E, Kopyra J 2011, *Phys. Rep.* **508** 1
[10] Mazevet S, Morrison M A, Boydstun O, Nesbet R K 1999 *Phys. Rev. A* **59** 477
[11] Feng H, Sun W G and Morrison M A 2003 *Phys. Rev. A* **68** 062709
[12] Greene C H and Jungen Ch 1985 *Phys. Rev. Lett.* **55** 1066
[13] Houfek K and Čížek M in *Low-Energy Electron Scattering from Molecules, Biomolecules and Surfaces* (P. Čársky and R. Čurík, Eds.), (CRC Press, New York 2012)
[14] Rescigno T N, Isaacs W A, Orel A F, Meyer H D and McCurdy C W 2002 *Phys. Rev. A* **65** 032716
[15] Haxton D J, Zhang Z, Meyer H D, Rescigno T N and McCurdy C W 2012 *Phys. Rev. A* **69** 062714
[16] Čurík R and Čársky P 2003 *J. Phys. B* **36** 2165
[17] Allan M and Andric L 1995 *J. Chem. Phys.* **105** 3559
[18] Čurík R and Gianturco F A 2002 *J. Phys. B* **35** 1235
[19] Joachain C J *Quantum Collision Theory*, pp 577-621, (Amsterdam: Elsevier Science Publishers B.V., 1983)
[20] Perdew J P and Zunger A 1981 *Phys. Rev. B* **23** 5048
[21] Čársky P and Čurík R, in *Computational Chemistry: Reviews of Current Trends* (J.Leszczynski, Ed.), Vol 10, World Scientific (2006)
[22] Itikawa Y 1969 *J. Phys. Soc. Jpn.* **27** 444
[23] Collins L A and Norcross D W 1978 *Phys. Rev. A* **18** 467
[24] Heitler W 1941 *Proc. Cambridge Phil. Soc.* **37** 291
[25] Nishimura H and Tawara H 1991 *J. Phys. B* **24** L363
[26] Szmytkowski C and Kwitnewski S 2002 *J. Phys. B* **35** 2613

[27] Makochekanwa C, Kato H, Hoshino M, Tanaka H, Kubo H, Lopes A R, Lima M A P and Ferreira L G 2006 *J. Chem. Phys.* **124** 024323
[28] R. Čurík R and Gianturco F A 2002 *J. Phys. B* **35** 717
[29] Rinkevicius Z, Tunnel I, Salek P, Vahtras O and Hans A 2003 *J. Chem. Phys.* **119** 34
[30] Sadlej J 1988 *Coll. Czech. Chem. Commun.* **53** 1955
[31] Krishtenko S V Maslov A I and Shevelko V *Molecules and Their Spectroscopic Properties*, (Springer, Berlin 1998)
[32] Milosavjević A R, Guliani A, Šević D, Hubin-Franskin M J and Marinković B P 2006 *Eur. Phys. J. D* **35** 411
[33] Maljkovič J B, Blanco F, Čurík R, García G, Marinković B P and Milosavljević A R 2012 *J. Chem. Phys.* **137** 064312

Measurements of differential cross sections for elastic electron scattering and electronic excitation of silver and lead atoms

S. D. Tošić
Laboratory for Atomic Collision Processes, Institute of Physics, University of Belgrade, Pregrevica 118, 11080 Belgrade, Serbia

E-mail: sanja.tosic@ipb.ac.rs

Abstract. This paper presents absolute differential cross sections (DCSs) and integrated cross sections (ICSs) for both elastic and inelastic electron scattering by lead and silver atoms in the energy range from 10 to 100 eV. The DCSs were measured as a function of scattering angle. Scattering angles are from 1° to 150° for the excitations of the unresolved $4d^{10}5p\ ^2P_{1/2,\ 3/2}$ silver line and $6p7s\ ^3P_{0,1}$ lead line, while for the elastic scattering they span from 10° to 150°. The measurements utilize crossed beam technique with effusive atomic beam being perpendicularly crossed by electron beam. Monoenergetic electron beam is obtained by means of hemispherical selector and it is focused by cylindrical electrostatic lenses while effusive atomic beam is formed by heating of Knudsen type oven. Absolute values for the resonance states are obtained by normalization of relative differential cross sections to the optical oscillator strengths, while the absolute values for the elastic scattering are obtained from the intensity ratios at particular scattering angles. Obtained absolute DCSs were extrapolated to 0° and 180° and numerically integrated to yield integral, momentum transfer and viscosity cross sections. The experimental results have been compared with the corresponding calculations.

1. Introduction

Electron interaction with metal atom vapours provides fundamental information on the structure and collisional dynamics of atomic system, as well as information on basic interaction between particles in scattering process. The main observable in these processes is differential cross section (DCS) which indicates the probability of specific interaction at certain electron energy and scattering angle and it is given by:

$$DCS = \frac{k_f}{k_i}\left|A_{fi}\right|^2 \tag{1}$$

where A_{fi} is a scattering amplitude and k_f and k_i are wave vectors of scattered and incident particles, respectively. Beside the general importance of both experimental and theoretical collisional electron cross section data for understanding and explanation of fundamental electron interaction with isolated metal atoms, there is also a particular relevance of these results for the biomedical radiation research as well as for many plasma diagnostic and modeling techniques which require electron impact cross sections as input data for the calculation of plasma parameters [1, 2]. Besides this, detailed electron-metal atom collision data are very important in astrophysics for spectra's analysis, for chemical composition determination of various astrophysical objects and also for its abundance analysis [3-12].

They are essential for determination of transport properties in astrophysical plasmas of some stellar atmospheres and also for modeling of some metal rich stellar atmospheres (atmospheres of some DZ white dwarfs) [13].

In metal atoms, DCS for resonant electron excitation is more pronounced than DCS for elastic scattering. To determine the absolute DCS experimentally, it is necessary to know several parameters [14]. Generally, experimental investigations are often limited by the high working temperature which is necessary to vaporize the metal sample in order to produce a well collimated effusive atomic beam. At the same time, at these temperatures, metal deposition can make contacts between electron lenses, which can induce difficulties during the measurements.

Here we present experimentally obtained absolute differential cross sections (DCSs) for both elastic and inelastic electron scattering by lead and silver atoms as well as corresponding integrated cross sections. The DCSs were measured at intermediate electron impact energies from 10 eV to 100 eV and scattering angles are from 1° to 150° for the combined excitations of the unresolved $4d^{10}5p\ ^2P_{1/2,\ 3/2}$ silver line and $6p7s\ ^3P_{0,1}$ lead line, while for the elastic scattering they span from 10° to 150°. The experimental results have been compared with the corresponding calculations.

2. Experimental technique and procedure

In the crossed beam arrangement where an electron beam has been perpendicularly crossed by the effusive atomic beam, the scattered electrons have been analyzed in the high resolution electron spectrometer ESMA which has been described in our previous papers dealing with electron scattering by other metal targets [15-20].

A hairpin thermoelectron source was used and electron beam was formed by the electron monochromator which consists of systems of cylindrical electrostatic lenses and hemispherical electrostatic energy selector. Scattered electrons were analyzed by the hemispherical electron energy analyzer which is of the same type as the monochromator. It can rotate around the atomic beam axis so we could analyze and detect scattered electrons in the scattering angle range from -30° to 150°. A channel electron multiplier was used as a detector.

A metal vapor beam was produced by heating an oven crucible containing metal by two separate heaters. In this way we separately heated the top and bottom of the oven and provided a variable temperature difference at approximately 100 K. The working temperature was about 1170 K for Pb and 1300 K for Ag and background pressure was of the order of 10^{-5} Pa. The overall energy resolution (FWHM) was typically 120 meV for Pb and 160 meV for Ag while the angular resolution was 1.5°. The position of true zero scattering angle was determined before each angular distribution measurement by checking the symmetry of the inelastically scattered electrons at negative and positive scattering angles (usually from -10° to +10°). The measured angular distributions were converted into relative DCSs by using the appropriate effective length correction factors [21]. The absolute values of differential cross sections for excitation are obtained through the normalization procedure to the optical oscillator strengths (OOS) for the observed transition. Following the method described by Felfli and Msezane [22], we normalized the relative DCSs by using the forward scattering function (FSF) introduced by Avdonina *et al.* [23]. Due to this procedure which is based on Lassettre limiting theorem and the fact that GOS tends to OOS as momentum transfer squared (K^2) tends to zero, relative DCSs were converted to the generalized oscillator strengths according to formula:

$$GOS(K,E) = \frac{\omega}{2}\frac{k_i}{k_f}K^2 DCS(K,\theta) \tag{2}$$

where ω is the excitation energy, k_i and k_f are the electron momenta before and after the collision, respectively, and the momentum transfer K is defined by:

$$K^2 = 2E\left[2 - \frac{\omega}{E} - 2\sqrt{1 - \frac{\omega}{E}}\cos\theta\right] \tag{3}$$

where E is the impact energy. The obtained GOS values were fitted and extrapolated from the small values of K^2 obtained from zero scattering angle at each electron impact energy (Eq. 3) and normalized to the FSF. These normalization factors were then used to obtain absolute DCS values. FSF procedure was applied at all energies except at 10 eV for electron silver excitation where the necessary condition $E > 2.5\omega$ is not satisfied (ω is the excitation energy of the transition). A more detailed description of the normalization procedure has been described in details in our previous papers [15, 24]. Briefly, in the case of lead we generate FSF using the OOS value of 0.21 for the $6p7s$ $^3P_1$ state ($6p7s$ $^3P_0$ is optically forbidden). In the case of silver, contrary to the case of lead, both unresolved states are optically allowed and have appropriate OOS values so we used the OOS values of 0.452 for the $^2P_{3/2}$ level and 0.223 for the $^2P_{1/2}$ level. The absolute values for the elastic scattering are obtained from the intensity ratios at particular scattering angles.

Obtained absolute DCSs were extrapolated to 0° and 180° using the measured values at small scattering angles and corresponding calculations, relativistic distorted wave calculations (RDW) for excitation processes [15, 16] and optical potential calculations for elastic electron scattering [17, 19]. After extrapolation, numerical integration was applied and integral, momentum transfer and viscosity cross sections were derived using formulas

$$Q_I = 2\pi\int_0^\pi \sigma(\theta)\sin\theta \, d\theta \tag{4}$$

$$Q_M = 2\pi\int_0^\pi \sigma(\theta)\left[1 - \left(1 - \frac{\omega}{E_0}\right)^{1/2}\cos\theta\right]\sin\theta \, d\theta \tag{5}$$

$$Q_V = 2\pi\int_0^\pi \sigma(\theta)\left[1 - \left(1 - \frac{\omega}{E_0}\right)\cos^2\theta\right]\sin\theta \, d\theta \tag{6}$$

where σ is the differential cross section (DCS) and ω is the excitation energy.

3. Results

3.1 Electron scattering by lead atom

Both the elastic electron scattering and electron-impact excitation of the unresolved $6p7s$ $^3P_{0,1}$ levels of Pb atom have been investigated. The measurements at small scattering angles for the resonant $6p^2$ $^3P_0$ $\rightarrow$ $6p7s$ $^3P_1$ transition in Pb are shown in Figure 1. One can see the normalized generalized oscillator strengths (GOSs) determined through normalizations to the optical oscillator strength using the forward scattering function (FSF) method [24]. It is evident that absolute GOS values for zero scattering angles lie on the FSF curve and also that the normalized GOSs decrease with increasing momentum transfer. At 20 eV we have linear dependence of K^2 but this linearity decreases at higher energies.

The absolute DCSs values for the $6p7s$ $^3P_{0,1}$ excitation at 60, 80 and 100 eV electron impact energies are presented in Figure 2 and compared with the relativistic distorted-wave (RDW) calculations using multi-configuration (MCGS) and single-configuration (SCGS) wave functions for the atomic states [16]. The comparison between our experimental data and theoretical results shows that there is excellent agreement in both the shape and magnitude except at 100 eV and scattering angles between

70° and 120° where the experiment gives somewhat smaller values. The good agreement is also evident at small scattering angles (see inset in Figure 2).

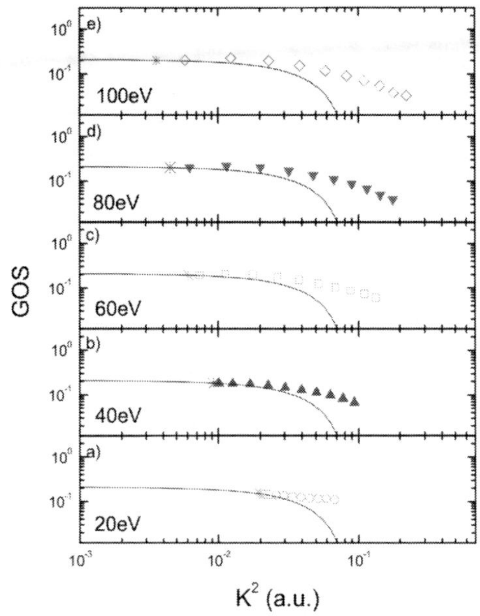

Figure 1. Generalized oscillator strengths (GOS) for the $6p7s\ ^3P_1$ state of lead atoms versus momentum transfer squared (K^2) at 20, 40, 60, 80 and 100 eV electron-impact energies. The stars show the appropriate minimal values of K^2 for zero scattering angle and the solid line represents the forward scattering function (FSF) generated using the optical oscillator strength (OOS) value of 0.21 [24].

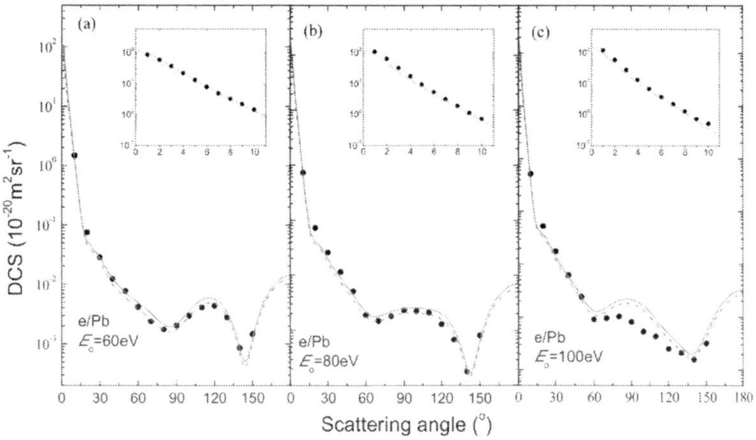

Figure 2. Differential cross sections for the $6p7s\ ^3P_{0,1}$ excitation of lead at (a) 60 eV, (b) 80 eV and (c) 100 eV electron impact-energies. Filled circles with error bars denote the present experimental results. The solid line shows DCSs calculated by the MCGS approximation and the dashed line shows the results obtained using the SCGS approximation [16]. In the insets, the low angle parts of DCS are zoomed out.

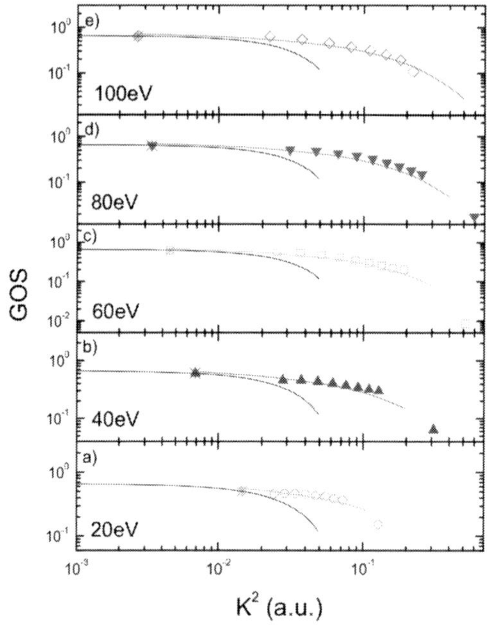

Figure 3. Generalized oscillator strengths (GOS) for the $4d^{10}5p\ ^2P_{1/2,\ 3/2}$ state of silver atoms versus momentum transfer squared (K^2) at 20, 40, 60, 80 and 100 eV electron-impact energies. Labels are the same as for the Fig. 1. The colour solid lines represent corresponding RDW calculations [15].

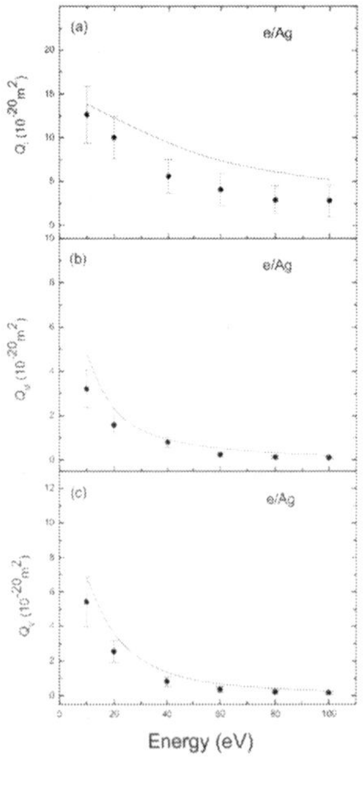

Figure 4. (a) Integral Q_I, (b) momentum transfer Q_M, and (c) viscosity cross sections Q_V for electron impact excitation of the $4d^{10}5p\ ^2P_{1/2,3/2}$ state of silver atom. Filled circles with error bars denote the present experimental results while the solid line shows the combined ICSs calculated by the RDW method.

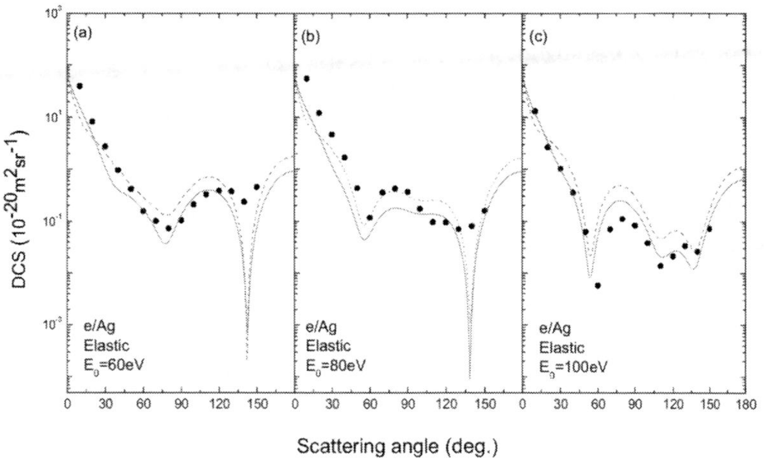

Figure 5. Differential cross sections for the elastic electron scattering by silver atoms at (a) 60 eV, (b) 80 eV and (c) 100 eV electron impact-energies. Filled circles with error bars denote the present experimental results. The dashed line shows the results obtained using the SEPSo calculation while solid line shows DCSs calculated by the SEPASo approximation [19].

3.2 Electron scattering by silver atom

We have also investigated the electron excitation of the first combined resonant $4d^{10}5p$ state of silver atom as well as the elastic electron scattering by this target [15, 19]. The obtained results for generalized oscillator strengths (GOS) are presented in Figure 3 together with corresponding relativistic distorted wave (RDW) calculations [15] where the atomic orbitals and wave functions for both the ground and excited states (fine-structures levels) were calculated using the multiconfiguration Dirac–Fock (MCDF) program [25]. Since both unresolved states of silver are optically allowed and therefore they both have the OOS values, we generated FSF using OOS value of 0.675 which was obtained as a sum of the OOS values of the $^2P_{1/2}$ (0.223) and $^2P_{3/2}$ (0.452) levels of silver [19]. As expected [22], the region of linearity of the GOS values decreases as E increases from threshold. Good agreement between experiment and theoretically obtained GOS results for combined the $4d^{10}5p^2\mathrm{P}_{1/2,3/2}$ excitation using RDW calculations verify that our normalization is reliable. At 10 eV electron impact energy where the necessary condition for normalization defined as $E > 2.5\omega$ is not satisfied, we normalized our experimental results to the calculated RDW data.

In Figure 4 we present results for integrated cross sections. As one can see, both experiment and theory predict decreasing of Q_I, Q_M and Q_V with increasing incident electron energy and there is a good agrrement in shape of ICS curves. At 10 eV and 20 eV theory gives slightly higher ICS values but since the calculated DCSs are higher than the measured ones over the whole angular range this is not surprising.

Figure 5 shows present measured DCSs for elastic electron scattering by silver together with optical potential calculations [19]. There are two sets of calculated results, with (SEPASo) and without (SEPSo) taking into account absorption effects (potential). It is evident that experimental data exhibit the same shape as the theories and that both approximations give very similar results.

4. Conclusion

The elastic and inelastic electron collisions with Pb and Ag atoms have been investigated. We have obtained absolute values for generalized oscillator strengths (GOS), differential cross sections (DCS) and integrated cross sections (ICS) for electron impact excitation of the $4d^{10}5p\ ^2P_{1/2,\,3/2}$ silver line and the $6p7s\ ^3P_{0,1}$ lead line, as well as DCSs and ICSs for elastic electron scattering by these targets. Measurements were performed at electron energies of 10, 20, 40, 60, 80 and 100 eV and at scattering angles up to 150°. The measured results were put on the absolute scale using FSF method. Comparisons with corresponding relativistic distorted wave (RDW) calculations for excitation [15, 16, 24] and optical potential calculations for elastic scattering [17, 19] were made. We found reasonably good agreement between experiment and theory.

It is known that the data about electron metal atom cross sections are of particularly interest for different applications in many fields such as the astrophysics, biomedicine, plasma physics, surface and material science etc. In order to contribute to research in these areas by getting new reliable results, our future work will be concentrated to the electron interaction with other metal atoms.

5. References

[1] Marinković B P, Pejčev V, Filipović D M, Šević D, Milosavljević A R, Milisavljević S, Rabasović M S, Pavlović D and Maljković J B 2007 *Journal of Physics: Conf. Ser.* **86** 012006

[2] Fortov V E , Iakubov I T and Khrapak A G 2007 Chapter Electrical conductivity of partially ionized plasma *Physics of Strongly Coupled Plasma* (Oxford University Press)

[3] Van Eck S, Goriely S, Jorissen A and Plez B 2001 *Nature* **412** 793

[4] Cowan J J, Sneden C, Truran J W and Burris D L 1996 *ApJ* **460** L115

[5] Goswami A, Aoki W, Beers T C, Christlieb N, Norris J E, Ryan S G and Tsangarides S 2006 *MNRAS* **372**, 343

[6] Cardelli J A 1994 *Science* **265**, 209

[7] Smith G 1988 *J. Phys. B: At. Mol. Opt. Phys.* **21** 2827

[8] Dimitrijević M S and Sahal-Brechot S 1999 *Astron. Astrophys. Suppl. Ser.* **140** 191

[9] Liang Gui-Yun, Bian Xia and Zhao Gang 2004 *Chinese Physics* **13** 0891

[10] Holweger H 1972 *Solar. Phys.* **25** 14

[11] Thoren P 2000 *Astron. Astrophys.* **358** L21

[12] Chmielewski Y 2000 *Astron. Astrophys.* **353** 666

[13] Sreckovic V A, Ignjatovic Lj M, Mihajlov A A and Dimitrijevic M S 2010 *Monthly Notices of the Royal Astronomical Society* **406** 590

[14] Rabasović M S, Tošić S D, Pejčev V, Šević D, Filipović D M and Marinković B P 2008 *Facta Universitatis, Series Phys. Chem. Technol.* **6** 119

[15] Tošić S D, Pejčev V, Šević D, McEachran R P, Stauffer A D and Marinković B P 2012 *Nucl. Instr. Meth. B*, **279** 53

[16] Milisavljević S, Rabasović M S, Šević D , Pejčev V, Filipović D M, Lalita Sharma, Rajesh Srivastava, Stauffer A D and Marinković B P 2008 *Phys. Rev. A*, **76** 022714

[17] Tošić S D, Rabasović M S, Šević D, Pejčev V, Filipović D M, Lalita Sharma, Triphati A N, Rajesh Srivastava and Marinković B P 2008 *Phys. Rev. A*, **77** 012725

[18] Rabasović M S, Tošić S D, Šević D, Pejčev V, Filipović D M and Marinković B P 2009 *Nucl. Instr. Meth. B*, **267** 279

[19] Tošić S D, Kelemen V I, Šević D, Pejčev V, Filipović D M, Remeta E Y and Marinković B P 2009 *Nucl. Instr. Meth. B*, **267** 283

[20] Predojević B 2006 CP876 *The Physics of Ionized Gases: 23rd Summer School and International Symposium* ed Lj Hadžijevski, B P Marinković and N Simonović (American Institute of Physics) p 88

[21] Brinkman R T and Trajmar S 1981 *J. Phys. E,* **14** 245
[22] Felfli Z and Msezane A Z 1998 *J. Phys. B: At. Mol. Opt. Phys.* **31** L165
[23] Avdonina N B, Felfli Z and Msezane A Z 1997 *J. Phys. B: At. Mol. Opt. Phys.* **30** 2591
[24] Milisavljević S, Rabasović M S, Šević D, Pejčev V, Filipović D M, Lalita Sharma, Rajesh Srivastava, Stauffer A D and Marinković B P 2008 *Phys. Rev. A*, **75** 052713
[25] Grant I P, McKenzie B J, Norrington P B and Pyper N C 1980 *Comput. Phys. Commun.* **21** 207

Acknowledgments

I am very grateful to Dr B P Marinković, Dr V Pejčev and other colleagues from the Laboratory of Atomic Collision Processes. I am also grateful to Dr R Srivastava, Dr A Stauffer, Dr R Mc Eachran, Dr V Kelemen and Dr E Yu Remeta for performing and sending calculated cross sections results. The work was supported by the Ministry of Education, Science and Technological Development of Republic of Serbia (Project No. 171020) and ESF/COST Action MP1002 Nano-scale insights in ion beam cancer therapy (Nano-IBCT).

Journal of Physics: Conference Series **399** (2012) 012005 doi:10.1088/1742-6596/399/1/012005

Preliminary results from recent experiments and future roadmap to Shock Ignition of Fusion Targets

D Batani[1,6], G Malka[1], G Schurtz[1], X Ribeyre[1], E Lebel[1], L Giuffrida[1], V Tikhonchuk[1], L Volpe[2], A Patria[2], P Koester[3], L Labate[3], L A Gizzi[3], L Antonelli[4], M Richetta[4], J Nejdl[5], M Sawicka[5], D Margarone[5], M Krus[5], E Krousky[5], J Skala[5], R Dudzak[5], A Velyhan[5], J Ullshmied[5], O Renner[6], M Smid[6], O Klimo[6], S Atzeni[7], A Marocchino[7], A Schiavi,[7] C Spindloe[8], T O'Dell[8], T Vinci[9], J Wolowski[10], J Badziak[10], T Pysarcizck[10], M Rosinski[10], Z Kalinowska[10], and T Chodukowski[10]

[1] Université de Bordeaux, CNRS, CELIA, UMR 5107, F-33405 Talence, France
[2] Università di Milano-Bicocca, Pza della Scienza 3, 20126 Milano, Italy
[3] INO-CNR, Pisa, Italy
[4] Università di Roma "Tor Vergata", Roma, Italy
[5] Institute of Physics ASCR /PALS Research Centre, Za Slovankou 3, 18221 Prague 8, Czech Republic
[6] Czech Technical University, Prague, Czech Republic
[7] Università di Roma "La Sapienza", Roma, Italy
[8] Scitech Precision, Rutherford Appleton Laboratory, UK
[9] LULI, Ecole Polytechnique, France
[10] IPPLM, Warsaw, Poland

Abstract. Shock ignition (SI) is a new approach to Inertial Confinement Fusion (ICF) based on decoupling the compression and ignition phase. The last one relies on launching a strong shock through a high intensity laser spike ($\leq 10^{16}$ W/cm^2) at the end of compression. In this paper, first we described an experiment performed using the PALS iodine laser to study laser-target coupling and laser-plasma interaction in an intensity regime relevant for SI. A first beam with wavelength $\lambda = 1.33$ μm and low intensity was used to create an extended preformed plasma, and a second one with $\lambda = 0.44$ μm to create a strong shock. Several diagnostics characterized the preformed plasma and the interaction of the main pulse. Pressure up to 90 Mbar was inferred. In the last paper of the paper, we discuss the relevant steps, which can be followed in order to approach the demonstration of SI on laser facilities like LMJ.

1. Introduction

Shock ignition (SI) [1-5] is a novel approach to Inertial Confinement Fusion (ICF) [6-8], based on the separation of the compression and ignition phases The first phase implies compression of a thermonuclear DT pellet by ns laser beams at $I < 10^{15}$ W/cm^2. The second relies on a laser pulse

[6] To whom any correspondence should be addressed. E-mail: batani@celia.u-bordeaux1.fr

with intensities I ≈ 10^{15} - 10^{16} W/cm^2, driving a very strong shock (P ≈ several 100 Mbar), by, which generates the hot spot required for ignition. SI has potentials for high gain and could also allow achieving ignition using relatively small laser energy for compression (a few hundred kJ).

Several experiments have shown that compression to a regime of interest for ICF is possible [9-12] Ignition by the standard approach of creating a central hot spot by hydrodynamic cumulation at the end of the implosion [1] is the objective of the current National Ignition Campaign [12] at the National Ignition Facility, NIF [13]. It requires, *inter-alia*, the achievement of high implosion velocity (about 360 km/s for the NIF baseline target). Separation of compression and ignition allows lower implosion velocity [1], thus reducing risks associated to hydrodynamic instabilities. Such a separation is also common to fast ignition [14,15]. However, SI has the advantage that it is compatible with present-day "NIF like" laser technology [1]. Also, it does not rely on such an extreme physics as fast ignition (i.e. generation of an intense beam of relativistic electrons, its propagation in dense matter and deposition of energy in the compressed core). Preliminary SI experiments in spherical geometry [16,17] are encouraging and demonstration experiments are possible within the next decade on NIF or LMJ [18].

The physics related to SI is also largely unexplored. Laser-plasma interactions at intensities above 10^{14} W/cm^2 are characterized by non-linearity. Strong parametric instabilities (SBS: Stimulated Brillouin Scattering, SRS: Stimulated Raman Scattering, TPD: Two-Plasmon Decay) may arise with the unwanted effect of reflecting a substantial amount of incident laser light and generating fast electrons These may preheat the fuel of ICF targets, making compression more difficult. Also, laser beam filamentation may produce strong inhomogeneities, which alter the uniformity of target compression and enhance parametric instabilities. We notice that the generation of fast electrons, provided they are not too energetic, may not be dangerous in SI, since they may not be able to penetrate the large areal density <ρr> achieved at late stages of compression.

While, of course, SI "demonstration" experiments need spherical geometry, many related issues can be addressed using planar targets. Indeed, 1D planar geometry offers the advantage of a simpler scheme and easier access of diagnostics. In this context, some results have been reported in [19], for intensities ≤ 10^{15} W/cm^2.

In this paper, first we report results of an experiment in the intensity range up 10^{16}W/cm^2, the upper limit that is considered for SI. The experiment was performed using two beams of the Prague Asterix Laser System, PALS [20]. The goals of the experiment were to study: 1) the coupling of the high-intensity beam to the payload through an extended plasma corona, and the generation of a strong shock, 2) the effect of laser-plasma instabilities at I ≈ 10^{16} W/cm^2, and the amount of light which is reflected, and 3) the generation of hot electrons and their impact on laser-payload coupling.

In the last paper of the paper, we discuss the relevant steps, which can be followed in order to approach the demonstration of SI on laser facilities like LMJ.

2. Experiment at PALS

The PALS Iodine Laser delivers pulses at the fundamental wavelength l = 1.3 μm, with duration τ = 300 ps [20]. In the experiment we used two pulses, namely an auxiliary one delivering about 30 J, and the main one delivering up to 300 J, with time delay of 0 - 1.2 ns with respect to the auxiliary beam (see later). This pulse is preceded (by about 8 ns) by a small prepulse due to leakage in the mode-locker, with relative power about 10^{-3}. The auxiliary beam was used to create an extended plasma. It was operating at the fundamental frequency and focused to an extended spot (diameter ≈ 900 μm, nearly flat top, intensity I ≈ 10^{13} W/cm^2) so to create an approximately 1D expanding plasma. The beam was smoothed using a random phase plate (RPP) to produce a uniform irradiation. The fundamental wavelength was used in order to generate a plasma with relatively high temperature at the critical surface, despite the low intensity, thereby allowing better approaching the conditions of the plasma corona in SI. The main beam was used in the first phase of the experiment, to create the XRL beam for diagnostics [21]. In the second phase, it was converted to 3w (l =438 nm and E ≤ 250 J) and focused with a lens of diameter 30 cm and focal length f = 60 cm to a spot of diameter about 100 μm and intensity I = 10^{16} W/cm^2 to create a strong shock.

In most of the shots we used a target with a 25-μm plastic layer (parylene-C; C_8H_7Cl, with Cl to allow for X-ray spectroscopy) on the laser side, and 25 μm Al on the rear. The low-Z material on the front mimicked the typical ICF ablator material. However, in some targets we used an additional 10 μm Al step on the back and an intermediate Cu layer (5 μm thick) between plastics and Al. These respectively allowed for shock chronometry (Al being a standard material for this kind of measurements [22]) and for hot electrons characterization. To the same goal, we used targets with two layers Ti (20 μm) / Cu (20 μm). Also, we used targets with different plastic thickness in front of the Cu layer: these allowed estimating the average energy of hot electrons by looking at the signal reduction vs. overcoat thickness. Finally, thicker targets were used for crater measurements.

Several diagnostics were used. In phase 1, the extended plasma was characterized by X-ray deflectometry, using the PALS X-ray laser to get density profiles [16,17]. In phase 2, addressing intense interaction and shock generation, we used: i) X-ray CCD working in a single-hit mode to measure plasma extension and characterize its emission [18, 19], ii) ion collectors for cross-checking of plasma temperature [20,21], iii) shock chronometry (measuring the self emission from target rear side with a streak camera) [22], iv) optical spectroscopy, and calorimetry of the radiation reflected within the cone of the focusing lens (f/2), to evaluate the onset and amount of back reflected light from parametric instabilities, v) a set of mini-calorimeters to measure the light scattered outside the lens cone, and vi) Kα imaging to evaluate hot electron generation [23].

In addition, optical interferometry was used in both phases [24] because, unlike X-ray deflectometry, it did not require using the main beam, even if of course the accessible electron density was smaller. Finally post-mortem analysis of the crater size in thick targets was used as a diagnostics of the total deposited energy. X-ray pin-hole cameras, were used in both phases to obtain the transversal size of preformed plasma. Finally, X-ray spectroscopy was also used in both phases to obtain plasma temperature [25]. The X–ray spectra were measured using a spherically–bent mica crystal spectrometer, which was aligned to provide a spatial resolution along the laser axis. In the 4th order, it covered the wavelengths 4.17 ÷ 4.52 Å, to image the chlorine Heα and Lyα lines. The Heγ ÷ Heη and Lyβ lines were observed in the 5th order.

The copper Kα emission was measured using a spherically–bent quartz (211) crystal, which was set up as a monochromator in imaging mode (Bragg angle θ = 88.7°) to provide a quasi–monochromatic distribution of Kα intensity, 2D-spatially resolved along the target surface. Another X-ray keV spectrometer used a flat ADP (ammonium di-hydrogen phosphate) crystal with 2d ≈ 10.659 Å placed at ≈ 20 cm from the source and with a Bragg angle ≈ 19°. It recorded X-ray spectra with an observed range from 2600 to 3600 eV. We detected X-ray lines from Li-, He- and H-like Cl ions, however the spectral resolution was poor due to large source size (≈ 1mm implying a spectral resolution $\Delta E \approx 40$ eV) [26]. The 3 diagnostics used the Kodak AA400 film to detect the signal.

2D plasma density profiles were obtained with XRL deflectometry, a technique based on the deformation of Talbot pattern of 2D grating caused by gradients of index of refraction (plasma electron density), assuming cylindrical symmetry and using Abel transform. The diagnostics is described in [27]. A Ne-like zinc X-ray laser was used, emitting at 21.2 nm, operated in single pass providing 150-ps pulses of 200 μJ. A Mo-Si multilayered spherical mirror with f = 250mm was used to image the plasma on back-illuminated X-ray CCD with magnification M = 8.2. Fig. 1 shows the obtained 2D density profiles 0.3 ns and 0.9 ns after the arrival driving pulse.

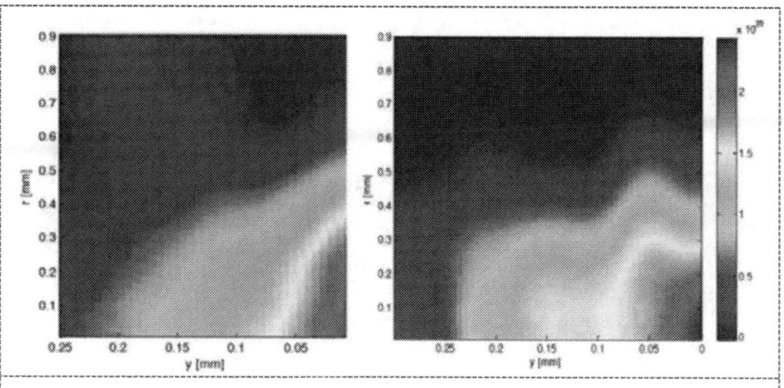

Figure 1. Experimental results from XRL deflectometry: 2D density electron density profiles (in cm^{-3}) at 0.3 ns (left) and 0.9 ns (right) after the arrival driving pulse.

The plasma with $n_e > 10^{20}$ cm^{-3} extends over 200 μm perpendicularly to target surface and over 800 μm radially (comparable to the spot size ≈ 1 mm), in agreement with X-ray PHC images. The density along the axis of the focal spot is characterized by an exponentially decreasing profile. The low-density portion of such a profile superimposes to the one obtained by optical interferometry and it is well reproduced by 1D hydro simulations performed with the code MULTI [28]. The coronal plasma temperature predicted by such simulations is ≈ 500 eV and ≈ 200 eV in the overcritical region. X-ray keV spectra analyzed with the help of the codes RATION [29] and SPECT3D-A [30], yielded similar temperatures.

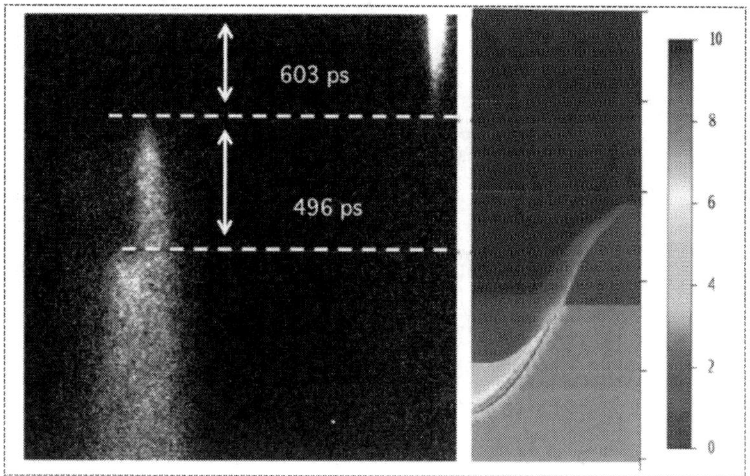

Figure 2. Left) Shock chronometry image for a stepped target with E(3ω) = 245 J, E pre-pulse = 29 J, and delay between auxiliary and main beam 500 ps. Right) Simulations of shock breakout with the code DUED showing the material density in the color scale (Al from 0 to 35 μm, plastics 35 to 55 μm).

The main pulse was fired in the preformed plasma with delays Δt = 0, …, 1200 ps. We also made shots without the creation beam (main only). Fig. 2 (a) shows a streak camera image of a shock breakout on a stepped target, while (b) shows a simulation of shock propagation (we performed 2D simulations with the hydro codes DUED [31, 32] and MULTI-2D [33]). Shock pressure is rapidly decreasing partly due to 2D effects (the total target thickness being comparable to the focal spot radius) but mainly, due to the short duration of the laser pulse (a relaxation wave is generated on the front side at the end of the pulse and rapidly catches up with the shock front). Data analysis and simulations indicate that the peak shock pressure is ≈ 90 MBar on front side during pulse interaction, and decreases to less than 10 MBar at shock breakout.

Results from the X-ray pin-hole camera allowed confirming the focalization of the main laser beam to 100 μm. Also, by measuring the energy deposited in each pixel, we could obtain emission spectra and get monochromatic images. Spectra obtained with CH/Cu/Al (and Ti/Cu) showed clear K-a lines from Cu (or Ti and Cu), confirming the presence of hot electrons. The penetration depth in CH was estimated by comparing results with different plastic thickness as ≈ 27 μm. By using the online database ESTAR of NIST (with the table of Mylar which as a similar density to the plastic used in our experiment), this corresponds to an average hot electron energy ≈ 50 keV, in agreement with predictions from scaling laws [34] and available data [35]. Data obtained from Ti/Cu targets are also compatible with this result. The Kα imager showed images with size ≈ 200 μm, larger than the focal spot but compatible with the hot electron range (taking into account that at such low energies, hot electron propagation is practically isotropic). By measuring the total flux of Ka photons (with the CCD and the Kα imager) we could estimate the total number of hot electrons, finding a conversion efficiency from laser to hot electrons of the order of 1%, in agreement with what estimated in [36].

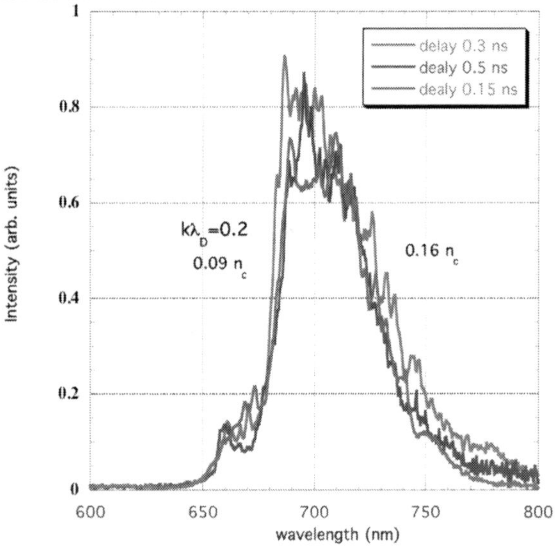

Figure 3. Backreflected Raman spectra at different prepulse delays

Finally, the analysis of the backscattered light showed a little amount of reflection within the focusing lens cone (< 5% in all cases) and mainly dominated by SBS. The scattered light outside the lens cone was of the same order of that within the lens cone, bringing the total reflectivity to ≤ 10%. We also observed a substantial stability of SRS spectra vs. pulse delay Δt (see Fig.3). SRS spectra are characterized by a Landau cut-off at short wavelengths (λ ≈ 670 nm), while at long wavelengths extend up to λ ≈ 720 nm (value at 0.5 of the maximum). No sign of SRS generation at ≈ $n_c/4$ is present. This is probably the signature of strong delocalized absorption in the extended plasma corona. Alternatively, such low density could be the sign of cavitation taking place in the plasma, as shown in [37].

Our experimental results show that during interaction we can obtain shock pressure P ≈ 90 Mbar. This is indeed the highest pressure measured so far in this kind of experiment.

3. Preliminary Roadmap to Shock Ignition

Shock Ignition has been selected as the main approach to demonstration of ignition via ICF by the HiPER Project [38]. The goals of the HiPER Project are to study the follow up to NIF ignition and prepare the way to future Fusion reactors. The preparatory phase of the project covers the period 2009-2013 and it is mainly funded by the EU, UK, France, and the Czech Republic. The work includes both scientific and "more technical" issues like: i) Study of high-energy high-repetition laser drivers, ii) Study of target mass production, injection, tracking and positioning at high repetition frequency, iii) Studies on chamber design, material resistance, material activation at high radiation fluxes, iv) Study of advanced ignition concepts.

Shock Ignition has been selected as the main route to IFE being promising and, unlike fast ignition, compatible with present day's laser technology, allowing demonstration on LMJ within the next decade

The Laser Megajoule (LMJ) in France is a Nd:glass facility, becoming operational in 2015, which will finally deliver 2 MJ of laser energy in ≈ 10 ns using 160 beams (see Fig. 4). It has been constructed in the framework of the French Programme for the Stewardship of Nuclear Stockpiles but thanks to the agreement between region Aquitaine and CEA, ≥ 20% of shots will be allocated for civilian academic research oriented towards fusion for energy.

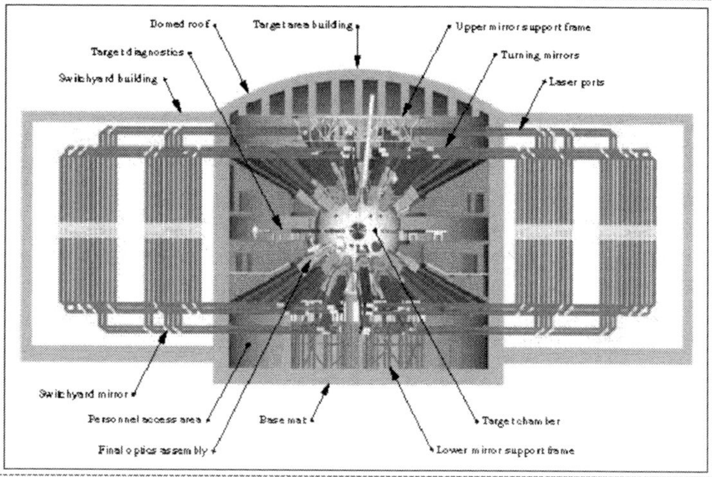

Figure 4. In-principle scheme of LMJ laser

In order to realize the demonstration of feasibility of shock ignition on LMJ, one must first be able to compress a target using the Polar Direct drive options. Indeed LMJ, like NIF, has been constructed to work with indirect drive, which means that the beams are not uniformly distributed over the whole surface of the spherical targets, but they come in two poles (corresponding to the entrance holes of the laser holhraum). In order to perform uniform direct drive, one must repoint the laser beams in order to cover the whole target surface. After that, the energy of the beam must be balanced: the beams, which have been repointed, will be incident at some angle on the target surface, thereby having a larger focal spot and a lower intensity. Therefore the energy of the beams, which have not been repointed, needs to be lowered in order to get uniform laser intensity everywhere. When this is done, it turns out that about only 0.5 MJ of the initially available 2MJ can be used. Luckily, such energy would still be sufficient for performing a direct-drive implosion.

The next step would then be performing shock ignition demonstration experiments. This requires dedicating a part of the beams of LMJ to drive the final laser spike.

For instance a 40 quads pattern could split the quads, defocus and repoint the beams with 80 beams used for compression (PDD) and the generation of the spike, and 80 beams for spike only.

Before getting o the demonstration of PDD and shock ignition on LMJ, several issues can and need to be investigated on intermediate facilities. These include physical issues (Capability of generating 300 Mbar shocks, Absorption at oblique incidence, Impact of parametric instabilities on laser absorption and hot electron generation, impact of hot electrons ion shock formation, smoothing, timing and dynamics of shock collision, shock isotropization and smoothing by electron conduction). The physical results from all such points should also be used to validate code platforms in order to get predictive capability

Finally, also some laser issues need to be investigate, for instance LMJ beams are focused on target using gratings (unlike NIF). In principle, by putting some programmed chirp on the laser beam, this should allow to do a dynamic pointing of the laser beams in order to follow the laser implosion in "real time".

In Europe there is a possibility to use a whole Laser "Rainbow" to study such topics, moving from the less big lasers (LULI, Vulcan, PALS, Phelix) to larger ones (Orion, LIL) and finally the LMJ/PETAL laser facility. This offers a unique opportunity for European Science and for ICF studies.

Conclusions

Shock ignition appears to be a promising approach to demonstration of nuclear fusion ignition by laser direct drive. In order to contribute to investigate the physical unknowns of shock ignition we have performed an experiment at PALS. Our experiment has shown that we can couple a laser beam to a payload and generate a rather strong shock (90 MBar) even in presence of an extended plasma corona. This is indeed the 1 highest pressure measured so far in this kind of experiments, showing a clear progress in approaching a shock ignition relevant regime.

Unlike in experiments performed in spherical geometry [13,14], we measured little back-reflection due to parametric instabilities. Although this sounds as good news for SI, nevertheless one should be cautious, since the growth rate and saturation of parametric instabilities critically depends on plasma conditions, which in our experiment are quite different from a real SI experiment in spherical geometry, especially concerning plasma extension and plasma temperature. Clearly a more detailed analysis is required here. Our results also point out to the need of improving the modelization of energy transport in SI experiments, for instance taking fully into account the effect of parametric instabilities, filamentation, hot electrons generation, magnetic fields, etc.

The final part of the paper was devoted to briefly explain the steps (physical roadmap), which are needed in order to progress from present physical experiment, to the future demonstration of Polar Direct Drive and Shock Ignition on the LMJ/PETAL laser facility.

Acknowledgments
We acknowledge the support of HiPER project and Preparatory Phase Funding Agencies (EC, MSMT and STFC). The work was also supported by the European Union under the "LaserLab" program of the 7th FP, and by the Czech Ministry of Education, Youth and Sports, projects LC528 and LM2010014. Scitech Precision and Rutherford Appleton Laboratory Target Fabrication Group supplied targets. Finally, we thank the technical staff of PALS for help in running the experiments.

References

[1] L.J.Perkins, et al., Phys. Rev. Lett. **103**, 045004 (2009)
[2] R. Betti, et al., Phys. Rev. Lett. **98**, 155001 (2007)
[3] X Ribeyre, et al., Plasma Phys. Control. Fusion **51**, 015013 (2009)

Journal of Physics: Conference Series **399** (2012) 012005 doi:10.1088/1742-6596/399/1/012005

[4] S.Atzeni, et al., Plasma Phys. Control. Fusion **53**, 035010 (2011)
[5] S. Atzeni and G. Schurtz, Proc. SPIE **8080**, 808022 (2011)
[6] J.Nuckolls, et al., Nature **239**, 139 (1972)
[7] J. Lindl, Phys. Plasmas **2**, 3933 (1995)
[8] S.Atzeni and J.Meyer-ter-Vehn, The Physics of Inertial Fusion, Clarendon-Oxford (2004)
[9] H. Azechi, et al. Las. Particle Beams, **9**, 193 (1991)
[10] T. C. Sangster, et al., Phys. Plasmas **17**, 056312 (2010)
[11] C. D. Zhou, et al., Phys. Rev. Lett. **98**, 025004 (2007)
[12] N. B. Meezan, L. J. Atherton, et al. «National Ignition Campaign Hohlraum energetics» Phys. Plasmas **17**, 056304 (2010)
[13] E.I. Moses, C.R Wuest «The National Ignition Facility: Laser Performance and First Experiments» Fusion Science and Technology, **47**, 314 (2005)
[14] M.Tabak, et al., Phys. Plasmas **15**, 056306 (2008)
[15] W. Theobald, et al., Phys. Plasmas **18**, 056305 (2011)
[16] W. Theobald, et al., Phys. Plasmas **15**, 056306 (2008)
[17] P.B.Radha, et al., IEEE Trans. Plasma Science, **39**, 1007 (2011)
[18] N. Fleurot, C. Cavailler, J.L. Bourgade «The LaserMégajoule (LMJ) Project dedicated to inertial confinement fusion: Development and construction status » Fusion Engineering and Design, 74, 147 (2005)
[19] S. D. Baton, M. Koenig, et al. "Experiment in Planar Geometry for Shock Ignition Studies" Phys. Rev. Lett. 108, 195002 (2012)
[20] K. Jungwirth, et al., Phys. Plasmas **8**, 2495 (2001)
[21] B. Rus, et al., Phys. Rev. A **66**, 063806 (2002)
[22] M. Koenig, et al., Phys. Rev. Lett., **74**, 2260 (1995)
[23] A. Morace, D. Batani "Spherically bent crystal for X-ray imaging of laser produced plasmas" Nuclear Instruments and Methods in Physics Research A, 623:22, 797- 800, Elsevier BV, 11 (2010)
[24] A. Kasperczuk, T. Pisarczyk, et al. «Plasma jet generation by flyer disk collision with massive target» Optica Applicata, Vol. XXXVII, No. 1–2, 2007
[25] M. Smıd, L. Antonelli, O. Renner «X–ray spectroscopic characterization of shock–ignition–relevant plasmas» Acta Polytechnica in press (2012)
[26] L. Antonelli, et al., Acta Tecnica, **56**, T57 (2011)
[27] J. Nejdl, et al. Proc. SPIE **7451**, 745117 (2009)
[26] R. Ramis, et al., Comput. Phys. Commun. **49**, 475 (1988)
[27] W. Lee, User Manual for RATION (Lawrence Livermore National Laboratory, 1990)
[30] J.J.MacFarlane, et.al. SPECT3D-A Multi- Dimensional Collisional-Radiative Code for Generating Diagnostic Signatures Based on Hydrodynamics and PIC Simulation Output. High Energy Density Physics, 3:181, 2007
[31] S.Atzeni, Computer Phys. Commun. **143**, 107 (1986)
[32] S. Atzeni, et al., Computer Phys. Commun. **169**, 153 (2005)
[33] R. Ramis, et al.,Computer Phys. Commun. **180**, 977 (2009)
[34] Estar database: http://physics.nist.gov/ Phys- RefData/Star/Text/ESTAR.html
[35] F.N. Beg, et al., Phys. Plasmas **4**, 2 (1997)
[36] Guskov
[37] O Klimo, et al., Plasma Phys. Control. Fusion **52**, 055013 (2010)
[38] High Power Laser Energy Research Facility, http://www.hiper-laser.org/index.asp

Journal of Physics: Conference Series **399** (2012) 012006 doi:10.1088/1742-6596/399/1/012006

Properties of plasmas produced by short double pulse laser ablation of metals

J Hermann[1], L Mercadier[1], E Axente[1,2], S Noël[1]

[1] LP3, CNRS - Aix-Marseille University, Marseille, France
[2] Lasers Department, INFLPR, PO Box MG 36, 77125 Magurele, Ilfov, Romania

E-mail: hermann@lp3.univ-mrs.fr

Abstract. We investigate the composition of plasmas produced by laser ablation of metals with two time-delayed short laser pulses using fast imaging and time- and space-resolved optical emission spectroscopy. The ablated material is deposited on mica substrates and analyzed by atomic force microscopy. The laser-produced craters are inspected by optical microscopy to evaluate the ablated material quantity. It is shown that the fraction of nanoparticles in the ablation plume is strongly altered when a second laser pulse of sufficiently large delay is applied. Comparing the results obtained for different metals, we observe a significant nanoparticle reduction for interpulse delays of the order of the characteristic time of electron-lattice thermalization. More detailed analyses show that the plume changes occur on two different characteristic times, indicating two different mechanisms at its origin. Here, we discuss the involved processes and we propose a simple and efficient technique for the measurement of electron-lattice thermalization times based on plume observations during double pulse laser ablation.

1. Introduction

Pulsed laser ablation of solid materials with intense short pulses has attracted growing interest in the last decade, since reliable pico- and femtosecond laser systems are available. With respect to nanosecond laser pulses, the interaction of short laser pulses with condensed matter is significantly different as the pulse duration is shorter than the characteristic time of energy relaxation [1, 2]. Thus, ablation occurs after the laser irradiation if short pulses are employed whereas nanosecond laser pulses interact with the plume [3-5]. The heating of the plume in the nanosecond regime leads to the generation of a strong Bremsstrahlung continuum that is undesired in analytic applications by laser-induced breakdown spectroscopy (LIBS) [6, 7]. In addition, the re-heated plume exerts a pressure on the melted layer and thus generates microscopic droplets that present the main drawback in pulsed laser deposition (PLD) of thin films using nanosecond laser sources [8]. The main advantages of material processing by short laser pulses are due to the reduction of the heat-affected zone and multiphoton absorption induced by the high laser intensity. They ensure an increased precision for micro- and nanostructuring of all materials [9, 10]. According to the minimization of thermal effects and in particular the reduction of the melting depth, short pulse laser ablation is expected to be stoechiometric and therefore promising for material analysis by LIBS [11].

Recent experimental and theoretical investigations show that small particles in the nanometer size-range present a large fraction of the ablated material [12-14]. This particularity of short pulse laser ablation was attributed to the fast heating and energy relaxation. Recent theoretical investigations [15,

16] propose two principal mechanisms for the nanoparticle generation during the quasi-adiabatic expansion of the plume. One consists in the transformation of matter into a liquid-gas mixture that favours the nanoparticle generation. This mechanism is so-called *"phase explosion"* or *"thermal decomposition"* [17, 18]. The *"mechanical decomposition"* of the metastable melt due to shock- and rarefaction waves was foreseen to present another source of efficient nanoparticle generation [19]. The particular role of pressure relaxation was stressed by Chimier and co-workers [20] and Norman *et al.* [21]. Nanoparticle formation by condensation is supposed to play a minor role during short pulse ablation under vacuum [22].

The precise theoretical description of short pulse laser ablation and nanoparticle generation is difficult as the knowledge of material properties is incomplete in the extreme non-equilibrium conditions that are characteristic of short pulse laser heating. Two main approaches are applied for theoretical modelling. Both are based on the two-temperature (TTM) model [23] that takes into account that electrons and lattice have different temperatures under material excitation by short laser pulses. The combination with a hydrodynamic model was employed in several numerical studies [12, 16, 24, 25] whereas the microscopic description using molecular dynamics was alternatively used in other numerical investigations [18, 19, 26, 27, 28]. The latter approach benefits from the strong enhancement of computational performance in the past years. The difficulty of the molecular dynamics calculations is due to the lack of knowledge of the interaction potential between the individual particles. In the hydrodynamic approach, the main problem comes from the incomplete knowledge of the equation of state. The improvement of both theoretical approaches requires therefore reliable experimental data characterizing the ablation of materials in the short pulse regime.

In this context, the here presented experimental investigations were mainly motivated by a deeper understanding of short pulse laser ablation and nanoparticle generation from metal targets. A particular interest was focused on the reduction of nanoparticles in the ablation plume. Nanoparticles radiate a blackbody-like continuum [22, 29, 30] and their presence is therefore not suitable for material analysis via LIBS. The elimination of nanoparticles may be also suitable for the deposition of ultra-smooth thin films via PLD [31] or for high-precision micromachining [32].

We have shown in previous papers that the fraction of nanoparticles generated during metal ablation only weakly depends on the laser fluence [14] but is strongly influenced by the strength of electron-lattice coupling [33]. This behaviour was attributed to the competition between the fast electron heat transport and the energy transfer to the lattice that governs the metals heat regime for irradiation with laser pulses shorter than the characteristic time of electron-lattice thermalization τ_{el}. An increase of laser fluence favours the heat transport towards the bulk, increasing thus the ablation depth while the average thermal energy of the ablated material remains almost constant. The influence of the electron-lattice coupling was evidenced by comparing short pulse laser ablation of gold and copper. Indeed, strong coupling in the case of copper limits the heat transport and the ablation depth is accordingly small. The average energy of the ablated material is large as evidenced by the large atomization degree of the plume [33]. Contrarily, the more efficient heat transport in the weakly coupling gold leads to a larger amount of ablated material having lower average thermal energy as evidenced by the large fraction of nanoparticles within the plume.

In the present paper, we investigate the short pulse ablation of metals with two time-delayed femtosecond laser pulses.

Recently, the so-called *"double pulse"* technique became popular, in particular in material analysis by LIBS, where an increased analytical signal and a better signal-to-noise ratio are requested [34]. In this application, mainly performed with nanosecond laser sources, the application of the second delayed laser pulse may have different roles. In most cases, an enhanced excitation due to plasma re-heating by the second pulse is desired [35]. In some cases, the first laser pulse is applied to induce a shock wave in the surrounding air. Thus, the second laser pulse ablates the material under reduced pressure, minimizing thus the contribution of ambient air to the plasma emission [36]. Semerok and Dutouquet [37] ablated copper and aluminium with double pulses of durations ranging from 50 fs to 2 ps. They observed an increase of the plume brightness and a decrease of crater depth for delays in the range from 1 to 10 ps. Both effects were attributed to plasma re-heating by the second laser pulse. Ablating the same metals with laser pulses of 5 ps duration, Le Harzic *et al.* [38] observed a reduced ablation depth and a smoother crater bottom when applying a second laser pulse with a delay of about 10^2 ps. Performing plume analysis during laser ablation of titanium with double pulses of 100-fs duration, Scuderi *et al.* [39] observed that the amount of nanoparticles decreased with the delay between both pulses whereas the number of atoms and ions increased. They attributed the changes of plume composition to the fragmentation of nanoparticles. The double pulse ablation of silicon has been investigated by time-of-flight mass spectrometry [40] and optical emission spectroscopy [41, 42], demonstrating that the ion yield and the plume emission intensity significantly increase for delays in the range of some tens of picoseconds. In addition, the laser-produced crater becomes smoother [40]. These changes were attributed to the alteration of the optical properties that take place when the material melts. With respect to solid silicon, the melt has a much smaller optical penetration depth leading thus to a strongly increased laser energy coupling.

In metals, strong absorption occurs at any temperature due to the large number of free electrons [43]. In that case, as mentioned above, the ablation regime and thus the plume composition are determined by the interplay between electron heat diffusion and electron-lattice coupling [33]. The theoretical description requires thus the knowledge of the electron heat conductivity k_e and the electron-lattice coupling parameter G as functions of electron temperature T_e and lattice temperature T_l. As rough estimations, the electron heat conductivity is given by $k_e = k_0\, T_e / T_l$, whereas G is taken constant. Here, k_0 is the electron heat conductivity at ambient temperature. More precise evaluations of both parameters were recently proposed by Lin *et al.* [44]. In the present paper, we will limit us to qualitative interpretations of the experimental results based on the simple theoretical considerations of the electron-lattice coupling.

2. Experimental arrangement

A schematic presentation of the experimental setup is given in figure 1. The ablation experiments were carried out with a Ti:Sapphire laser (Spectraphysics, model Hurricane) delivering pulses of 100-fs duration at a repetition rate of 1 kHz. The laser was operated at 800-nm wavelength. An aperture of 2-mm diameter was employed to select the centre part of the Gaussian beam which was imaged with an achromatic lens of 50-mm focal length onto the target surface. The two delayed laser pulses were obtained by turning the beam polarization with the aid of a half-wave plate and splitting the laser beam with the aid of a polarized large band prism.

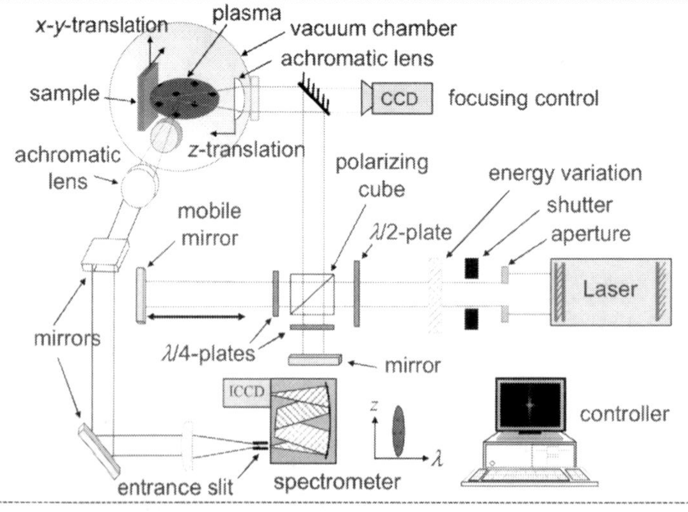

Figure 1. Schematic presentation of the experiment.

Each beam crosses a quarter-wave plate before and after reflection on a mirror at 90° incidence. After recombination by the prism, two laser pulses of equal energy and orthogonal polarization were obtained. The interpulse delay was varied from 0 to 300 ps by adjusting the length of one beam path. According to the total laser pulse energy of 2×25 μJ incident onto the samples surface and a spot diameter of 35 μm, a maximum laser fluence of 2×2 J cm^{-2} was obtained. A mechanical shutter was used to control the number of applied double pulses.

The metal targets of 0.5-mm thickness, 10×10-mm^2 area and 99.95% purity were placed in a vacuum chamber of 10^{-4}-Pa residual pressure. Inside the chamber, target holder and focusing lens were mounted on motorized translation axes. A glass plate was placed between target and lens to prevent the latter from deposition of ablated material. The glass plate was replaced regularly to minimize the laser beam extinction by the deposit.Fast imaging of the ablation plume was performed with the aid of a focusing objective and an intensified charge-coupled device (ICCD) (Andor, model iStar). The time delay between laser pulse and observation gate was adjusted to 400 ns with the aid of a delayed pulse generator. During the experiments, 20 laser shots were applied at maximum to each irradiation site. For measurements that require data acquisition over more than 20 ablation events, several sites were irradiated. The target was translated 50 μm perpendicularly to the laser beam to separate the irradiation sites. The laser-produced craters were inspected by optical microscopy after the ablation experiments. The ablated material was deposited on mica substrates, placed parallel to the target at a distance of 20 mm from the target surface. It was analyzed by atomic force microscopy (AFM) in the semi-contact mode. Several areas of 2×2-μm^2 dimension were scanned to characterize each deposit in order to count a number of particles sufficiently large for statistical analysis of the particle size distribution.

3. Results and discussion

3.1. Influence of interpulse delay on plume composition and ablation depth

In the present section, we evaluate the influence of the interpulse delay on the composition of the laser-produced plasma and on the quantity of ablated material. In addition, the plume excitation temperature is measured for copper ablation using optical emission spectroscopy.

3.1.1. Analysis of plume composition. Two plume images recorded during ablation of copper using a single pulse (a) and two delayed laser pulses (b) are shown in figure 2.

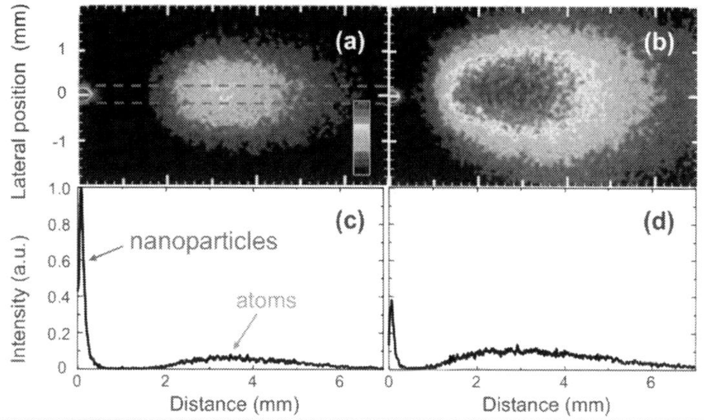

Figure 2. Plume images recorded during ablation of copper with a single pulse of 4–J cm^{-2} laser fluence (a) and with two pulses of 2 J cm^{-2} delayed by 33 ps (b). The intensity profiles (c) and (d) were obtained from the images (a) and (b), respectively. The delay between observation gate and laser pulses was 400 ns.

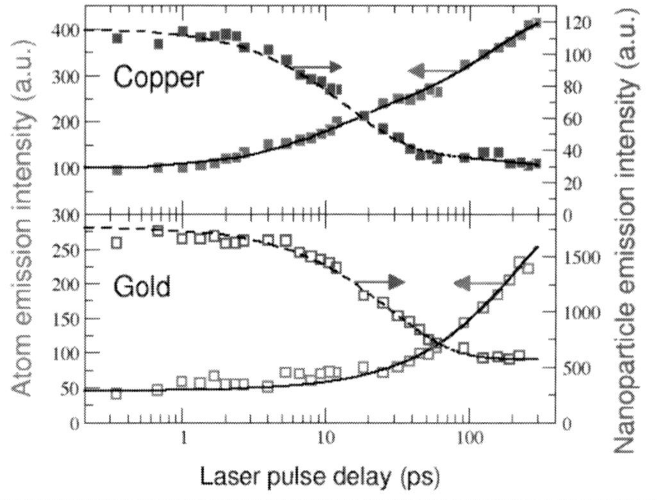

Figure 3. Emission intensities of atoms (red colour) and nanoparticles (blue colour) deduced from plume images (see figure 2) for various delays between two laser pulses of 2 J cm^{-2}. The continuous lines present the intensity increase and decrease according to formula (1), suggesting two characteristic times for the observed plume changes. With respect to ablation of gold, the changes of emission intensity occur at shorter times for ablation of copper.

To enhance the presentation quality, the intensity matrix was smoothened by using an appropriate procedure. Both images reveal the presence of two plume components. A "slow" component of high intensity is located close to the target whereas a "fast" component of lower emission intensity is observed at larger distance. The splitting into two main plume components was observed for short pulse laser ablation of several metals [14, 22, 29, 30]. Analysis by optical emission spectroscopy revealed that the fast component mainly contains neutral atoms, whereas nanoparticles dominate the slow component. The ten-level colour palette is adjusted to the emission intensity of nanoparticles.

With respect to the single-shot experiment (a), a large increase of the emission intensity of the atomized plume component is observed when two delayed pulses were applied (b). At the same time, the nanoparticles plume intensity decreases as shown by the intensity profiles (c) and (d) that were obtained from the plume images (a) and (b), respectively, by averaging the signal over several pixel rows around the plasma symmetry axis as indicated by the dashed lines in figure 2 (a).

The emission intensities of atoms and nanoparticles deduced from the plume images are presented in figure 3 as functions of interpulse delay for copper (a) and gold (b). It is shown that strong changes of the emission intensities of atoms and nanoparticles occur for both metals with increasing interpulse delay. Indeed, the emission intensity of atoms increases by a factor of four whereas the nanoparticle intensity decreases by the same factor. With respect to ablation of gold (b), the intensity changes occur at a shorter delay for double pulse ablation of copper (a). The evolution of the emission intensity changes of both components versus interpulse duration t can be approximated by the following biexponential function [45]

$$I(t) = I_0 + I_1\left(1 - e^{-t/\tau_1}\right) + I_2\left(1 - e^{-t/\tau_2}\right). \quad (1)$$

Here, I_0 is the intensity for an interpulse delay t = 0, I_1 and I_2 are the amplitudes of intensity changes that occur on different time scales, characterized by τ_1 and τ_2, respectively. The continuous and dashed lines present the emission intensities of atoms and the nanoparticles, respectively, calculated with formula (1) by inserting the parameters listed in table 1.

Table 1. Intensity values I_0, I_1 and I_2 and characteristic times τ_1 and τ_2 used to describe the emission intensity evolution of atoms and nanoparticles as a function of interpulse delay using formula (1).

Metal	Plume component	I_0 (a.u.)	I_1 (a.u.)	I_2 (a.u.)	τ_1 (ps)	τ_2 (ps)
Cu	atoms	95	105	240	10	140
Cu	nanoparticles	117	-78	-12	13	200
Au	atoms	45	0	240	-	200
Au	nanoparticules	1740	-1170	0	30	-

The approximation by a biexponential function suggests two characteristic times for the observed plume changes. The shorter time τ_1 scales about 10 ps for copper and 30 ps for gold. These values are close to the characteristic times of electron-lattice relaxation $\tau_{el} \approx 1$–10 ps and 30–100 ps for copper and gold, respectively [46, 47]. The time τ_2 is about one order of magnitude larger than τ_1. It characterizes a slower process that strongly influences the atomic emission intensity increase as shown by the relative large I_2-value. Contrarily, we have $I_2 / I_1 << 1$ for the nanoparticle emission intensity evolution, indicating thus that the slow process has no significant influence on this component of the plume.

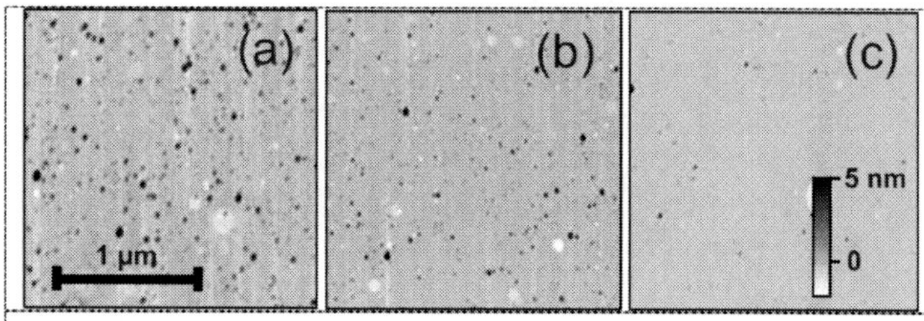

Figure 4. AFM images of copper nanoparticles deposited on mica substrates with two laser pulses of 2 J cm^{-2}, delayed by 0, 6 and 90 ps, respectively for (a), (b) and (c).

The reduction of the amount of nanoparticles with increasing interpulse delay was confirmed by atomic force microscopic analyses of the ablated material deposited on mica substrates. To illustrate the strong decrease of the number of deposited nanoparticles with t, we present in figure 4 AFM images of mica substrates after deposition with 500 laser double pulses and three different interpulse delays. With respect to ablation with $t = 0$ ps (a), the number of nanoparticles is strongly reduced when increasing the interpulse delay to 6 ps (b). A further decrease occurs for long t-values, as illustrated in figure 4(c), where only a few nanoparticles are observed. The number of nanoparticles deduced from the AFM analysis is presented in figure 5 as a function of interpulse delay and compared to the biexponential function [equation (1)] that was used to characterize the nanoparticle emission intensity during copper ablation (see figure 3).

3.1.2. Ablation depth measurements. The ablation depth was measured in the following way: for each interpulse delay, a series of 10 craters was drilled with an increasing number of double pulses as shown in figure 5. The depth of each crater was measured using optical microscopy by focussing on the samples surface and the crater bottom [48]. According to the linear increase of crater depth z with the laser pulse number n_{las}, the ablation depth was deduced from the slope $\Delta z / \Delta n_{las}$.

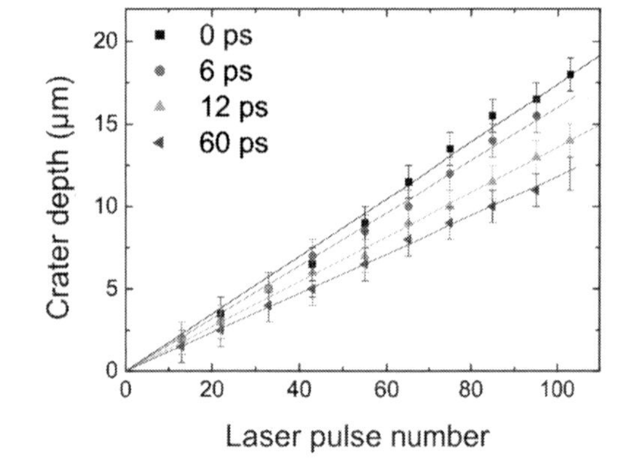

Figure 5. Crater depth as a function of laser pulse number for ablation of copper with interpulse delays of 0, 6, 12 and 60 ps. The ablation depth (per double pulse) was deduced from the slope.

The ablation depth as function of interpulse delay is presented in figure 6 for ablation of copper and gold. For an interpulse delay $t = 0$, the ablation depth equals the value obtained with a single pulse of 4-J cm^{-2} fluence, whereas the depth measured for large delays is slightly smaller than the value corresponding to one pulse of 2-J cm^{-2} fluence. This behaviour is in agreement with the ablation depth measurements performed by Semerok and Dutouquet [36] after laser ablation of copper with short laser pulses of durations ranging from 50 fs to 2 ps. Similar to the characterization of the plume emission intensities, we approximate the ablation depth by a biexponential function by replacing the intensity values I in equation (1) by the appropriate depths z (continuous and dashed lines for copper and gold, respectively). It is noted, that the times τ_1 and τ_2 equal the characteristic times of the plume intensity evolution. The main contribution of the observed ablation depth decrease is attributed to the change of the samples heat regime as we have $z_1 >> z_2$, similar to the evolution of nanoparticle emission intensity. This indicates that nanoparticles present the major part of the ablated mass.

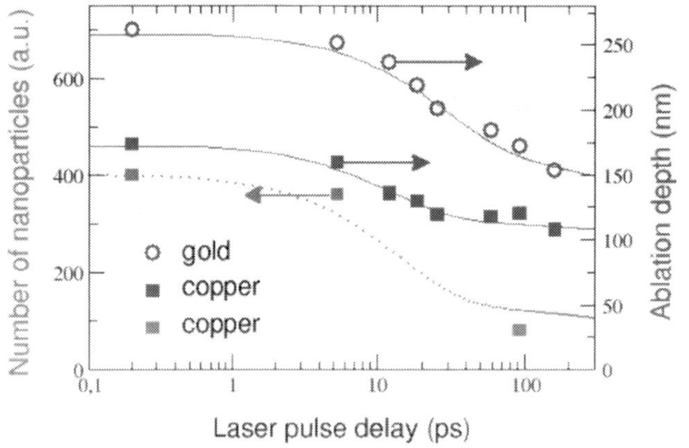

Figure 6. Number of Cu nanoparticles (red colour) deposited on mica substrates with double pulses of 2 × 2 J cm^{-2}. The decrease vs interpulse delay is similar to that of the nanoparticle emission intensity presented by the dotted line (taken from figure 3). The ablation depths of Cu (squares) and Au (circles) are also shown. The continuous and dashed lines were obtained using equation (1) for $\tau_1 = 12$ and 30 ps, respectively, replacing the characteristic intensities by the appropriate depths.

Excitation temperature measurements. The excitation temperature of atoms was determined for double pulse laser ablation of copper from the intensity ratio of the Cu I 510.55 and 515.32 nm transitions having significantly different upper level energies (see table 2). Both lines are observed in the spectra presented in figure 7 for different interpulse delays. It is shown that the Cu I 515.32 nm-intensity significantly increases with t, whereas the Cu I 510.55 nm-intensity is almost constant.

Table 2. Radiative decay A, upper level energy E and statistical weight g of the spectral lines used for temperature measurements.

Transition	A (s^{-1})	E (cm^{-1})	g
Cu I 510.55 nm	2.0×10^6	30784	4
Cu I 515.32 nm	6.0×10^7	49935	4

Assuming a Boltzmann distribution of the excited level population densities, the temperature can be deduced from the intensity ratio of spectral lines emitted from the same species using

$$T = \frac{E_2 - E_1}{k_B}\left[\log\left(\frac{I_1\,\lambda_1\,A_2\,g_2}{I_2\,\lambda_2\,A_1\,g_2}\right)\right]^{-1}. \tag{2}$$

Here, k_B is the Boltzmann constant, λ and I are the resonance wavelength and the measured intensity of the transitions, respectively. The indexes 1 and 2 stand for the transition of larger and smaller upper level energy (see table 2). The derived excitation temperature is presented in figure 8 versus interpulse delay. It is shown that the temperature is almost constant up to an interpulse delay of about 20 ps and a significant increase is only observed for larger t-values. The late temperature increase is in agreement with the observed intensity rise of the atomized plume component (section 3.1.1) that was described by the larger characteristic time τ_2 in the biexponential function [equation (1)]. The delay of the increase supports the assumption that plasma heating due to the interaction of the second laser pulse with the vaporized matter is an additional mechanism responsible of the plume alteration, different from processes that cause the nanoparticle reduction on the timescale of electron lattice relaxation.

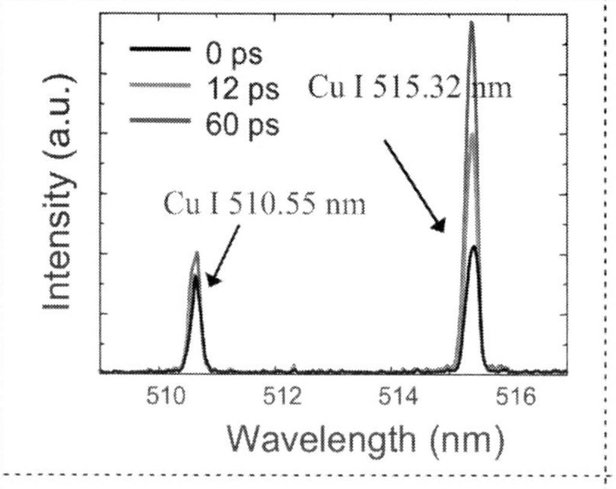

Figure 7. Spectral line emission of Cu I 510.55 and 515.32 nm recorded during laser ablation of Cu for different interpulse delays at a distance of 0.5 mm from the samples surface. The gate was delayed by 60 ns with respect to the laser pulses.

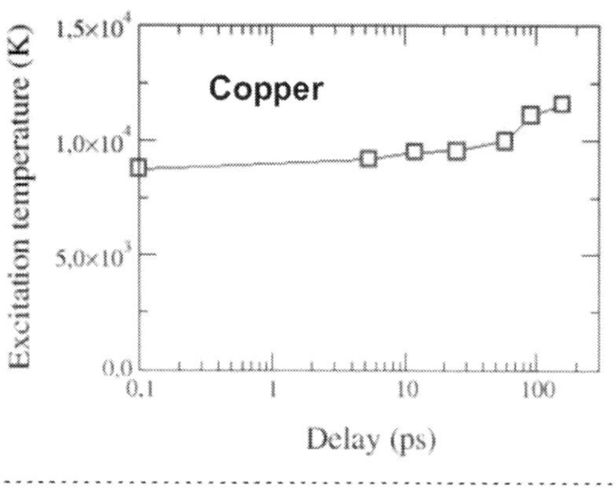

Figure 8. Excitation temperature as a function of interpulse delay deduced from the intensity ratio of the Cu I 510.55 and 515.32 nm transitions. The experimental conditions are identical to those of figure 7.

3.2. Probing electron-lattice coupling via plume observations

In the previous section, we demonstrated that the plume changes may be characterized by two different times. The shorter time equals the characteristic time of electron-lattice thermalization and describes the most significant change of plume atomization. In the present section, we take advantage of the dependence of the plume changes versus interpulse delay in order to deduce the characteristic time of electron-lattice thermalization of transition metals. Thus, we focus our attention on the plume changes that occur on the shorter timescale characterized by τ_l. At this timescale, the emission intensity variation versus interpulse delay is characterized by a monoexponential function $I(t)=I_0+I_1\left(1-e^{-t/\tau_1}\right)$. Here, I_0 is the intensity at $t=0$, I_0+I_l the intensity at $t=\infty$, and τ_l the characteristic time of plume intensity change that equals the electron lattice thermalization time. After the evaluation of I_0 and I_1 and assuming $\tau_{e\text{-}l}=\tau_l$, the characteristic time of electron lattice thermalization can be derived from the linearized function [49]

$$\log[1-(I-I_0)/I_1]=t/\tau_{e-l}\,. \tag{3}$$

Table 3. Atomic mass M, characteristic time electron-lattice relaxation $\tau_{e\text{-}l}$ deduced from the atomic emission intensity changes versus interpulse delay according to equation (3) for double pulses with two different laser fluences.

Metal	M (amu)	$\tau_{e\text{-}l}$ (ps) 2×0.5 J cm^{-1}	$\tau_{e\text{-}l}$ (ps) 2×2 J cm^{-1}
Ti	47.9	50	60
Zr	91.2	90	110
Hf	178.5	200	240

Applying equation (3) to the atomic emission intensities measured for Ti, Zr and Hf, a linear evolution is observed for the three metals with different slopes as shown in figure 9. It is noted that the inferred $\tau_{e\text{-}l}$-values follow the atomic mass ratio 1:2:4 of the three metals (see table 3). The linear increase of $\tau_{e\text{-}l}$ with the atomic mass M is expected. At large electron temperature, the metal is expected to behave like a hot plasma and the electron-lattice relaxation scales as $\tau_{ei}\propto M\,T_e^{3/2}$ [5].

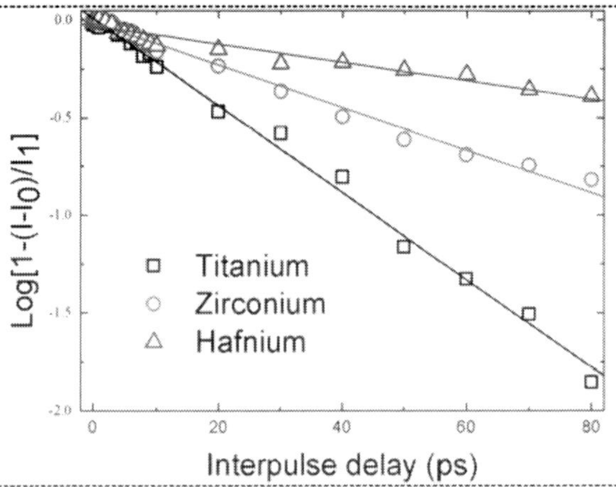

Figure 9. Analysis of the atomic emission intensity dependence versus interpulse according to equation (3) for ablation of Ti, Zr and Hf with double pulses of 2×0.5 J cm^{-2}.

An almost linear dependence of $\tau_{e\text{-}l}$ on M was also predicted by the calculations of Eidmann *et al.* for Cu, Ag, and Au [24].

Beside the $\tau_{e\text{-}l} \propto M$ dependence, we observe a slight influence of the laser fluence on the electron-lattice thermalization. The electron-lattice thermalization times listed in table 3 show that an increase of laser fluence by a factor of 4 (from 2 × 0.5 to 2 × 2 J cm^{-2}) leads to a $\tau_{e\text{-}l}$-augmentation of about 20% only. The formula presented above predicts a strong increase of the relaxation time in $T_e^{3/2}$ that is a consequence of the lowering of the energy exchange rate with increasing electron velocity. We deduce that T_e only weakly increases with laser fluence, and that most additional laser energy leads to an enhanced energy transport into the bulk [14, 33].

A broad spread in the values of electron lattice relaxation time and coupling parameter can be observed in literature. In case of nickel, for instance, G-values ranging from 3 to 10 × 10^{17} W m^{-3}K^{-1} have been published [50, 51]. The data were obtained using different values of laser fluence. Most $\tau_{e\text{-}l}$ and G data were obtained in pump-probe experiments by measuring changes in surface reflectivity [52, 53]. These experiments are limited to relatively weak laser excitation to avoid nonlinear saturation processes occurring in very thin metal films (~ 200 nm) used as targets. Relaxation times ranging from 1 to 4 ps and larger coupling parameters of about 10^{17} m^{-3} K^{-1} were reported for copper [52]. Corkum *et al.* [54], however, studied multipulse laser-induced damage to deduce the surface temperature of bulk metal targets. These authors obtained G-values one order of magnitude smaller than the values reported in references [52] and [55] for the same material. The $\tau_{e\text{-}l}$-values we present in table 3 are therefore consistent with the values reported in literature.

4. Summary and conclusion

The present experimental investigations of metal ablation with two time-delayed short laser pulses demonstrate that the plume composition is strongly altered if the interpulse delay is of the order of the electron-lattice thermalization time. The changes of the emission intensities of atoms and nanoparticles with the interpulse delay reveal two characteristic times, suggesting different mechanisms to be responsible for the plume alteration. The samples heating regime is expected to be responsible for changes on the timescale of electron-lattice thermalization. With increasing lattice temperature, the electron heat conductivity decreases and the laser energy deposited by the second laser pulse is confined in a smaller volume. According to the decrease of the heated depth and the rise of the average temperature of the ablated matter, the atomization degree of the plume increases. Effects of pressure relaxation are also expected to cause plume changes on the timescale of electron-lattice thermalization [20, 21]. For times about one order of magnitude larger, plasma re-heating due to interaction of the second delayed laser pulse with the vaporized material is expected. The observations are supported by measurements of the ablation crater depth as well as AFM analysis of the nanoparticles deposited on mica substrates. The present experiments indicate that the change of the samples heat regime is the dominant mechanisms for the reduction of nanoparticles within the plume, whereas the increase of the atom emission intensity is mainly due to plasma re-heating by the second delayed laser pulse.

In addition, we found for Ti, Zr and Hf that the electron-lattice thermalization time deduced from the plume changes increases linearly with the atomic mass as expected for the electron-ion thermalization in a hot plasma. Furthermore, $\tau_{e\text{-}l}$ is shown to increase slightly with laser fluence as a result of the change of the heat regime. The present investigations demonstrate the possibility of characterizing the electron-lattice coupling of metals by simply observing the ablation plumes produced laser ablation with two delayed short laser pulses. With respect to conventional pump-probe reflectivity measurements, the presented results demonstrate that multipulse laser irradiation can be applied for measurements of the electron-phonon coupling.

Finally, the efficient reduction of the amount of nanoparticles within the plume on one hand, and the increase of the atomic emission intensity on the other hand, improve the performances of material analysis via LIBS. The double-pulse technique may be also promising for other applications such as micromachining or pulsed laser deposition.

Acknowledgments
The authors acknowledge the support from LASERLAB-EUROPE through the CNRS-LP3 001848 project. E.Axente acknowledges the support by the Romanian minister of research through the TE98 84/02.08.2010 project.

References
[1] B.N. Chichkov B N, Momma C, Nolte S, von Alvensleben F and Tünnermann A 1996 *Appl. Phys.* A **63** 109
[2] Bonn M, Denzler D N, Funk S, Wolf M, Wellershoff S S and Hohlfeld J 2000 *Phys. Rev.* B **61** 1101
[3] Von der Linde D and Sokolowski-Tinten K 2000 *Appl. Surf. Sci.* **154–155** 1
[4] Singh R K and Narayan J 1990 *Phys. Rev.* B **41** 8843
[5] Zeldovich Y B and Raizer Yu P 1966 *Physics of Shock Waves and High Temperature Hydrodynamic Phenomena* (London: Academic Press)
[6] Hermann J, Boulmer-Leborgne C and Hong D 1998 *J. Appl. Phys.* **83** 691
[7] Liu H C, Mao X L, Yoo J H and Russo R E 1999 *Spectrochim. Acta Part* B **54** 1607
[8] Chrisey D B and Hubler G K 1994 *Pulsed-Laser Deposition of Thin Films* (New York: Wiley)
[9] Pronko P P, Dutta S K, Squier J, Rudd J V, Du D and Mourou G, 1995 *Opt. Commun.* **114** 106
[10] Nolte S, Momma C, Jacobs H, Tünnermann A, Chichkov B N, Wellegehausen B and Welling H 1997 *J. Opt. Soc. Am.* B **14** 2716
[11] Russo R E, Mao X L, Liu C and Gonzalez J 2004 *J. Anal. At. Spectrom.* **19** 1084
[12] Eliezer S, Eliaz N, Grossman E, Fisher D, Gouzman I, Henis Z, Pecker S, Horovitz Y, Fraenkel M, Maman S and Lereah Y 2004 *Phys. Rev.* B **69**, 144119
[13] Amoruso S, Ausanio G, Bruzzese R, Vitiello M and Wang X 2005 *Phys. Rev.* B **71**, 033406
[14] Noël S, Hermann J and Itina T E 2007 *Appl. Surf. Sci.* **253** 6310
[15] Cheng C and Xu X, 2005 *Phys. Rev.* B **72** 165415
[16] Povarnitsyn M E, Itina T E, Sentis M, Khishchenko K V and Levashov P R 2007 *Phys. Rev.* B **75** 235414
[17] Miotello A and Kelly R 1995 *Appl. Phys. Lett.* **67** 3535
[18] Zhigilei L V 2003 *Appl. Phys.* A **76** 339
[19] Perez D and Lewis L 2002 *Phys. Rev. Lett.* **89** 255504
[20] Chimier B, Tikhonchuk V T and Hallo L 2008 *Appl. Phys. A* **92** 843
[21] Norman G E, Starikov S V and Stegailov V V 2012 *Sov. Phys. JETP* **114** 792
[22] Amoruso S, Bruzzese R, Spinelli N, Velotta R, Vitiello M, Wang X, Ausanio G, Iannotti V and Lanotte L 2004 *Appl. Phys. Lett.* **84** 4502
[23] Anisimov S I, Kapeliovich B L and Perel'man T L 1974 *Sov. Phys. JETP* **39** 375
[24] Eidmann K, Meyer-ter-Vehn J, Schlegel T and Hüller S 2000 *Phys. Rev.* E **62** 1202
[25] Laville S, Vidal F, Johnston T W, Barthélemy O, Chaker M, Le Drogoff B, Margot J and Sabsabi M 2002 *Phys. Rev.* E **66** 066415
[26] Zickfeld K, Garcia M E and Bennemann K H 1999 *Phys. Rev.* B **59** 13422
[27] Itina T E, Gouriet K, Zhigilei L V, Noël S, Hermann J and Sentis M 2007 *Appl. Surf. Sci.* **253** 7656
[28] Ivanov D S and Zhigilei L V 2007 *Phys. Rev. Lett.* **98** 195701
[29] Albert O, Roger S, Glinec Y, Loulergue J C, Etchepare J, Boulmer-Leborgne C, Perrière J and Millon E 2003 *Appl. Phys.* A **76** 319
[30] Grojo D, Hermann J and Perrone A 2005 *J. Appl. Phys.* **97** 063306
[31] Sasaki A, Liu J, Hara W, Akiba S, Saito K, Yodo T and Yoshimoto M 2004 *J. Mat. Research* **19** 2725
[32] Bruneau S, Hermann J, Dumitru G, Sentis M and Axente E 2005 *Appl. Surf. Sci.* **248** 299

[33] Hermann J, Noël S, Itina T E, Axente E and Povarnitsyn M E 2008 *Laser Physics* **18** 374
[34] Babushok V I, DeLucia F C Jr, Gottfried J L, Munson C A and Miziolek A W 2006 *Spectrochim. Acta Part* B **61** 999
[35] Uebbing J, Brust J, Sdorra W, Leis F and Niemax K 1991 *Appl. Spectrosc.* **45** 1419
[36] Cristoforetti G, Legnaioli S, Palleschi V, Salvetti A and Tognoni E 2004 *Spectrochim. Acta Part* B **59** 1907
[37] Semerok A and Dutouquet C 2004 *Thin Solid Films* **453** 501
[38] Le Harzic R, Breitling D, Sommer S, Föhl C, König K, Dausinger F and Audouard E 2005 *Appl. Phys.* A **81** 1121
[39] Scuderi D, Albert O, Moreau D, Pronko P P, Etchepare J 2005 *Appl. Phys. Lett.* **86**, 071502
[40] Spyridaki M, Koudoumas E, Tzanetakis P, Fotakis C, Stoian R, Rosenfeld A and Hertel I V 2003 *Appl. Phys. Lett.* **83** 1474
[41] Hu Z, Singha S, Liu Y and Gordon R J 2007 *Appl. Phys. Lett.* **90** 131910
[42] Schiffern J T, Doerr D W and Alexander D R 2007 *Spectrochim. Acta Part* B **62**, 1412
[43] Prokhorov A M, Konov V I, Ursu I and Mihailescu I N 1990 *Laser Heating of Metals* (Bristol: Adam Hilger)
[44] Lin Z, Zhigilei L V and Celli V 2008 *Phys. Rev.* B **77** 075133
[45] Noël S and Hermann J 2009 *Appl. Phys. Lett.* **94** 053120
[46] Hohlfeld J, Wellershoff S S, Grüdde J, Conrad U, Jänke V and Matthias E 2000 *Chem. Phys.* **251** 237
[47] Schäfer C, Urbassek H M, Zighilei L V 2002 *Phys. Rev.* B **66** 115404
[48] Noël S and Hermann J 2007 *SPIE - ROMOPTO 2006 proceedings* **6785**, 67850F
[49] Axente E, Mihailescu I N, Hermann J, Itina T E 2011 *Appl. Phys. Lett.* **99** 081502
[50] Elsayed-Ali H E, Norris T B, Pessot M A and Mourou G A 1987 *Phys. Rev. Lett.* **58** 1212
[51] Caffrey A P, Hopkins P E, Klopf J M and Norris P M 2005 *Microscale Thermophys. Eng.* **9** 365
[52] Beaurepaire E, Merle J-C, Daunois A and Bigot J-Y 1996 *Phys. Rev. Lett.* **76** 4250
[53] Groeneveld R H M, Sprik R and Lagendijk A 1990 *Phys.Rev.Lett.* **64** 784
[54] Corkum P B, Brunel F, Sherman N K and Srinivasan-Rao T 1988 *Phys.Rev. Lett.* **61** 2886
[55] Allen P B 1987 *Phys. Rev. Lett.* **59** 1460

26th Summer School and International Symposium on the Physics of Ionized Gases (SPIG 2012) IOP Publishing
Journal of Physics: Conference Series **399** (2012) 012007 doi:10.1088/1742-6596/399/1/012007

Application of keV and MeV ion microbeams through tapered glass capillaries

T Ikeda, T M Kojima, T Kobayashi, W Meissl[1], V Mäckel, Y Kanai and Y Yamazaki

Atomic Physics Laboratory, RIKEN, 2-1 Hirosawa, Wako, Saitama, 351-0198, Japan

E-mail: tokihiro@riken.jp

Abstract. We have developed a method to produce micrometer-sized beams of keV energy highly charged ions (HCIs) and MeV energy protons/helium ions with tapered glass capillary optics for the applications of micrometer sized surface modifications and a biological tool, respectively. The transmission experiments of keV HCIs through the glass capillaries show a density enhancement of about 10, beam guiding up to 5°, and the extracted beam keeping the initial charge-state. The combination of MeV ion beams and the capillary with a thin end window at its outlet was used for the irradiation of a part of nucleus of a HeLa cell in culture solution. *Escherichia coli* cells are irradiated by MeV proton microbeam to determine the minimum dose to stop the single flagellar motor. Scanning irradiation of polymer surface by the beam extracted from the capillary in solution containing acrylic acid was found to provide a deposition layer with large affinity with water.

1. Introduction

Ion beam irradiation has been a common technique in not only physics experiments but also other fields as tools for surface modification, material analysis, investigation of radiation effect in biology, and so on. Ions with keV energy can affect only the surface layer of a sample in vacuum, and ions with MeV energy can be extracted to air or liquid. In the case of the keV energy beam, highly charged ions (HCIs) accelerated by 100 V- 10 kV (slow HCIs) are effective for the above purposes. Slow HCI can transfer its large potential energy to only the surface of the sample because it cannot be implanted deeply into the sample. The potential energy is sufficiently large to affect the surface, e.g., that of an Ar^{8+} is about 600 eV. Therefore slow HCIs have high ability to modify surfaces and cause efficient sputtering. Once a micrometer-sized beam (microbeam) is available, these functions specific to slow HCIs can be used to realize, e.g., micro-patterning of modifications and element-sensitive micro-imaging. However, microbeam of slow HCIs is not yet practically available. The use of collimator implies not only the drastic reduction of the beam intensity but also the charge transfer and the energy loss of the ions due to the touching the inner edge of collimating aperture. If the lenses are added, a good emittance of the initial beam is also required. In order to overcome the problems, a different type of optics is needed.

In the case of MeV energy beam, a few MeV protons or He ions are suitable for obtaining the range from 1 μm to 100 μm in liquid, which is almost the same size of a living cell or an organelle in a cell.

[1] Present address: Vienna University of Technology, Wiedner Hauptstrasse 8-10/134, A-1040 Vienna, Austria

The other advantages of this energy region are that the accelerators are relatively smaller than those for higher energy physics and the radiation level in the experimental area is lower. However, it was difficult to deliver the ions with a few MeV to a living cell in culture solution. Therefore, a delivery method of microbeam to a target in a liquid is required.

For the requirements above, we have developed a method to provide microbeams of keV and MeV ions employing tapered insulator tube whose outlet size is in the order of micrometer. When a beam is transmitted through the tapered insulator tube, the outgoing beam size will be the same as the outlet size of the tube. We selected glass as insulating material. In the present article, we denote the tapered glass tube "single tapered glass capillary" or just "glass capillary". In the second section, the single tapered glass capillary is briefly explained. In the third and fourth sections, some experiments using slow HCIs and MeV energy ions are described, respectively.

2. Glass capillary

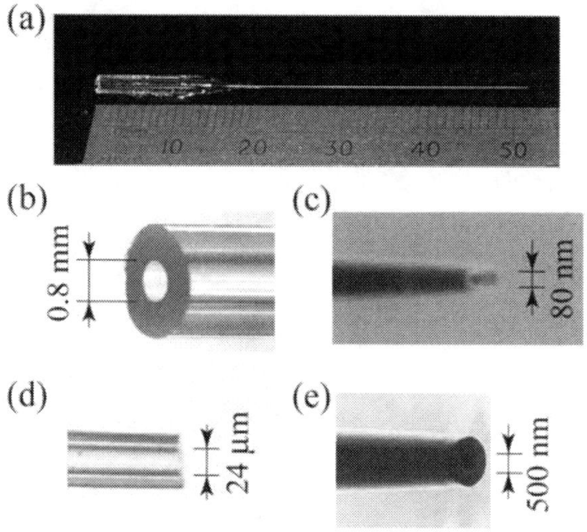

Figure 1. (a) Photograph of the single tapered glass capillary, (b) beam inlet with the outer diameter of 2 mmϕ, (c) original beam outlet, (d) and (e) outlets cut at the inner diameters of 24 μmϕ and 500 nmϕ, respectively.

Glass capillaries are sometimes known as glass pipettes for microinjection or glass electrodes for measuring the changing voltage potential within the neurons in biological and medical fields. A number of glass capillary sizes and several kinds of the glass materials are commercially available for these purposes. The first use of the glass capillary in a beam transmission experiment was reported by Narusawa's group in Kochi University of Technology. They found that a He^+ microbeam extracted from a glass capillary with the outlet size of 0.3 μmϕ kept its initial kinetic energy of 2 MeV [1]. In the case of keV energy ion beams, the transmission through a glass capillary with the outlet size of 24 μmϕ was observed for the first time in RIKEN [2] by the present authors. The glass capillaries we use are made of borosilicate, which is typically composed of SiO_2: 80.9 %, Al_2O_3: 2.3 %, B_2O_3: 12.7 %, Na_2O: 4.0 %, K_2O: 0.04 %. Its density of 2.23, which is almost the same as silica glass, is hard enough to prevent the slow HCI beam from penetrating a capillary wall with a thickness of less than 1 μm under the condition of oblique incidence. Soda lime glass capillaries are also used in our experiments. The material consists of SiO_2: 70%, Na_2O: 11%, CaO: 7.0%, and so on. The softening temperatures are 821°C and 700°C for borosilicate and soda lime glass, respectively. The tapered capillary is prepared by heating a straight glass tube, whose dimension is 2 or 3 mmϕ in outer diameter, around 1 mmϕ in inner diameter and 90 mm in length, and then stretched by pulling both ends with a constant force to have an outlet size of from a few tens μm down to submicron in diameter D_{out} as shown in figure 1. Any size of D_{out} larger than 1 μmϕ (figure 1(d)) is obtained by a device "microforge" and the capillary taper angle is adjusted by heater temperature and pulling force of another device "puller".

Both devices are commercially provided by NARISHIGE, Inc. Figure 1(e) is a photo of an outlet modified by a focused ion beam device.

3. Microbeams of slow HCI with glass capillaries and Teflon tube optics

Transmission of slow HCI beams through a single tapered glass capillary can be explained as follows: The incident ions entering the capillary hit the inner wall and consequently cause it to become charged, although the ions stop at the hitting point. When the accumulated charge becomes large enough to prevent the following incident ions from touching the inner wall, the ions travel more or less parallel to the wall, and then some of the ions can exit from the outlet. Although some entering ions do not transmit, this has the following advantages. (1) Extracted beam keeps its initial charge state and kinetic energy. (2) The taper can enhance the density of the output beam. (3) The size of the output beam is the same as that of the outlet diameter. Considering these points, single tapered glass capillaries are one of the feasible techniques to produce a microbeam of slow HCIs. This technique is based on a self-organized charge up process of the inner glass wall mentioned above, which was experimentally demonstrated by Stolterfoht *et al.* [3] using insulator capillaries with a 3 keV Ne^{7+} beam. In the experiments, polyethylene terephthalate (PET) foils with multi-capillaries of 100 - 200 nmϕ in diameter and 10 μm in length were employed. In order to explain the phenomenon, some model calculations have been carried out [4,5].

3.1. Transmission properties for slow HCIs

Using the guiding process, the first microbeam by a single tapered glass capillary was obtained with 8 keV Ar^{8+} as mentioned in the previous section [2]. Figure 2 shows the number of the transmitted ions as a function of time using an injected beam of 0.2 pA (~1.5×10^5 Ar^{8+} ions/s) with a stability of ~10%, where the vacuum level of the experimental chamber was ~10^{-5} Pa. The capillary with D_{out} = 24 μmϕ was used in this experiment. The number of the transmitted ions grew slowly with a time constant of several tens of seconds (charge-up time), and then became more or less stable for more than 1200 s. However, no transmission was obtained in the first a few seconds as shown in inset. There was a dip around t = 9 s, which has been sometimes observed in experiments. In order to understand the structure, theoretical works have been carried out using the measured shapes of capillaries [6].

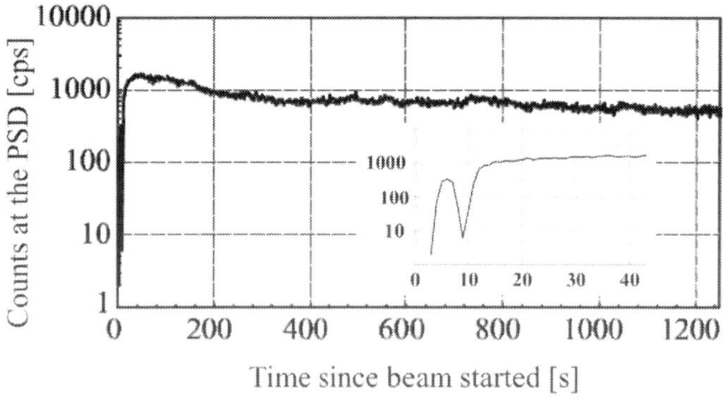

Figure 2. Transmitted ion intensity for 8 keV Ar^{8+} as a function of time. The transmission intensity grew slowly with a time constant of several tens of seconds (charge-up time) as shown in inset, and then became more or less stable for more than 1200 s. The maximum intensity was about 1600 cps for the incident current of 0.2 pA.

The maximum intensity in figure 2 was about 1600 cps (counts per second). The corresponding transmission efficiency was about 1%. The density enhancement factor defined by the ratio of N_{out}/S_{out} to N_{in}/S_{in} is estimated to be ~10, where N_{out} is the number of transmitted ions, N_{in} the number of injected ions into the capillary, and S_{out} and S_{in} are the geometrical outlet and inlet cross sections of the capillary, respectively. The factor more than 1 means that the optics makes the beam density higher (focusing effect). This factor depends on the taper angle, the outlet size and the injected current. The angular divergence of the transmitted beam was estimated to be ~5 mrad. The beam transmission was

examined when the capillary was tilted stepwise by 1° from −5° to +5° (= 87 mrad) relative to the axis of the injected beam of ~0.01 pA keeping the capillary inlet position. The extracted beams were obtained for all the tilting angles by using a position sensitive detector (PSD) downstream of the capillary. This proves that the beam was well guided in the direction of the capillary tilted by as large as 100 mrad (guiding effect). It was noted that the deflection angle is an order of magnitude larger than the half opening angle (~8 mrad) of the tapered capillary when the cross section of the capillary along its axis is assumed to be a trapezoid. And the divergence of the injected beam was a few mrad. The charge-state distribution of the transmitted HCIs through the capillary without tilting was also measured by a combination of the PSD and a deflector between the capillary outlet and the PSD. There was no indication of charge-changed ions on the PSD but some background events, i.e., the ions were transmitted along the capillary without changing the incident charge state. To realize a sub-micron beam, extraction from the outlet of D_{out} = 900 nmϕ was obtained for an Ar^{8+} beam accelerated by 8 kV (corresponding to the energy of 64 keV). The vertical and horizontal divergences of the transmitted beam were 2.0 mrad and 2.7 mrad, respectively [7]. There have been some reports for slow HCI transmission experiments with Ar^{8+} of 80 keV and Xe^{23+} of 230 keV [8], I^{q+} (q = 10 - 50) of 3 keV/q [9], Ar^{9+} of 4.5 keV [10], Ar^{8+} of 8-60 keV [11], and so on.
Not only the stable transmission reported above but also unstable behaviour has been sometimes observed especially with higher incident currents. Figure 3 shows transmitted ion intensity through a 4 μmϕ outlet for 64 keV Ar^{8+} as a function of time. The incident current was kept around 20 pA. The plot is an extracted view of 5 minutes (t = 600 - 900 s) from a 25-minute measurement. Several spikes were observed with approximately constant transmission of 4,000 cps, whose corresponding density enhancement factor is ~10. This fluctuating behaviour is possibly due to the higher incident current comparing with that of figure 2. Although the transmission is supported by a self-organized charge up, discharge process must contribute to the charge distribution. Considering the almost constant input charge onto the surface, the spikes in transmission are expected to be due to a dynamically resistive change of the glass.

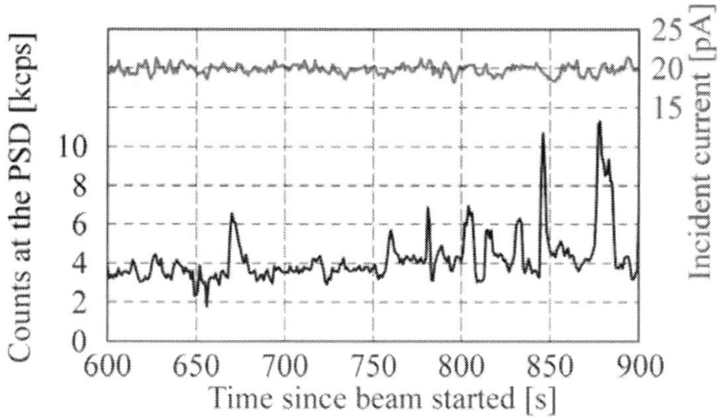

Figure 3. Transmitted ion intensity through a 4 μmϕ outlet for 64 keV Ar^{8+} as a function of time. The incident current was kept around 20 pA. Several spikes were observed with approximately constant transmission of 4,000 cps, whose corresponding density enhancement factor is ~10.

In order to investigate the resistive switching of bulky glass under irradiation of slow HCI, a guiding experiment using an Ar^{8+} of 104 keV beam through a gap (0.1 mm) of a paired glass plates was performed and compared with a hysteretic resistive switching model [12]. We observed a slowly and regularly oscillating transmission current through the gap during an injection of a steady beam. The oscillation frequency was almost proportional to the injection current intensities. This phenomenon can be attributed to the charge–discharge iteration cycle with dynamic resistive change of the glass plate, which includes switching like insulator-metal transition according to the electric filed induced by the deposit charge from the input beam. The model calculation was applied to balky insulator sample for the first time and well reproduces the experimental observation.

3.2. Beam guiding with flexible Teflon tubes

Based on the guiding effect of the glass capillary, a guiding experiment by larger angles for Ar^{8+} of 8 keV beam through flexile Teflon tubes is in progress [13]. Figure 4(a) shows a schematic drawing of the sample setup. Teflon tubes with outer diameter of 2 mm and inner diameter of 1 mm were used. The sample tube was placed inside a groove having a fixed curvature machined in the holder. The definition of the bending angle ϕ is given in figure 4(a). The tube holder had three grooves of different curvatures and one for a straight tube. The tested bending angles were $\phi = 9.6^{\circ}$, 17.5°, 26.7°, and 0°, which correspond to the radii of curvature R = 270, 150, 100 mm, and straight tube, respectively. The holder was made of aluminium and the grooves were filled with an electrically conductive paste while inserting the Teflon tube. Both ends of each tube, i.e. the faces of inlet and outlet but not inside, were also coated with the conductive paste. The transmitted ion current, I_t , was measured with an aluminium plate set just after the outlet of the tube, and its time dependence was recorded every second by a PC data acquisition system. The beam current on the holder, I_h, was also monitored during the measurement. The total current, I_{tot} , namely the beam current passing through the 2 mm diameter aperture of the shield, was determined as the sum of on-holder and transmitted currents, i.e. $I_{tot} = I_h + I_t$. Figure 4(b) presents the ratio of transmitted current to total current. The region from 0° to 7° in horizontal axis corresponds to the transmission ratio for the straight tube (in dashed-line region). Open squares denote the measurement with tilt angle (from -7° to $+5^{\circ}$). The bars denote the statistical error of one standard deviation. Assuming a certain value for the beam fraction injected into the tube, one can convert the ratio I_t/I_{tot} into a transmission efficiency. If 32% of the total current, which is estimated from a Gaussian beam profile with FWHM=2 mm, is injected, $I_t/I_{tot} = 0.28$ at 0° (open square)

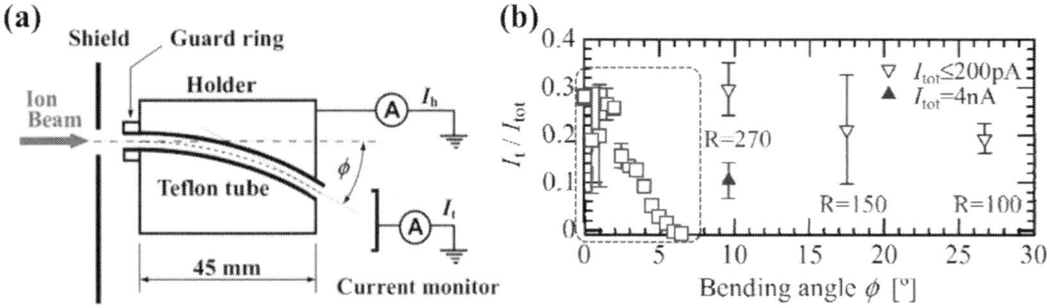

Figure 4. (a) Schematic diagram of the experimental setup. The definition of bending angle is denoted as ϕ. (b) Summary of I_t/I_{tot} for curved tubes in comparison with those for tilted straight tubes. The beam-guidable bending angles of the curved tubes turned out to be much larger than the limit of guiding with a tilted straight tube.

corresponds to a transmission efficiency of 87.5%. The transmission efficiency became zero when the tube was tilted more than 7° from the centre. The ratio I_t/I_{tot} for the curved tubes is also summarized in figure 4(b). When the primary beam intensity was of the order of a few nA, the maximum transmitted current was high, but unstable. In these three cases, the lower the incident current, the more stable the ion transmission. This suggests that there are appropriate intensities for a good balance of charge and discharge to guide the ion beams stably. The results demonstrate that these curved tubes can guide 8 keV Ar^{8+} ions with transmission efficiencies of several tens of per cent. The beam-guidable bending angles of the curved tubes turned out to be indeed much larger than the limit of guiding with a straight tube of the same size. This may open a way to develop a new scheme of flexible ion beam deflector systems like fibre optics guides.

4. Microbeams of MeV ions with glass capillaries

4.1. Glass capillary with thin end-window

Several research groups have been intensively working on the preparation of microbeams for biological applications. In a conventional scheme, a well-focused energetic ion beam is extracted in air via a vacuum isolation window, then injected into a biological cell in culture solution. Another straightforward scheme involves passing energetic ions through a micrometre-sized aperture with or without a thin window at the end. Drawbacks of these schemes are (1) a relatively large cylindrical volume is damaged along the beam trajectory in addition to the targeted point, (2) serious energy and angular stragglings are induced during passage through the vacuum isolation window and air, which deteriorates the beam quality, and as a result determines the lower limit of the beam size at the target, and (3) real-time control/monitoring of the bombarding point is not easy even when a micrometre beam is prepared. In order to overcome such technical but serious problems, we have developed a scheme using a tapered glass capillary. The transport mechanism for MeV ions is governed by small angle scattering with the inner wall, which is different from that for slow HCIs. The point for the break-through for cell irradiation is that the capillary has a thin window at its outlet [14]. This scheme can realize pinpoint energy deposition and three-dimensional selection of the bombarding point by observing the outlet through a microscope with a precision of a micron or better in an arbitrary position of a living cell or in any liquid object. As the material of the end window, borosilicate glass [14,15] or plastic is available.

4.2. Irradiation to human cells

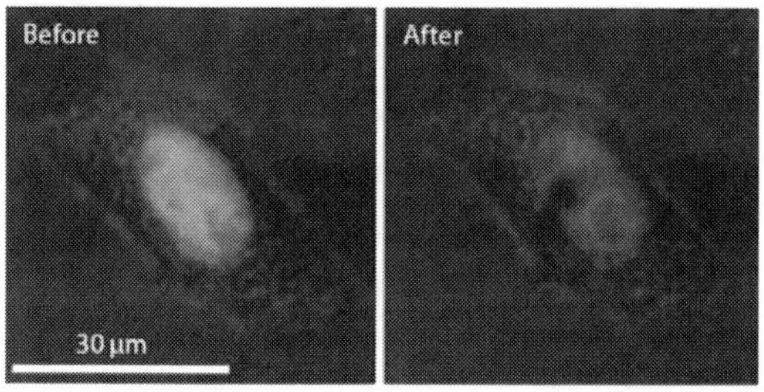

Figure 5. The target of the irradiation was a nucleus of HeLa cell in culture solution, whose nucleus was labelled with green fluorescent protein (histone H2B-GFP). The left and right panels are fluorescent images before and after the irradiation combined with the phase contrast images, respectively.

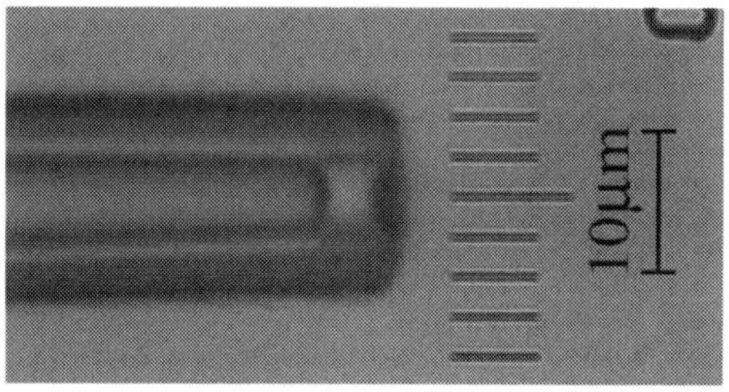

Figure 6. Plastic scintillator end-window with diameter and thickness are 5 µmϕ and ~4 µm, respectively. When an ion passes through the plastic scintillator end-window, the window yields scintillation photons due to the energy loss at the window.

Figure 5 shows the demonstration of a cell irradiation using a glass capillary with a borosilicate end window. The target was a HeLa cell in culture solution, whose nucleus was labelled with green fluorescent protein (histone H2B-GFP). The left panel of figure 5 is a fluorescent image before irradiation combined with the phase contrast image of the same sample. A proton beam of 1 MeV was generated by Pelletron Accelerator at RIKEN, then extracted with a current of ~10 pA from the end window whose diameter and thickness were 5 µmϕ and 8 µm, respectively. After 20 s irradiation, only the irradiated area was bleached, as shown in the right panel. The total dose of this irradiation was relatively high so that the bleaching of GFP was demonstrated at a 3-dimensionally confined volume. However, in some cases, irradiation by only one ion is sufficient to create a change or damage inside the target cell. Therefore, single ion irradiation is required for accurate dose control. To reduce the beam intensity, we have already installed a beam chopping system which generates short pulsed beams of the order of hundreds ns. The number of ions in a pulse can be adjusted to be about 1. The number of ions in a pulse follows a Poisson distribution, i.e., some pulses are empty or have more than one ion. If the ion energy is larger than ~10 MeV, a detector behind the target cell would be available for counting the ions. However, in the case of our a few MeV ions, plastic scintillator end-window has been introduced to confirm the single ion irradiation before stopping at the target [16]. Figure 6 is a photo of the window whose diameter and thickness are 5 μmϕ and ~4 μm, respectively. The end-window diameters from about 1 to 100 μmϕ are available. The thickness is similar to the diameter. When a few MeV ion passes through the plastic scintillator end-window just before the target, the end-window yields scintillation photons due to the energy loss at least 100 keV for several μm thickness (for protons). The intensity of the scintillation will be monitored by a photo multiplier. The method for the dose control is in progress.

4.3. Irradiation to Escherichia coli cells

In contrast, some targets need a higher dose to change. A microbeam extracted from the glass capillary with end-window was applied to a single cell of *Escherichia coli* in liquid media to evaluate the effect of irradiation on cellular physiology [17]. As *E. coli* cells can move in the liquid media by rotating a bundle of their peritrichous flagella, the rotational behaviour of the single flagellar motor of the cell tethered to the substrate was observed during irradiation and the growth profiles of the cells attached to the substrate were monitored after irradiation. Then the irradiation dose required to stop the flagellar motor was evaluated. A proton microbeam of 2 MeV (before end window) with an intensity of 7.2 x 10^4 particles /s was extracted from the plastic end window whose diameter and thickness were 11 μmϕ

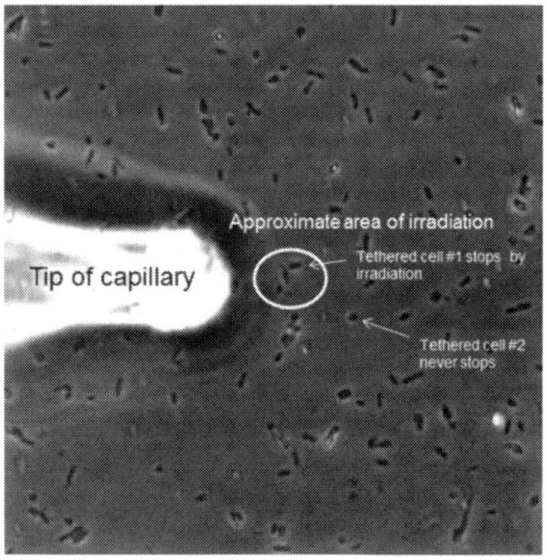

Figure 7. A bottom view of the specimen. The area targeted by the microbeam is indicated with a circle. Actively rotating tethered cells, marked with arrows, exist both inside (the cell #1) and outside (the cell #2) the target area. The cell #1 stopped rotating, while the cell #2 continued to rotate, indicating that the beam was thin enough to irradiate the specific area.

and less than 10 μm, respectively. Figure 7 is a bottom view of the specimen. The area targeted by the microbeam is indicated with a circle. Actively rotating tethered cells, marked with arrows, exist inside (the cell #1) and outside (the cell #2) the target area. The tethered cell within the target area stopped rotating, while another tethered cell outside the target area continued to rotate, indicating that the beam was thin enough to irradiate the specific area. In addition to the determination of the minimum dose to affect the activity, the growth and division of cells adhering to the substrate were also reported. This methodology of irradiating cells *in situ* enables us to observe both the targets and the unirradiated controls simultaneously, and it is quite useful to evaluate the effect of ionizing radiation on various cellular processes in the future works, e.g., damage to cell membrane, effects on membrane potential and respiratory chain enzymes.

4.4. Surface modification in liquid

The following application is not for biology but for surface modification. The range of the 3 MeV protons in water after passing through a Kapton end window with a thickness of 7 μm was calculated to be 140 μm using the SRIM software package. The surfaces of polyethylene (PE, $-[CH_2CH_2]_n-$) and polytetrafluoroethylene (PTFE, $-[CF_2CF_2]_n-$) sheets were selected as the samples, and aqueous acrylic acid (AAc, CH_2CHCO_2H) in concentrations between 0 and 10 wt.% was used as solution. Acrylic acid readily combines with itself or other monomers by reacting at its double bond site to form hydrophilic polymers. A scanning irradiation was performed with the distance of about 100 μm between the capillary outlet and the sample surface. After the irradiation, deposition on only the scanned area due to grafting and polymerization of AAc monomers was found to have a large affinity with water [18]. The deposited layer was stable and showed good adhesion to the base polymer. This method can provide techniques of coating or surface modification with fine pattern.

Other activities employing the tapered single glass capillaries with MeV ions were reported for PIXE [19-20] and NRA [21], as well as fundamental studies of the beam transport mechanism [22-25]. This glass made optics has been applied to not only ion beams but also electron [26-28], positron [29,30], and muon beams [31,32].

5. Summary

A method to produce microbeams of keV energy HCIs with single tapered glass capillaries based on a self-organized charge-up process has been developed. The transmission in this energy showed focusing effect up to the enhancement of 10, the guiding effect, and preservation of its initial charge state. Employing a flexibly curved Teflon tube, a beam guiding by large angles up to 26.7° was obtained. Microbeams of a few MeV protons/He ions provided by glass capillaries with end-windows have been used for the irradiation of HeLa cells and *Escherichia coli* cells. A method for dose estimation is in progress using plastic scintillator end-window combined with a pulsed beam system. A scanning surface modification experiment by the irradiation of proton beam was performed. The deposition on only the scanned area due to grafting and polymerization of AAc monomers was found to have a large affinity with water.

Acknowledgments

This work was partly supported by Human Frontier Science Program (Walter Meissl), JSPS KAKENHI Grant Numbers 17654079, 20510119, 23656225, 22604015, 22740269, Grant of "Developments and Applications of Exotic Quantum Beams" from RIKEN, and FY2006 and FY2008 Incentive Research Grants from RIKEN.

References

[1] Nebiki T, Yamamoto T, Narusawa T, Breese M B H, Teo E J and Watt F 2003 *J. Vac. Sci. Technol.* A **21** 5 1671-74

[2] Ikeda T, Kanai Y, Kojima T M, Iwai Y, Kambara T, Hoshino M, Nebiki T, Narusawa T and Yamazaki Y 2006 *Appl. Phys. Lett.* **89** 163502

[3] Stolterfoht N, Bremer J-H, Hoffmann V, Hellhammer R, Fink D, Petrov A and Sulik B 2002 *Phys. Rev. Lett.* **88** 133201

[4] Schiessl K, Palfinger W, Tökési K, Nowotny H, Lemell C and Burgdörfer J 2005 *Phys. Rev.* A **72** 062902

[5] Schiessl K, Lemell C, Tőkési K and Burgdörfer J 2009 *J. Phys. Conf. Ser.* **194** 012069
[6] Schweigler T, Lemell C and Burgdörfer J 2011 *Nucl. Instrum. Methods Phys. Res.* B **269** 1253-56
[7] Ikeda T, Kojima T M, Iwai Y, Kanai Y, Kambara T, Nebiki T, Narusawa T and Yamazaki Y 2007 *J. Phys. Conf. Ser.* **58** 68-73
[8] Cassimi A *et al.* 2008 *Int. J. Nanotechnol.* **5** 809-17
[9] Nakayama R, Tona M, Nakamura N, Watanabe H, Yoshiyasu N, Yamada C, Yamazaki A, Ohtani S and Sakurai M 2009 *Nucl. Instrum. Methods Phys. Res.* B **267** 2381-84
[10] Bereczky R J, Kowarik G, Aumayr F and Tőkési K 2009 *Nucl. Instrum. Methods Phys. Res.* B **267** 317-20
[11] Kreller M, Zschornack G and Kentsch U 2011 *Nucl. Instrum. Methods Phys. Res.* B **269** 1032-35
[12] Ikeda T, Iwai Y, Kojima T M, Onoda S, Kanai Y and Yamazaki Y 2012 *Nucl. Instrum. Methods Phys. Res.* B **287** 31-34
[13] Kojima T M, Ikeda T, Kanai Y, Yamazaki Y and Esaulov V A 2011 *J. Phys. D: Appl. Phys.* **44** 355201
[14] Iwai Y, Ikeda T, Kojima T M, Yamazaki Y, Maeshima K, Imamoto N, Kobayashi T, Nebiki T, Narusawa T and Pokhil G P 2008 *Appl. Phys. Lett.* **92** 023509
[15] Ikeda T, Kanai Y, Kojima T M, Iwai Y, Kanazawa Y, Hoshino M, Kobayashi T, Pokhil G P and Yamazaki Y 2007 *J. Phys. Conf.Ser.* **88** 012031
[16] Ikeda T 2011 *Japan Patent* No. 2011-161568
[17] Kato M, Meissl W, Umezawa K, Ikeda T and Yamazaki Y 2012 *Appl. Phys. Lett.* **100** 193702
[18] Kobayashi T, Miyamoto S, Ikeda T, Kojima T M., Ogiwara K and Yamazaki Y 2012 *Nucl. Instrum. Methods Phys. Res.* B **272** 405-8
[19] Nebiki T, Kabir M H and Narusawa T 2006 *Nucl. Instrum. Methods Phys. Res.* B **249** 226-29
[20] Simon M J, Döbeli M, Müller A M and Synal H-A 2012 *Nucl. Instrum. Methods Phys. Res.* B **273** 237-40
[21] Sekiba D *et al.* 2008 *Nucl. Instrum. Methods Phys. Res.* B **266** 4027-36
[22] Fujita N, Ishii K and Ogawa H 2011 *Phys. Scr.* T **144** 014033
[23] Hasegawa J, Jaiyen S, Polee C, Chankow N and Oguri Y 2011 *J. Appl. Phys.* **110** 044913
[24] Jaiyen S, Chankow N, Hasegawa J, and Oguri Y 2012 *Nucl. Instrum. Methods Phys. Res.* B **271** 13-18
[25] Gong Z, Yan S, Ma H, Nie R, Xue J and Wang Y 2012 *Nucl. Instrum. Methods Phys. Res.* B **272** 370-73
[26] Dassanayake B S, Bereczky R J, Das S, Ayyad A, Tőkési K and Tanis J A 2010 *Phys. Rev.* A **81** 020701(R)
[27] Wickramarachchi S J, Dassanayake B S, Keerthisinghe D, Ayyad A and Tanis J A 2011 *Nucl. Instrum. Methods Phys. Res.* B **269** 1248-52
[28] Wang W, Chen J, Yu D Y, Yang B, Wu Y H, Zhang M W, Ruan F F and Cai X H 2011 *Phys. Scr.* T **144** 014023
[29] Oshima N, Iwai Y, Kojima T M, Ikeda T, Kanazawa Y, Hoshino M, Suzuki R and Yamazaki Y 2009 *Mater. Sci. Forum* **607**, 263-65
[30] DuBois R D and Tőkési K 2012 *Nucl. Instrum. Methods Phys. Res.* B **279** 186-89
[31] Kojima T M, Tomono D, Ikeda T, Ishida K, Iwai Y, Iwasaki M, Matsuda Y, Matsuzaki T and Yamazaki Y 2007 *J. Phys. Soc. Jpn.* **76** 9 093501
[32] Tomono D, Kojima T M, Ishida K, Ikeda T, Iwai Y, Tokuda M, Kanazawa Y, Matsuda Y, Matsuzaki T, Iwasaki M and Yamazaki Y 2011 *J. Phys. Soc. Jpn.* **80** 044501

The non-resonant neutralization dynamics of the multiply charged Rydberg ions escaping solid surfaces

S. M. D. Galijaš, N. N. Nedeljković and M. D. Majkić
University of Belgrade, Faculty of Physics, P.O. Box 368, Belgrade, Serbia
E-mail: hekata@ff.bg.ac.rs

Abstract. We investigate the intermediate and final population of the Rydberg states ($n_A \gg 1, l_A = 0-3, m_A$) of multiply charged ions escaping solid surfaces by using the two-state vector model. All our calculations have been carried out for intermediate projectile velocities ($v \approx 1$ a.u.). The electron capture is non-resonant and characterized by the selective population of the ionic Rydberg states. The final population probabilities enable us to elucidate the role of the ionic core polarization and to analyze the n_A, l_A and v probability distributions. We consider the ions SVI, ClVII and ArVIII with core charges $Z = 6, 7$ and 8, respectively, and the ions KrVIII and XeVIII with $Z = 8$. The model is applied on two different ion-surface systems in order to emphasize the influence of the solid work function. We discuss the appearance of resonances (pronounced maxima) in the probability distributions. The resonances are explained by means of the electron tunneling in the very vicinity of the ion-surface potential barrier top. It is demonstrated that the resonances are characteristic only for some combination of the surface and ionic characteristics (argon anomaly), in agreement with available beam-foil experimental data.

1. Introduction

The electron exchange during the interaction of ions with solid surfaces has been intensively studied in the last two decades both theoretically [1] and experimentally [2]. This kind of problem, concerning the neutralization dynamics of multiply charged ions, has been also considered within the framework of two-state vector model (TVM) [3, 4, 5, 6, 7, 8, 9, 10, 11], for wide range of projectile velocities v.

In the present article we report the TVM of the non-resonant electron capture from the solid conducting band into the Rydberg states (n_A, l_A, m_A) of multiply charged ionic projectiles ($v \approx 1$ a.u.) [4, 5], by taking into account the ionic core polarization [10, 11]. The state of a single active electron is simultaneously described by two state vectors $|\Psi_1(t)\rangle = \hat{U}_1(t_{in}, t)|\Psi_1(t_{in})\rangle$ and $|\Psi_2(t)\rangle = \hat{U}_2(t_{fin}, t)|\Psi_2(t_{fin})\rangle$, within the first and the second scenarios, respectively, where $\hat{U}_1$ and $\hat{U}_2$ are the corresponding evolution operators. An entanglement of these states at intermediate stages of the ion-surface interaction is described by a mixed electron flux through the moving Firsov plane S_F used to define the ionic region. Model is almost independent of the form of the near-surface potential, and a dynamical response of the surface can be neglected, so that the classical electrostatic image potentials can be used outside the surface. The interaction of the active electron with polarized ionic core in the first scenario is the Coulomb interaction.

However, in the second scenario the active electron moves closer to the charge cloud of the electrons already bound to the nucleus. In that case, the Simons-Bloch potential [12] can be used: $\hat{U} = -Z/r_{\mathrm{A}} + \sum_{l'=0}^{\infty}(c_{l'}/r_{\mathrm{A}}^2)\hat{P}_{l'}$, where $\hat{P}_{l'} = |l'\rangle\langle l'|$ is the projection operator onto the subspace of a given angular momentum l'. This effective potential accounts for the experimentally observed quantum defects of the energy spectra $\tilde{E}_{\mathrm{A}} = -\tilde{\gamma}_{\mathrm{A}}^2/2$ through the constants $c_{l'}$.

2. The non-resonant Rydberg state population

2.1. Intermediate stages of the population dynamics

Our attention is focused on the electron capture into the Rydberg states of multiply charged Rydberg ions (core charge Z) escaping solid surfaces at intermediate velocities $v = \mathrm{d}R/\mathrm{d}t \approx 1$ a.u., where R is the ion-surface distance. In order to take into account the ionic core polarization, the TVM developed in [4, 5] is appropriately modified [10, 11], following the idea of [8]. Within the framework of the TVM, the neutralization probability density $T_{\mu_{\mathrm{M}},\nu_{\mathrm{A}}}(t) = |A_{\mu_{\mathrm{M}},\nu_{\mathrm{A}}}(t)|^2$ (per unit energy parameter γ_{M} of the energy $E_M = -\gamma_{\mathrm{M}}^2/2$) is defined using the two-amplitude $A_{\mu_{\mathrm{M}},\nu_{\mathrm{A}}} = \langle\Psi_2(t)|\hat{P}_{\mathrm{A}}|\Psi_1(t)\rangle$; the quantity $\hat{P}_{\mathrm{A}}$ represents the projection operator onto the ionic region. For description of the intermediate stages of the process we use the normalized probability density $\tilde{T}_{\mu_{\mathrm{M}},\nu_{\mathrm{A}}}(t) = T_{\mu_{\mathrm{M}},\nu_{\mathrm{A}}}(t)/T^{\mathrm{fin}}_{\mu_{\mathrm{M}},\nu_{\mathrm{A}}}$, where $T^{\mathrm{fin}}_{\mu_{\mathrm{M}},\nu_{\mathrm{A}}} = \lim_{t\to t_{\mathrm{fin}}} T_{\mu_{\mathrm{M}},\nu_{\mathrm{A}}}(t)$.

Appropriate analytical expression for the normalized probability in the considered intermediate velocity region is given by [13]

$$\tilde{T}_{\mu_{\mathrm{M}},\nu_{\mathrm{A}}}(t) \approx \left|1 - \mathcal{B}(R)\mathrm{e}^{-(\tilde{\beta}-\mathrm{i}\frac{w}{v})R}\right|^2, \tag{1}$$

where $\mathcal{B}(R) = 1 + (\beta R)^{\alpha+1}/\Gamma(\alpha+2)$, $\alpha = Z/\tilde{\gamma}_{\mathrm{A}} - 3/2 + 1/(4\gamma_{\mathrm{M}})$ and $\beta = \tilde{\beta} - \mathrm{i}w/v = \gamma_{\mathrm{M}} + (\tilde{\gamma}_A - \gamma_{\mathrm{M}})g - \mathrm{i}w/v$. The quantity w in equation (1) is defined by $w = (\gamma_{\mathrm{M}}^2 - \tilde{\gamma}_{\mathrm{A}}^2)/2 - v^2(1-2g)/2$, where $g = a/R$, whereas a represents the position of the Firsov plane S_{F} in respect to the ionic core [4]. The population of the Rydberg states is mainly localized at the ion-surface distances $R = R_c^N$; the distances R_c^N can be considered as neutralization distances under the fixed initial and final states of the active electron. Within the framework of TVM, the quantities $R_c^N = R_c^N(\mu_{\mathrm{M}}, \nu_{\mathrm{A}})$ are defined as a positions of maxima of the total rates $\tilde{\Gamma}_{\mu_{\mathrm{M}},\nu_{\mathrm{A}}} = d\tilde{T}_{\mu_{\mathrm{M}},\nu_{\mathrm{A}}}/dt$; in the explicit form (scaled by velocity v) we have

$$\begin{aligned}\tilde{\Gamma}_{\mu_{\mathrm{M}},\nu_{\mathrm{A}}}(t)/v &= \frac{2(\alpha+1)}{R}\mathrm{e}^{-\tilde{\beta}R}\cos\mathcal{K}(R) \\ &+ 2f(R)\left\{\left(\frac{\alpha+1}{R} - \tilde{\beta}\right)[f - \cos\mathcal{K}(R)] + \frac{w}{v}\sin\mathcal{K}(R)\right. \\ &\left. - \frac{(\alpha+1)}{R}\mathrm{e}^{-\tilde{\beta}R}\right\},\end{aligned} \tag{2}$$

where $\mathcal{K}(R) = wR/v + \arg\mathcal{B}(R)$ and

$$f(R) \approx \left(1 + \frac{|\beta|^{\alpha+1}}{\Gamma(\alpha+2)}R^{\alpha+1}\right)\mathrm{e}^{-\tilde{\beta}R}. \tag{3}$$

In figure 1 we present the normalized probability density $\tilde{T}_{\mu_{\mathrm{M}},\nu_{\mathrm{A}}}(t)$ and the corresponding scaled rates $\tilde{\Gamma}_{\mu_{\mathrm{M}},\nu_{\mathrm{A}}}(t)/v$ for the electron capture from the state with the energy parameter $\gamma_{\mathrm{M}} = \gamma_{\mathrm{M}}^{\star}$ (dominantly contributing to the transition) into the Rydberg states $\nu_{\mathrm{A}} = (n_{\mathrm{A}} = 6, 8, 10; l_{\mathrm{A}} = 1)$ of the ArVIII ion escaping the solid surface with velocity $v = 1$ a.u. At smaller ion-surface distances R we have a kind of instability and at larger R the neutralization process becomes stabilized. Therefore, the last peak can be used for calculation of the neutralization distances R_c^N. In figure 2 we analyzed the normalized intermediate rate $\tilde{\Gamma}_{\mu_M,\nu_A}$ for the population of the Rydberg states of the ArVIII ion with the $n_A = 8$, for $l_A = 0, 1$ and 2, and for $v = 1$ a.u. and $v = 2$ a.u. As we can see, the neutralization distances are slowly shifted toward larger ion-surface distances by increasing l_A. In the absence of core polarization, i.e. in the the pointlike core case (dashed curves) we obtain almost the same curve as for $l_A = 3$. Also, from figure 2 we conclude that the velocity dependence of R_c^N in the intermediate velocity region is practically negligible.

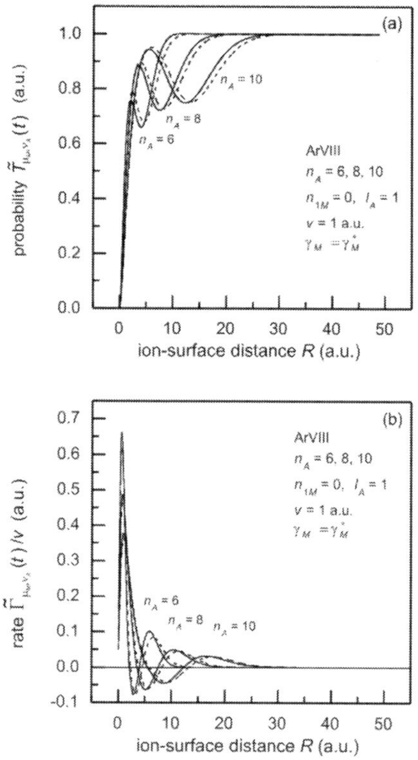

Figure 1. Normalized intermediate (a) probability density $\tilde{T}_{\mu_M,\nu_A}(t)$ and (b) rate $\tilde{\Gamma}_{\mu_M,\nu_A}(t)$ (scaled by v) via ion-surface distances R for the population of the Rydberg states $n_A = 6, 8, 10$, $l_A = 1$ of the ArVIII ion escaping the solid surface with velocity $v = 1$ a.u. Dashed curves correspond to the pointlike core case.

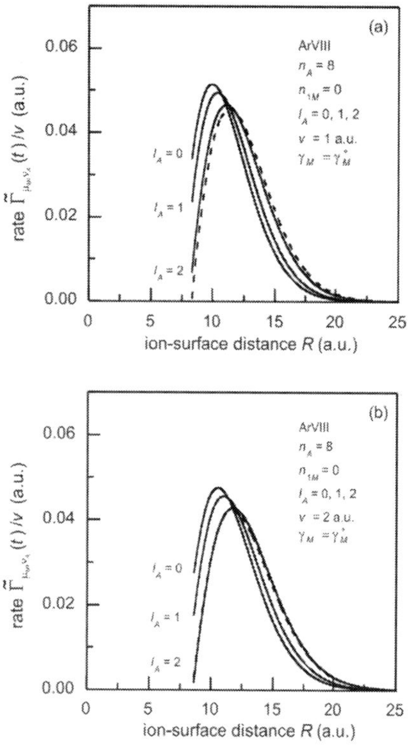

Figure 2. The last peak of the normalized intermediate rate $\tilde{\Gamma}_{\mu_M,\nu_A}(t)$ (scaled by v) via ion-surface distances R for the population of the Rydberg states $n_A = 8$, $l_A = 0, 1, 2$ of the ArVIII ion escaping the solid surface with velocity (a) $v = 1$ a.u. and (b) $v = 2$ a.u. Dashed curves correspond to the pointlike core case.

2.2. The regular population of the final Rydberg state

First we consider the population of the Rydberg states $n_A < n_{res}$, which occur via tunnelling mechanism. We shall consider the term "regular population" for this type of the population process to distinguish from the population of the Rydberg states $n_A = n_{res}$, populated in the very vicinity of the potential barrier top [4, 10]. By n_{res} we denoted the resonant quantum numbers. A specific feature of the TVM (independently of the type of electron transitions) is the use of the mixed flux $I_{\mu_{\rm M},\nu_{\rm A}}(t)$ through the Firsov plane defined by the following surface integral:

$$I_{\mu_{\rm M},\nu_{\rm A}}(t) = \frac{\rm i}{2}\int_{S_{\rm F}}\left[\frac{\nabla\Psi_1}{\Psi_1} - \frac{\nabla\Psi_2^*}{\Psi_2^*} - 2{\rm i}v\left(1-\frac{{\rm d}a}{{\rm d}R}\right)\boldsymbol{e}_z\right]\Psi_2^*(\boldsymbol{r},t)\Psi_1(\boldsymbol{r},t){\rm d}\boldsymbol{S}. \quad (4)$$

The kinematic factor g of the Firsov plane dynamics is given by equation (4.20) in reference [4]. The final transition probability follows directly from the mixed flux, i.e.

$$T_{\nu_{\rm A}}^{fin} = \sum_{\mu_{\rm M}} T_{\mu_{\rm M},\nu_{\rm A}}^{fin} = \int_{\gamma_{\rm M}}\sum_{n_{1M},m_M}\lim_{t\to t_{\rm fin}}\left|\int_{t_{\rm in}}^{t} I_{\mu_{\rm M},\nu_{\rm A}}(t){\rm d}t\right|^2 {\rm d}\gamma_{\rm M}. \quad (5)$$

The final population probability $P_{\nu_{\rm A}}^{fin}$ (that can be compared with experiments) for the principal quantum numbers $n_{\rm A} \leq n_{res}-1$ and for $l_{\rm A}=1$ and 2 is given by [4]

$$P_{\nu_{\rm A}}^{fin} = \left[1-\exp(-T_{\nu_{\rm A}}^{fin})\right]\exp\left(-\sum_{n'_{\rm A}\neq n_{\rm A}}^{n_{res}-1} T_{\nu'_{\rm A}}^{fin}\right), \quad (6)$$

where $\nu'_{\rm A} = (n'_{\rm A}, l'_{\rm A}=1, m'_{\rm A}=0)$.

In figure 3 we present the quantities $T_{\nu_{\rm A}}^{fin}$ and $P_{\nu_{\rm A}}^{fin}$ for the ions SVI, ClVII and ArVIII escaping the solid surface at velocities $v=1.94$ a.u., $v=2.50$ a.u. and $v=1.42$ a.u., respectively, relevant for the beam-foil experiments. We consider the surface with the solid work function $\phi = 3$ eV and with the bottom level of the solid $U_0 = 10$ eV. Comparing the $T_{\nu_{\rm A}}^{fin}$ and $P_{\nu_{\rm A}}^{fin}$ curves in figure 3, we recognize the importance of the multichannel description of the population process. Indeed, the distances between Rydberg levels are small and all channels participate simultaneously. This is of particular importance in the case of ArVIII ion for $v=1.42$ a.u., which is characterized by the higher population probabilities. Also, one can recognize that the core polarization induce the lowering of the probability distribution (compare the full curves with the dashed curves in figure 3).

The final population probabilities $P_{\nu_{\rm A}}^{fin}$ with core polarization taken into account (solid curves in figure 3) reflect the selective character of the Rydberg level population. That is, the Rydberg levels with $n_{\rm A} = n_0 = 6,7$ and 9 (for $l_{\rm A}=1$ and $m_{\rm A}=0$) of the ions SVI, ClVII and ArVIII, respectively, are populated with maximal probabilities. The selective population (with $n_0 = 6,7$ and 8) appears also in the point-like core approximation. For the ArVIII (which is, considering the electronic structure, close to the hydrogenlike ions) the probability distributions with the ionic core polarization taken into account and those obtained for the pointlike ionic cores are similar. The relatively close relation between the TVM theoretical final population probabilities for $n_{\rm A} < n_{res}$ and $l_{\rm A}=1$ and the corresponding experimental findings (dots) is also demonstrated in figure 3. However, due to the relative character of the experimental output, the results reveals the agreement of both probabilities (for the polarized and pointlike core cases) with beam-foil experiment (for SVI and ClVII [14] and for ArVIII [15]). The same agreement with experimental data can be obtained for other low-$l_{\rm A}$ values, $l_{\rm A}=0$ and $l_{\rm A}=2$ [15, 16].

In figure 4(a) we present the final population probabilities $P_{\nu_{\rm A}}^{fin}$ of the Rydberg states with $n_{\rm A} < n_{res}$ and $l_{\rm A}=1$ of the ions ArVIII, KrVIII and XeVIII escaping solid surface ($\phi=3$ eV and $U_0=10$ eV) at velocity $v=1.42$ a.u.. The theoretical results presented in figure 4(a) are compared with available beam-foil experimental data [15] for ArVIII ion (dots). Experimental data for KrVIII and XeVIII do not exist in the low-$l_{\rm A}$ region, and for the Rydberg states with $n_{\rm A}\approx Z$. The experimental data for KrVIII ionic projectiles exist only for the large-$l_{\rm A}$ Rydberg states ($l_{\rm A}\geq 5$) [15]. Also, the recently obtained data for ArVIII ion interacting with microcapillary foil are in the large-$l_{\rm A}$ region, and for somewhat lower ionic velocities ($v=0.2$

Journal of Physics: Conference Series **399** (2012) 012008 doi:10.1088/1742-6596/399/1/012008

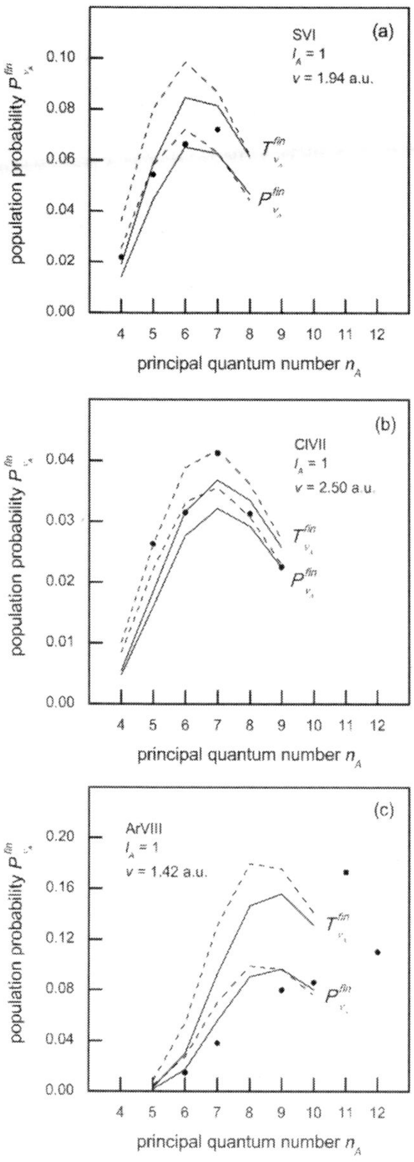

Figure 3. Final transition probabilities $T^{fin}_{\nu_A}$ and final population probabilities $P^{fin}_{\nu_A}$ via $n_A < n_{res}$ for the Rydberg states with $l_A = 1$ of the ions (a) SVI, (b) ClVII and (c) ArVIII escaping the solid surface ($\phi = 3$ eV, $U_0 = 10$ eV) with the velocities $v = 1.94$ a.u., $v = 2.50$ a.u. and $v = 1.42$ a.u., respectively. Dashed curves correspond to the pointlike core case. Dots are the results of the beam-foil experiments taken from [14] for SVI and ClVII ions and [15] for ArVIII ions.

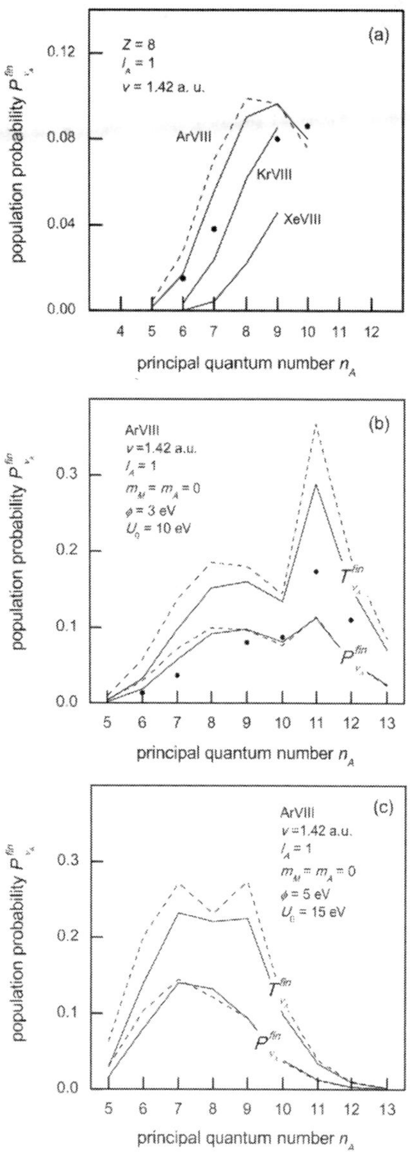

Figure 4. (a) Final population probabilities $P^{fin}_{\nu_A}$ via $n_A < n_{res}$ for the Rydberg states with $l_A = 1$ of the ions ArVIII, KrVIII and XeVIII escaping the solid surface ($\phi = 3$ eV and $U_0 = 10$ eV). In (b) and (c): probabilities $T^{fin}_{\nu_A}$ and $P^{fin}_{\nu_A}$ for $n_A < n_{res}$ and $n_A \geq n_{res}$ of the ArVIII ion, for surfaces with $\phi = 3$ eV, $U_0 = 10$ eV and $\phi = 5$ eV, $U_0 = 15$ eV, respectively. Dashed curves correspond to the pointlike core case. Dots represent the experimental results [15].

a.u.) [17]. From figure 4(a) we recognize that, due to the core polarization, the final population probabilities for KrVIII and XeVIII species are significantly lower in comparison to the ArVIII case. Because of the absence of the experimental data for KrVIII and XeVIII ions, for the considered Rydberg states, the obtained trends in the behavior of the population probabilities can not be directly experimentally verified.

2.3. Resonances in the final Rydberg state population

It is intriguing to discuss [5, 11] the population curves in the vicinity of the resonant quantum numbers $n_{res} > n_0$, i.e., for $n_A \in \mathcal{N}_r$, where the set $\mathcal{N}_r$ is complemental to the set $\mathcal{N}$ considered in section 2.2. We are looking for the existence of an extra pronounced maximum (resonance) in the final population distribution at $n_A = n_{res}$. Such a situation has been experimentally recognized [15] for the ArVIII ions, for $n_A = n_{res} = 11$ and $l_A = 1$ and $l_A = 2$.

In order to obtained appropriate analytical form for the final transition probabilities $T_{\nu_A}^{fin}$ and the final population probabilities $P_{\nu_A}^{fin}$ for $n_A \in \mathcal{N}_r$, we adapt the wave function Ψ_1 to the case of the tunneling in the very vicinity of the ion-surface potential barrier top [5]. The etalon equation method (EEM) for the calculation of the appropriate wave function Ψ_1 contains a scaling parameter α_s, through which one can express the fact that the corresponding electron transition are characterized by the close turning point configuration. In the considered case, the two-large-parameter asymptotic form of the wave function Ψ_1 is necessary. The case $n_A \in \mathcal{N}$, in which the one-large-parameter asymptotic form of the function $\Psi_1(t)$ is sufficiently accurate, follows in the limit $\alpha_s \to \infty$. The wave function Ψ_2, corresponding to the state $|\Psi_2(t)\rangle$ that evolves teleologically (in the second scenario) toward the final state $|\nu_A\rangle$ with $n_A \in \mathcal{N}_r$, can be obtained analogously as in the case of the regular population of the final Rydberg state. The explicit analytical form of these two functions is given in [11].

In figures 4(b) and 4(c) we present (full curves) the final transition probabilities T_{ν}^{fin} and the final population probabilities $P_{\nu_A}^{fin}$ for the Rydberg states of the ArVIII ion escaping the conducting solid surface at $v = 1.42$ a.u. We consider the population of the Rydberg states with principal quantum numbers $n_A \in \mathcal{N}$ and $n_A \in \mathcal{N}_r$, for angular momentum quantum number $l_A = 1$ and $m_M = m_A = 0$. The final distributions are discussed for two types of the surfaces. Dashed curves are the final probabilities obtained in the absence of core polarization, i.e., in the pointlike core approximation. Solid circles in Figure 4(b) are the properly normalized beam-foil experimental results [15]. The existence of the resonances is theoretically confirmed only for the ArVIII ion and the type of the surface considered in figure 4(b).

3. Concluding remarks

We analyzed the electron capture into Rydberg states of multiply charged Rydberg ions escaping solid surfaces at intermediate velocity, investigating the neutralization dynamics within the time interval between initial and final "measurement" within the framework of the TVM. At intermediate stages of the process, the main output of this analysis from the standpoint of an experiment is the neutralization distances R_c^N for the ions finally detected in a given Rydberg state. We also discussed the final low-l_A Rydberg state population distributions for $n_A \in \mathcal{N}$ of the SVI, ClVII, ArVIII, KrVIII and XeVIII multiply charged ions, escaping conduction solid surfaces at intermediate velocities, by using the nonresonant multichannel TVM. It was concluded that the core polarization (included via the Simons-Bloch potential [12]) systematically suppress the final population probabilities $P_{\nu_A}^{fin}$, comparing with the pointlike core approximation. This effect become essential when we consider the KrVIII and XeVIII ions. In addition, we find out a strong connections between the surface parameters (ϕ and U_0) and the position of maxima of the population distributions for a given ion and projectile velocity.

In order to elucidate the appearance of the extra pronounced maxima (resonances) in the population curves, the TVM has been adapted to the complemental case $n_A \in \mathcal{N}_r$. The resonances are treated by using the appropriate EEM accompanied by the scaling parameter α_s for the calculation of the wave function Ψ_1. The pronounced maxima in the final population distribution are obtained for the ArVIII ion interacting with solid surface with the lower valued of the work function and with angular momentum quantum numbers $l_A = 1$ and $l_A = 2$. We recognize that the resonances are observable only for the compatible combination of the ionic core structure and the solid surface band structure. We point out that the TVM population distributions (including resonances) are in agreement with available beam-foil experimental data.

Acknowledgments
This work was supported in part by the Ministry of Education and Science, Republic of Serbia (Project 171016).

References

[1] Burgdörfer J in *Review of Fundamental Processes and Applications of Atoms and Molecules* ed Lin C D (World Scientific: Singapore, 1993)
[2] Winter H 2002 *Phys. Rep.* **367** 387
[3] Nedeljković N N, Nedeljković Lj D, Janev R K and Mišković Z L 1991 *Nucl. Instrum. Methods Phys. Res.* B **58** 519
[4] Nedeljković N N, Nedeljković Lj D, Vojvodić S B and Mirković M A 1994 *Phys. Rev.* B **49** 5621
[5] Nedeljković Lj D and Nedeljković N N 1998 *Phys. Rev.* B **58** 16455
[6] Nedeljković N N, Nedeljković Lj D and Mirković M A 2003 *Phys. Rev.* A **68** 012721
[7] Nedeljković Lj D and Nedeljković N N 2003 *Phys. Rev.* A **67** 032709
[8] Nedeljković N N and Majkić M D 2007 *Phys. Rev.* A **76** 042902
[9] Nedeljković N N, Majkić M D, Galijaš S M D and Mitrović S B 2008 *App. Surf. Sci* **254** 7000
[10] Nedeljković N N, Galijaš S M D and Majkić M D 2009 *Surf. Sci.* **603** 2403
[11] Galijaš S M D, Nedeljković N N and Majkić M D 2011 *Surf. Sci.* **605** 723
[12] Simons G and Bloch A N 1973 *Phys. Rev.* B **7** 2754
[13] Galijaš S M D 2012 Ph.D. thesis University of Belgrade
[14] Veje E and Winter H 1988 *Z. Phys.* D **10** 457
[15] Bashkin S, Oona H and Veje E 1982 *Phys. Rev.* A **25** 417
[16] Hallin R, Leavitt J A, Lindgard A, Rathmann P W, Vach H and Veje E 1982 *Nucl. Instrum. Methods* **41** 202
[17] Morishita Y, Hutton R, Torri H, Komaki K, Brage T, Ando K, Ishii K, Kanai Y, Masuda H, Sekiguchi M, Rosmej F B and Yamazaki Y 2004 *Phys. Rev.* A **70** 012902

26th Summer School and International Symposium on the Physics of Ionized Gases (SPIG 2012) IOP Publishing
Journal of Physics: Conference Series **399** (2012) 012009 doi:10.1088/1742-6596/399/1/012009

Intermediate stages of the neutralization of multiply charged ions interacting with solid surfaces

M. D. Majkić, N. N. Nedeljković and S. M. D. Galijaš

University of Belgrade, Faculty of Physics, P.O. Box 368, Belgrade, Serbia

E-mail: hekata@ff.bg.ac.rs

Abstract.
We consider the electron capture (neutralization) into Rydberg states of multiply charged ions interacting with solid surface in the normal (escaping) and in the grazing incidence geometry. The time-symmetrized two-state vector model is used to investigate the intermediate stages of the population dynamics. For the fixed initial and final states of the active electron, the two wave functions which evolve simultaneously in two opposite directions of time (by two scenarios), are used to describe the transitional electron state. In both considered geometries, the main idea is to obtain the information about the position and the magnitude of the population process for the fixed initial and final states of the active electron. The results are compared with the classical overbarrier predictions and the measured kinetic energy gain due to the image acceleration of the ions. It is demonstrated that the ionic velocity influences the ion-surface distance at which the formation of the particular intermediate Rydberg state is mainly localized, as well as the probability of this formation.

1. Introduction

The interaction of ions/atoms with solid surfaces has been intensively studied both theoretically and experimentally [1,2]. The basic process in the system is the electron exchange between the atomic projectile and the solid surface, which can be described by several theoretical models. In order to elucidate the intermediate stages of the electron exchange dynamics, we developed the two-state vector model (TVM); the model has been introduced in [3] in which the proton neutralization was considered and has been developed ever since in our group.

In the present article we report the TVM analysis of the intermediate stages of the neutralization process for the multiply charged ions interacting with conducting solid surfaces. The transitional electron state is described simultaneously by two state vectors $|\Psi_1(t)\rangle$ and $|\Psi_2(t)\rangle$ evolving in two opposite directions in time. The first state vector evolves, in the first scenario, from the initial state (electron in the solid) "preselected" at the initial time $t = t_{\rm in}$ towards the future, while the second state vector evolves, in the second scenario, from the fixed final state (electron bound to the ion), "postselected" at the final time $t = t_{\rm fin}$.

The ions SVI, ClVII, ArVIII, KrVIII, and XeVIII, escaping solid surfaces in the normal direction, at low perpendicular velocity $v_\perp \ll 1$ a.u. [4,5], and the ions Ar^{Z+}, Kr^{Z+} and Xe^{Z+} for $Z \in [5, 35]$ (in a.u.) [6] colliding with solid surface at velocity v with parallel component $v_{||} < v_{\rm F}$, where $v_{\rm F}$ is the Fermi velocity of the solid, and with low perpendicular component

($v_\perp \ll 1$ a.u.), are considered as an example.

The TVM developed in [4,5] has been devoted to the normal emergency case. The neutralization probability $P_{\nu_A}(t)$, resulting in the formation of the Rydberg state $|\nu_A\rangle$, has been considered as a "sum" over the initial quantum numbers μ_M of the transition probability density $T_{\mu_M,\nu_A}(t)$; for description of the intermediate stages of the process we used the normalized probability $\tilde{P}_{\nu_A}(t) = P_{\nu_A}/P^{\mathrm{fin}}_{\nu_A}$ and the corresponding rate $\tilde{\Gamma}_{\nu_A}(t)$. The quantity $\tilde{P}_{\nu_A}(t)$ represents the neutralization probability at time t under the condition that the state $|\nu_A\rangle$ is populated at time $t = t_{\mathrm{fin}}$, i.e. $\tilde{P}_{\nu_A}(t_{\mathrm{fin}}) = 1$. In [6] we adapted the model to the scattering geometry. Under the grazing incidence geometrical conditions we calculated the intermediate transition probability density $T_{\mu'_M,\nu_A}(t)$ for shifted initial quantum numbers μ'_M, the velocity dependent intermediate transition probability $T_{\nu_A}(t) = \sum_{\mu'_M} \langle f\rangle_{\Omega'_k} T_{\mu'_M,\nu_A}(t)$, and the population probability $P^{\mathrm{P}}_{\nu_A}(t) = 1 - \exp[-T_{\nu_A}(t)]$. We note that, in the scattering geometry, the Fermi-Dirac distribution f of the electron momenta in solid is kinematically modified ($\boldsymbol{k} \to \boldsymbol{k}' = \boldsymbol{k} - \boldsymbol{v}_\parallel$), and averaged.

2. Calculation of the population probability $P^P_{\nu_A}(t)$

2.1. Formulation of the problem

The present report is mainly devoted to the scattering geometry. That is, we consider the intermediate stages of the Rydberg state population of multiply charged ions (core charge $Z \gg 1$) impinging a solid surface at velocity $\boldsymbol{v} = \boldsymbol{v}_\perp + \boldsymbol{v}_\parallel$, where $\boldsymbol{v}_\perp$ and $\boldsymbol{v}_\parallel$ are the perpendicular and parallel velocity components, respectively, see figure 1. At the initial time t_{in} the ion is on the ion-surface distance $R_{\mathrm{in}} \to \infty$, which decreases in time according to the law $v_\perp = -\mathrm{d}R/\mathrm{d}t$. The neutralization of the multiply charged ions interacting with solid surface in the escaping geometry can be considered as a special case $v_\parallel = 0$.

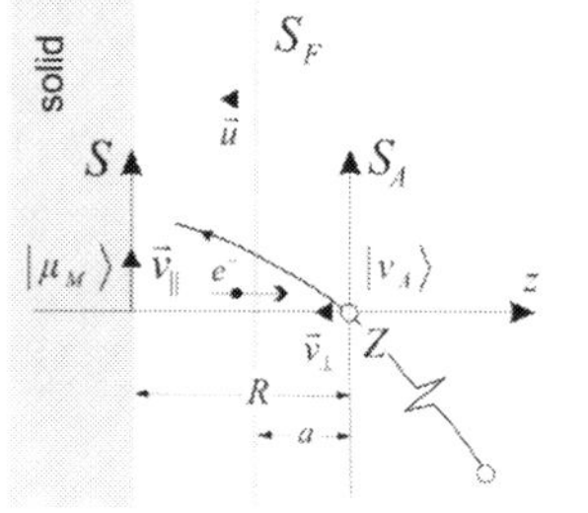

Figure 1. The TVM description of the population dynamics under the scattering conditions. At ion-surface distance R the electron can be resonantly captured by the ion (and recaptured by the solid) with probability $P^{\mathrm{P}}_{\nu_A}(t)$.

The TVM analysis of the population process in the scattering geometry requires some new elements in comparison to the normal emergency case. We formulate the problem in the system S which moves with velocity $\boldsymbol{v}_\parallel$ along the surface, with z axis along the instant ion-surface direction, see figure 1. At the time t the quantum behaviour of the active electron in the system S is simultaneously described by two state vectors $|\Psi_1(t)\rangle$ and $|\Psi_2(t)\rangle$. The kinematic modification can be used to connect the state $|\Psi_1(t)\rangle$ in the S and the state $|\Psi_1^{(0)}(t)\rangle$ in the system $S^{(0)}$ in rest (for $v_\parallel = 0$): $|\Psi_1(t)\rangle = |\Psi_1^{(0)}(t)\rangle/_{\boldsymbol{k}\to\boldsymbol{k}'=\boldsymbol{k}-\boldsymbol{v}_\parallel}$, where $\boldsymbol{k}$ is the momentum of the electron in solid. This transformation induces an effective change of the Fermi-Dirac distribution f of the vectors $\boldsymbol{k}$, which becomes dependent on the intensity and orientation of $\boldsymbol{k}'$; using the angle averaging $\langle f\rangle_{\Omega_{k'}}$ we obtain the necessary k'-distribution for our model. The state $|\Psi_2(t)\rangle$ in the S and the state $|\Psi_2^{\rm A}(t)\rangle$ in the rest frame $S_{\rm A}$ of the moving ion (which moves at velocity $\boldsymbol{v}_\perp$ in respect to S) are connected by the relation $|\Psi_2(t)\rangle = \exp(i\hat{G}_\perp)|\Psi_2^{\rm A}(t)\rangle$, where $\hat{G}_\perp = \boldsymbol{v}_\perp \cdot \boldsymbol{r} - v_\perp^2 t/2$ is the Galilei translation factor.

Within the framework of the TVM we take into account both the initial and the final conditions: $|\Psi_1^{(0)}(t_{\rm in})\rangle = |\mu_{\rm M}\rangle$ and $|\Psi_2^{\rm A}(t_{\rm fin})\rangle = |\nu_{\rm A}\rangle$. The parabolic state $|\mu_{\rm M}(R)\rangle = |\gamma_{\rm M}, n_{1\rm M}, m_{\rm M}\rangle$ is the eigenstate of the in-Hamiltonian $\hat{H}_1(R)$ of the first scenario and the spherical state $|\nu_{\rm A}(R)\rangle = |n_{\rm A}, l_{\rm A}, m_{\rm A}\rangle$ is the eigenstate of the out-Hamiltonian $\hat{H}_2(R)$ of the second scenario. For $R \to \infty$, the state $|\mu_{\rm M}(R)\rangle \to |\mu_{\rm M}\rangle$ describes the electron localized in solid, and for $R = R_{\rm fin}$ the state $|\nu_{\rm A}(R)\rangle \to |\nu_{\rm A}\rangle$ describes the electron that is completely localized around the ionic core. The energy spectrum of the Hamiltonian $\hat{H}_1(R)$ in the system $S^{(0)}$ is continuous $E_{\rm M} = -\gamma_{\rm M}^2/2$ (measured from the vacuum level of the ion-surface system). Note that the energy $\mathcal{E}_{\rm M} = k^2/2$ (measured from the bottom level U_0 of the solid) is connected with $E_{\rm M}$ by the relation $\gamma_{\rm M}^2/2 + k^2/2 = U_0$; the same relation holds for the shifted energies $E'_{\rm M} = -\gamma'^2_{\rm M}/2$ and $\mathcal{E}'_{\rm M} = k'^2/2$, relevant for $v_\parallel \neq 0$. The operator $\hat{H}_2(R)$ has a discrete spectrum $E_{\rm A}(R) = -\gamma_{\rm A}^2(R)/2$.

State $|\Psi_1^{(0)}(t)\rangle$ evolves by the first scenario according to the law $|\Psi_1^{(0)}(t)\rangle = \exp\left(-{\rm i}\int_{t_{\rm in}}^t \hat{H}_1(R){\rm d}t\right)|\mu_{\rm M}\rangle$. The second state $|\Psi_2^{\rm A}(t)\rangle$ evolves teleologically toward the fixed final state (by the second scenario). It means that the final state evolves backward in time: $|\Psi_2^{\rm A}(t)\rangle = \exp\left(-{\rm i}\int_{t_{\rm fin}}^t \hat{H}_2(R){\rm d}t\right)|\nu_{\rm A}\rangle$. The form of the Hamiltonians $\hat{H}_1(R)$ and $\hat{H}_2(R)$ is independent of the scattering geometry [4,5,6]. The ionic core polarization is taken into account using the Simons-Bloch potential.

2.2. *Basic expressions of the TVM*

The intermediate stages of the neutralization in the considered scattering geometry are described by two-state probability amplitude $A_{\mu'_{\rm M},\nu_{\rm A}}(t) = \langle\Psi_2(t)|\hat{P}_{\rm A}(t)|\Psi_1(t)\rangle$, where $\mu'_{\rm M} = (\gamma'_{\rm M}, n_{1\rm M}, m_{\rm M})$ and $\hat{P}_{\rm A}(t) = \int_{V_{\rm A}} |\boldsymbol{r}_{\rm A}\rangle\langle\boldsymbol{r}_{\rm A}|{\rm d}V$ is the projection operator onto the "ionic" region $V_{\rm A}$ (right from the Firsov plane $S_{\rm F}$ in figure 1). The intermediate probability per unit energy parameter γ'_M is given by relation: $T_{\mu'_{\rm M},\nu_{\rm A}}(t) = |A_{\mu'_{\rm M},\nu_{\rm A}}(t)|^2$. The corresponding intermediate transition probability $T_{\nu_{\rm A}}(t)$, in which the overall effectively extended conduction band participates, represents a "sum" over $\mu'_{\rm M}$:

$$T_{\nu_{\rm A}}(t) = \int \sum_{n_{1\rm M}, m_{\rm M}} \langle f\rangle_{\Omega_{\boldsymbol{k}'}} T_{\mu'_{\rm M},\nu_{\rm A}}(t) d\gamma'_{\rm M}. \tag{1}$$

The intermediate population probability $P^{\rm P}_{\nu_{\rm A}}(t)$ of the Rydberg state that will evolve into the final state $|\nu_{\rm A}\rangle$ is given by

$$P^{\rm P}_{\nu_{\rm A}}(t) = 1 - \exp\left[-T_{\nu_{\rm A}}(t)\right]. \tag{2}$$

In (1) the quantity $\langle f\rangle_{\Omega_{k'}} = \int_{4\pi} f {\rm d}\Omega_{\boldsymbol{k}'}/4\pi$ is the angle averaged (over all orientations of $\boldsymbol{k}'$ inside the solid) Fermi-Dirac distribution $f = I_{[0,E_F]}(k^2/2)$. By $E_F = v_{\rm F}^2/2$ we denoted the Fermi energy, where $v_{\rm F}$ is the Fermi velocity; in what follows we shall restrict our consideration to the case $v_\parallel < v_{\rm F}$. In that case we have: $\langle f\rangle_{\Omega_{k'}} = 1$ for $\phi_\uparrow < \gamma'^2_{\rm M}/2 < U_0$, and [7,8]

$$\langle f\rangle_{\Omega_{k'}} = \frac{v_{\rm F}^2 - \left(\sqrt{2U_0 - \gamma'^2_{\rm M}} - v_\parallel\right)^2}{4\sqrt{2U_0 - \gamma'^2_M v_\parallel}}, \phi_\downarrow < \gamma'^2_{\rm M}/2 < \phi_\uparrow. \tag{3}$$

In the expressions for $\langle f\rangle_{\Omega_{k'}}$ we use the notation $\phi_\downarrow = U_0 - (v_{\rm F} + v_\parallel)^2/2$ and $\phi_\uparrow = U_0 - (v_F - v_\parallel)^2/2$. In the case of $v_\parallel = 0$ we have $\phi_\downarrow = \phi_\uparrow = \phi$, where ϕ is the solid work function.

The TVM is based on the mixed flux $I_{\nu_A,\nu_A}(t)$ through the Firsov plane S_F, which represents a specific entanglement of the wave functions $\Psi_1(\boldsymbol{r},t)$ and $\Psi_2(\boldsymbol{r},t)$, i.e., we have $T_{\nu'_A,\nu_A}(t) = \int_{t_{in}}^{t} I_{\mu'_M,\nu_A}(t)dt$ The advantage of the model is that for the calculation of the mixed flux it is sufficient to know these wave functions on the Firsov plane only, which is, during the most intensive electron transition, positioned sufficiently far from both the surface and the ionic core. Therefore, taking into account that the electron transitions occur at large ion-surface distances R, we can apply the appropriate asymptotic methods. The wave functions $\Psi_1^{(0)}(\boldsymbol{r},t) = \Phi_{MA,\mu_M} \exp\left[f_M + i\left(\gamma_M^2/2\right)t\right]$ and $\Psi_2^A(\boldsymbol{r},t) = \Phi_{AM,\nu_A} \exp\left[f_A + i\left(\gamma_A^2/2\right)t\right]$ are expressed as space-time modifications of the eigenfunctions Φ_{MA,μ_M} and Φ_{AM,ν_A}, [9,10] of the Hamiltonians $\hat{H}_1$ and $\hat{H}_2$, respectively. For the function Φ_{MA,μ_M}, we use the expression analytically continued from the metallic region (through the barrier) into the ionic region [11] by the phase-integral method [12], using the appropriate JWKB connection formulas [13], obtained from the eigenfunctions $\tilde{\Phi}_{AM,\nu_A}$ of the atomic Hamiltonian $\tilde{\hat{H}}_2$.

The explicit expression for the mixed flux in the considered scattering geometry is given by [6]:

$$I_{\mu'_M,\nu_A}(t) = \frac{i}{2} e^{iw(R)t} \left[\gamma'_M + \gamma_A(R)\right] e^{f_A^* + f'_M} \int_{S_F} \Phi_{AM,\nu_A}^* \Phi_{MA,\mu'_M} dS, \tag{4}$$

where $f_{M'} = f_M/_{\gamma_M \to \gamma'_M}$ and

$$w(R) = \frac{\gamma'^2_M}{2} - \frac{\gamma_A^2(R)}{2} \tag{5}$$

is the energy difference considered from the system S. Using the expression for the mixed flux and taking into account that in the considered case the electron capture is determined by the resonant condition $\gamma'_M = \gamma_A(R)$, for the transition probability we get:

$$T_{\nu_A}(t) = T_{\nu_A}^{(0)} \langle f \rangle_{\Omega_{k'}} f_\gamma(\gamma'_M) \frac{v_\perp [\gamma'_M + \gamma_A(R)]^2}{\gamma'_M \tilde{\beta}} \left(1 + \frac{2\tilde{\alpha}}{\tilde{\beta}} \frac{1}{R}\right) R^{2\tilde{\alpha}} e^{-2\tilde{\beta}R}. \tag{6}$$

The quantities $T_{\nu_A}^{(0)}$ and $f_\gamma(\gamma'_M)$ and the parameters $\tilde{\alpha}$ and $\tilde{\beta}$ are explicitly given in reference [6]. The population probability $P_{\nu_A}^P(t)$ is given by (2). The population rates $\Gamma_{\nu_A}^P(t) = dP_{\nu_A}^P(t)/dt$, change sign during the ionic motion, which can be interpreted as interplay of the population and reionization processes. The position of the positive maximum of the rate determine the neutralization distances R_c^N, at which the population process is mainly localized.

3. RESULTS

3.1. Selectivity of the population process and the role of the projectile velocity

As an illustrative example, in figure 2 we present the quantities $P_{\nu_A}^P(t)$ for the population of some characteristic Rydberg states of the ions ArVIII and XeXIII slowly impinging the Al-surface. The most important fact that can be recognized from figure 2 is the selectivity of the population process: the larger n_A states are populated with lower $P_{\nu_A}^{\max}$. Moreover, the Rydberg states $n_A > n_c$ are populated with very low probabilities (considering the maxima); the values n_c are characteristic for the particular ion-solid system and the kinematic conditions. It means that, within the framework of the TVM, one can estimate the Rydberg states ($n_A = n_c$) that will be populated first during the grazing incidence of multiply charged ions on solid surface.

The role of parallel projectile velocity $v_\parallel$ is discussed in figure 3 in which we consider the population dynamics of the ion ArX. In figures 3(a) and 3(b) we present the intermediate population probability $P_{\nu_A}^P(t)$ via ion-surface distance R for the population of the Rydberg states with $n_A = 10$ and $n_A = 12$, respectively, for $l_A = 1$. We consider the population curves

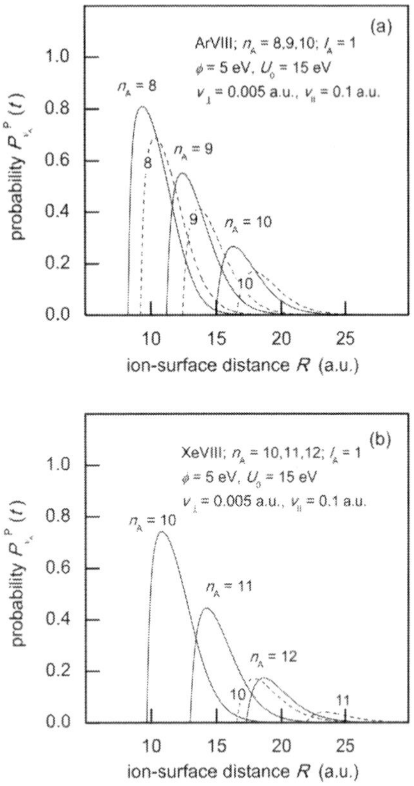

Figure 2. Population probabilities $P^{\rm P}_{\nu_{\rm A}}(t)$ of the Rydberg states with (solid curves) and without (dashed curves) core polarization taken into account.

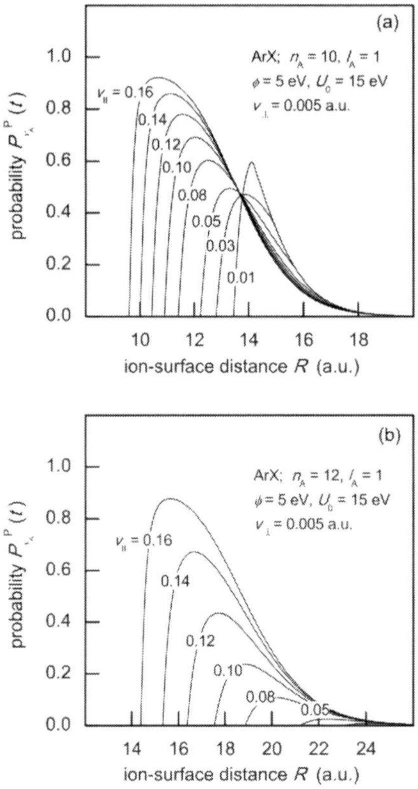

Figure 3. Population probabilities $P^{\rm P}_{\nu_{\rm A}}(t)$ of the Rydberg states of the ArX ion for some characteristic parallel velocities $v_\parallel$ presented in figure in a.u.

for some characteristic values of the projectile parallel velocity $v_\parallel \leq 0.16$ a.u. and for $v_\perp = 0.005$ a.u.

The essential difference between the results presented in these two figures is that the Rydberg state $n_{\rm A} = 10$ of the ArX ion considered in figure 3(a) could be populated in the normal incidence case ($v_\parallel = 0$), while the state $n_{\rm A} = 12$ presented in figure 3(b) could not. The common property of the presented population curves is the $v_\parallel$-behaviour: the distributions are shifted toward smaller ion-surface distances R with increasing $v_\parallel$. However, the behavior of the probability maxima is different. In the first case we recognize a decrease of $P^{\rm max}_{\nu_{\rm A}}$ with increasing $v_\parallel$ for $v_\parallel \leq 0.5$ a.u. and increase of probability maxima with further increase of $v_\parallel$, while in the second case the values $P^{\rm max}_{\nu_{\rm A}}$ increase with increasing $v_\parallel$ for all considered parallel velocities.

3.2. TVM and comparison with experiments

Direct observation of the neutralization distances $R^{\rm N}_{\rm c}$ has not yet been investigated experimentally in the case of multiply charged ions. However, an indirect insight into the neutralization dynamics is possible if we compare the measured projectile kinetic energy gain $\Delta E^{(Z)}$ [14,15] with the corresponding quantity calculated on the base of the TVM $R^{\rm N}_{\rm c}(n_{\rm c})$ values, where the $n_{\rm c}$ represents the principal quantum number of the "highest" Rydberg level that can be populated with probability $P^{\rm max}_{\nu_{\rm A}} \geq 0.2$.

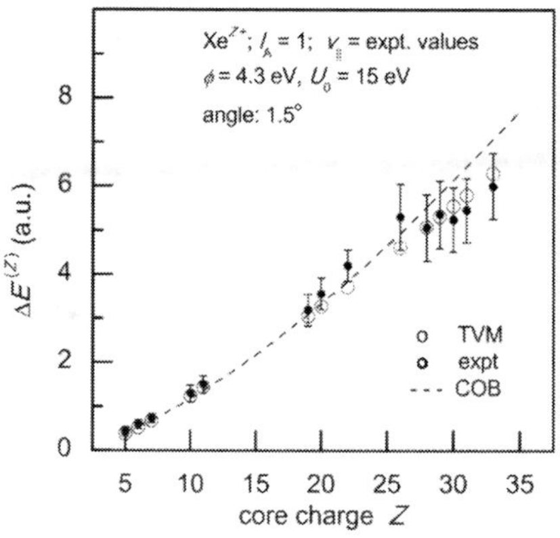

Figure 4. Kinetic energy gain $\Delta E^{(Z)}$ via ionic core charge Z of the ions Xe^{Z+}, under the experimental conditions. Solid circles represent the experimental data taken from [15] and open circles are the TVM values. All data are for $v_\perp = v \sin 1.5°$, for $\phi = 4.3$ eV, and for $U_0 = 15$ eV. Dashed curve is the COB quantity $\Delta E^{(Z)} \approx \phi Z^{3/2}/3\sqrt{2}$ for $\phi = 4.3$ eV [16].

Using the "staircase" model, we can calculate the energy gain of the ionic projectile $Z+$, considering the multielectron capture process as cascade neutralization $Z \to Z-1 \to ...1 \to 0$. According to the model, the charge Q of the projectile is instantaneously reduced to the charge $Q-1$ when the neutralization distance $R_c^{N,Q}$ for the population of the Rydberg state $n_A = n_c(Q)$ is reached. The image-force interaction of the projectile with the surface is changing accordingly: in each interval $R \in [R_c^{N,Q}, R_c^{N,Q-1}]$ the intensity of the image force is given by $F = (Q-1)^2/4R^2$. At the end of the neutralization process, due to the mentioned interaction of the projectile and the solid, we have $\Delta E^{(Z)} = \sum_{Q=1}^{Z}(2Q-1)/4R_c^{N,Q}$. We note that it is assumed that at $R_c^{N,Q}$ electron is captured into the field of the effective core charge $Q_c = Q$. In figure 4 we compare the TVM predictions (open circles) with the existing experimental results (full circles) [15], considering the ions Xe^{Z+}, where $Z \in [5, 35]$; we use the recursive relation $\Delta E^{(Z)} = \Delta E^{(Z-1)} + (2Z-1)/4R_c^{N,Z}$, taking for $\Delta E^{(4)}$ the experimental data.

For the surface parameters in figure 4 we use the values $\phi = 4.3$ eV and $U_0 = 15$ eV; the work function $\phi = 4.3$ eV is taken in accordance with experimental conditions. By dashed curve in figure 4 we present the corresponding COB expression $\Delta E^{(Z)}$ [16]. For $v_\parallel$ we take the experimental data $v_\parallel \approx v_{\rm in}$, where $v_{\rm in}$ is the initial projectile velocity; perpendicular ionic velocity is given by $v_\perp = v \sin \Phi$, where $\Phi = 1.5°$. We point out that, according to experimental conditions, the projectile velocity is 0.087 a.u. for the ions of low charge $Z \leq 11$; the ions of higher charge ($20 \leq Z \leq 33$) have been extracted through a fixed potential difference, and the corresponding projectile energies depended of the ionic charge: $E = 3.7Z$ keV, i.e. $v \in$ [0.15 a.u., 0.2 a.u.][15]. According to figure 4 we conclude that the TVM results are in accord with the experimental values including the experimentally observed saturation for $Z \gg 1$. The COB curve is in a good agreement with experimental values for low Z values, while in the large Z region the COB values are systematically larger in comparison to the experimental data. The obtained agreement of the TVM with available experimental results in the large Z case should be addressed to parallel velocity effect.

4. Concluding remarks

In this article we report the TVM study of the intermediate stages of the Rydberg state population of multiply charged ions impinging a conducting solid surface under the grazing incidence geometry. Within the framework of the model, the population probability distributions $P_{\nu_A}^{\rm P}(t)$ are represented by the the velocity dependent peak-shaped curves [6]. This is the specific feature of the scattering geometry; the corresponding physical picture of the population process is different in comparison to the normal incidence geometry case considered in [4,5]. The analytical form of the obtained population probabilities enables us to discuss the localization and the selectivity of the processes expressed via neutralization distances R_c^N and probability maxima $P_{\nu_A}^{max}$, for various relevant parameters of the ion-surface system.

The TVM results are compared indirectly with the existing theoretical (COB) and experimental results [15]. Due to the $v_{||}$-dependence of the quantity $R_c^N(n_c)$, the obtained kinetic energy gain also depends on the projectile parallel velocity in accordance with experimental findings.

Few additional comments may be relevant for further investigations of population dynamics of multiply charged ions interacting with solid surfaces, based on the TVM.

First, the TVM treatment of the present paper can be used to analyze the other ions as well as other solid surfaces in comparison to those considered in references [4,5,6]. Also, by varying the value of the parallel velocity $v_{||}$ one can obtain a more complete insight into the population dynamics; the extension of our model to the ionic velocities close to Fermi velocity v_F and the larger values is also possible. The influence of the angle of incidence on the population dynamics can be analyzed using the obtained $v_{\perp}$-dependence of the population probability.

Second, the TVM developed in the present paper for the grazing incidence of multiply charged ions with population of the low-l_A Rydberg states can be extended into the large-l_A region, by taking into account that the corresponding electron transitions do not take place in the vicinity of the z axis [17].

Acknowledgments
This work was supported in part by the Ministry of Education and Science, Republic of Serbia (Project 171016).

References

[1] Burgdörfer J in *Review of Fundamental Processes and Applications of Atoms and Molecules* ed Lin C D (World Scientific: Singapore, 1993)
[2] Winter H 2002 *Phys. Rep.* **367** 387
[3] Nedeljković N N, Nedeljković Lj D, Janev R K and Mišković Z L 1991 *Nucl. Instrum. Methods Phys. Res.* B **58** 519
[4] Nedeljković N N and Majkić M D 2007 *Phys. Rev.* A **76** 042902
[5] Nedeljković N N, Majkić M D, Galijaš S M D and Mitrović S B 2008 *App. Surf. Sci.* **254** 7000
[6] Nedeljković N N, Majkić M D and Galijaš S M D 2012 *J. Phys. B: At. Mol. Opt. Phys.* to be published
[7] Zimny R, Mišković Z L, Nedeljković N N and Nedeljković Lj D 1991 *Surf. Sci* **255** 135
[8] Mišković Z L and Janev R K 1989 *Surf. Sci.* **221** 317
[9] Nedeljković N N, Nedeljković Lj D, Vojvodić S B and Mirković M A 1994 *Phys. Rev.* B **49** 5621
[10] Nedeljković Lj D and Nedeljković N N 1998 *Phys. Rev.* B **58** 16455
[11] Nedeljković N N 1980 *Fizika* (Zagreb) **12** 275
[12] Fröman N and Fröman P O *Physical Probles Solved by the Phase Integral Method* (Cambridge: Cambridge University Press, 2002)
[13] Fröman N and Fröman P O 1965 *JWKB Aproximation, Contribution to the Theory*, Amsterdam: North-Holand
[14] Winter H 1992 *Europhy. Lett.* **18** 207
[15] Winter H, Auth C, Schuch R and Beebe E 1993 *Phys. Rev. Lett.* **71** 1939
[16] Lemell C, Winter H P, Aumayer F, Burgdörfer J and Mezer F 1996 *Phys.Rev.* A **53** 880
[17] Nedeljković N N, Nedeljković Lj D and Mirković M A 2003 *Phys. Rev.* A **68** 012721

Electron emission from a metal nano-tip by ultrashort laser pulses

G Wachter[1], C Lemell[1] and J Burgdörfer[1]

[1]Institute for Theoretical Physics, Vienna University of Technology, Wiedner Hauptstr. 8-10, A-1040 Vienna, Austria, EU

E-mail: georg.wachter@tuwien.ac.at

Abstract. We theoretically investigate the interaction of near-infrared few cycle laser pulses of moderate intensity with nano-scale metal tips. Macroscopic field enhancement leads to coherent electron emission from the tip apex. Electron spectra are simulated with time-dependent density functional theory (TDDFT). We investigate the dependence of the simulated electron spectra on the choice of exchange-correlation potential and atomic core pseudo-potential.

1. Introduction

Recent experimental and theoretical studies have shown [1–3] that a nano-tip illuminated by near-infrared moderate intensity few cycle laser pulses affords the possibility to study the effect of a strong time-dependent field acting on a solid-state system. The quantum dynamics near the surface can be inferred from analyzing the electrons which are photo-emitted by the laser pulse. The measured electron energy spectra show clear signs of strong field physics even at low nominal laser intensities which can be traced back to the field enhancement near the apex of the tip due the so-called lightning-rod effect. A sub-wavelength sized region near the tip apex experiences an enhanced laser field (field enhancement factor ~ 5 for tip radius ~ 10 nm). This allows for the observation of strong field effects at low nominal laser intensities and also lies at the heart of the observability of coherence effects as the emitted electrons originate from a tightly confined region in both space and time. The coherence of the emission process is reflected in the electron energy spectra, their carrier-envelope phase dependence and their angular distribution. Within a semi-classical description, the observed multi-photon peaks can be traced back to the periodic emission of coherent electron wave packets equi-spaced in time given by the laser period, leading to interference modulation of the energy spectra with the period of the photon energy [4]. Another feature typically associated with strong-field physics is found in the high-energy part of the electron spectra showing a plateau region followed by a cut-off. The latter stem from electrons undergoing a process known as "rescattering" in strong field physics for atoms: first, the electron tunnels out from its parent core near a maximum of the laser field. Subsequently, it is driven back towards the core as the laser field changes sign attaining energies of $\sim 3U_{\mathrm{p}}$ with the ponderomotive potential $U_{\mathrm{p}} = F_0^2/4\omega^2$ for laser amplitude F_0 and laser frequency ω. Upon rescattering in backward direction from the potential gradient near the core, the electron gains additional energy up to a final energy of $10\,U_{\mathrm{p}}$ determining the position of the observed cut-off [5]. The high-energy part of electron spectra from nano-tips has been observed [2] and its origin can unambiguously be traced to such a process [3, 4].

The plan for this paper is as follows: first, we review our quantum mechanical model for electron emission from a nano-tip based on time-dependent density functional theory (TDDFT). We show a sample simulation result where the rescattering process can be directly observed in the time-dependent electron density. We then focus on the dependence of the simulated electron spectra on two parameters entering the TDDFT description of the process, the exchange-correlation potential and the atomic core pseudo-potential used to simulate the outermost layer of atoms. Atomic units are used unless stated otherwise.

2. Time-dependent density functional theory for electron emission from nano-tips

We briefly review our microscopic description of electron emission based on time-dependent density functional theory (TDDFT) [6–9]. Since in the experiment the tip radius ($\sim$ 10 nm) is much larger than the length scale of the electronic system (Fermi wavelength $\sim$ 0.1 nm), translational symmetry in the surface plane is approximately conserved on an atomic scale. We hence restrict ourselves to a one-dimensional model, treating the coordinate along the surface normal z as reaction coordinate along which the macroscopic field enhancement is observed. A detailed description is given in [3]. In short, the time-dependent electron density $n(z,t)$ is expanded into Kohn-Sham orbitals $\psi_k(z,t)$

$$n(z,t) = \sum_{k=1}^{n_{\text{occ}}} c_k |\psi_k(z,t)|^2 \quad , \tag{1}$$

where n_{occ} is the number of occupied orbits up to the Fermi energy E_{F}. We use a metal slab of $\sim$200 a.u. width, yielding $n_{\text{occ}} \sim 50$ orbitals representing the conduction band. The weight coefficients c_k are derived from the projection of the three-dimensional Fermi sphere onto the tip axis [10] such that eq. 1 gives the ground state density for $t \to -\infty$. The electron density is expressed in terms of the Wigner-Seitz radius $r_s = 2.334$ a.u. according to the s-electron density of tungsten giving a Fermi energy of $E_{\text{F}} = 9.2$ eV.

The time evolution of the electron density is given by the time-dependent coupled Kohn-Sham equations

$$i\partial_t \psi_k(z,t) = \left\{ -\frac{1}{2}\partial_z^2 + V[n(z,t)] + V_{\text{ext}}(z,t) \right\} \psi_k(z,t) \, , \tag{2}$$

where $V[n(z,t)]$ contains the ground state, electrostatic, and exchange-correlation potentials. For the ground state we use a smooth jellium effective single particle potential of the form [11]

$$V_{\text{jellium}}(z) = \begin{cases} -\frac{V_0}{A\exp(Bz)+1} \, , z < z_{im} \\ \frac{1-\exp(-bz)}{4z} \, , z \geq z_{im} \end{cases} \quad , \tag{3}$$

where the depth of the potential well is adjusted to $V_0 = E_{\text{F}} + W$ with the work function W, the position of the image plane relative to the jellium edge is taken as $z_{im} = -0.2r_s + 1.25$, the parameter governing the transition to the asymptotically correct $1/(4z)$ image potential is chosen as $b = k_F$, and consequently $A = 4V_0/b - 1$ and $B = V_0/(4V_0/b - 1)$. This ground state potential has the advantage that the work function can be continuously adjusted. A clean tungsten (310) surface has a work function of $W_{(310)} = 4.35$ eV, which is, however, sensitive to surface adsorbates [12] and can serve only as a first estimate. Best agreement to experiment was found for a work function of $W = 6.2$ eV [3] which is also used throughout this work. Apart from the value of the work function, the exact shape of the ground state potential is of minor importance. We have tested other potentials including potentials accurately reproducing the electronic surface structure for a number of materials [13] yielding similar results.

To account for rescattering, a localized potential simulating the atomic cores of the first atomic layer is added to the jellium potential (see section 5). It is parametrized by a screened soft-core Coulomb potential

$$V_{\rm atom}(z) = -\frac{1}{1+|z|}e^{-|z|/\lambda_{\rm TF}} \tag{4}$$

with the Thomas-Fermi screening length $\lambda_{\rm TF} \approx 1$ a.u. for the electron gas and is centered at the position of the first atom layer, i.e. one half of the effective lattice position $a \times \frac{\sqrt{3}}{4} \approx 2.6$ a.u. inside the jellium edge for a tungsten (310) surface.

The electrostatic potential is derived from the Poisson equation in one dimension, $\frac{d^2}{dz^2}V_{es}(z,t) = -4\pi n(z,t)$. For the exchange-correlation potential, we assume the local density approximation (LDA) in the Perdew-Zunger parametrization [14] by default. The dependence of simulation results on this parameter is studied in section 4. The external potential from the laser pulse is given in dipole approximation $V_{\rm ext}(z,t) = zF_{\rm eff}(t) + zF_{\rm dc}$ where the enhancement due to the lightning rod effect is already included in $F_{\rm eff}$ and $F_{\rm dc}$ [3]. As in experiment, a small static extraction field $F_{\rm dc}$ is included which is adiabatically switched on before the laser pulse starts.

The Kohn-Sham equations (2) are integrated in real space by the Crank-Nicolson method with a constant time step of 0.05 a.u., extrapolating the time-dependent Hamiltonian at $t+\Delta t/2$ to second order in time. The simulation time is typically 120 fs ($\sim$ 5000 a.u.) which is chosen large enough so that electrons of low energy have passed the detection point (see below). The size of the simulation box was set to 1425 a.u. (9500 grid points) including absorbing potentials near the borders to avoid unphysical reflections. Electron energy spectra are determined [15] by a temporal Fourier transform of the wave functions at a detection point far from the surface ($\sim$ 900 a.u.) ensuring that the laser pulse has terminated at the time of arrival of the wavepacket. More information on calculating electron spectra from TDDFT can be found in [16, 17]. The calculated spectra are broadened by 0.5 eV, corresponding to a typical experimental spectrometer resolution. The electron energy is measured from the Fermi energy since the vacuum level is not well-defined in the presence of a static field.

3. Sample simulation

A typical simulation run is shown in fig. 1 for a laser intensity close to the experimental destruction threshold with a laser amplitude of $F_{\rm eff} = 0.023$ a.u.$=$ 9.77 GV/m ($I_{\rm eff} = 1.9 \times 10^{13}$ W/cm^2). Other parameters are chosen in accordance with experiment [3]. The pulse length is 6.4 fs (FWHM of the intensity envelope) with a photon energy of $\hbar\omega = 0.06$ a.u. corresponding to a near-infrared wavelength of $\lambda = 760$ nm. The carrier-envelope phase is chosen as $\phi_{CEP} = +\pi$ (a $(-\cos)$-pulse for the force), and a small static extraction field of $F_{\rm dc} = 0.00019$ a.u. $= 0.1$ GV/m $\approx 1\%$ $F_{\rm eff}$ is applied.

Electron density is emitted at the field maxima from the tip ($\lesssim 0$) into vacuum ($\gtrsim 0$) (grey vertical arrows in fig. 1). As the laser field changes sign, electrons are driven back towards the surface. They are reflected from the surface approximately at the zero crossings of the electric field (maxima of the vector potential), leading to interference patterns in the real-space density as the fast rescattered electrons overtake the slow direct electrons. Even though the pulse is several optical cycles long and the Keldish parameter $\gamma_{\rm eff} = 2.1 > 1$ is at the border to the multi-photon regime, the density plot is dominated by the two central field maxima where charge is emitted. This double slit in time leads to interferences in energy (multi-photon peaks), which are visible in the position-space density as fringes with increasing slope (examples marked with grey tilted arrows).

Inside the tip ($z \lesssim 0$), the induced density fluctuations act to screen the external laser field. Notably, they are several orders of magnitude larger than the emitted part of the density, but are still small in comparison to the bulk electron density $n_0 = 1.88 \times 10^{-2}$ a.u. The conduction band electrons can thus adjust adiabatically to the slowly varying perturbation of the external field (laser period $\approx$ 100 a.u.). The electron gas responds similar to a harmonic oscillator driven at frequencies far below its resonance since the photon energy of $\hbar\omega = 1.55$ eV is much smaller than the plasmon energy of $\hbar\omega_p = \sqrt{4\pi n_0} = 13.2$ eV. Fine scale Friedel oscillations (wavelength $\sim \lambda_F/2 = 3.8$ a.u.) are visible near the surface, similar to the response to a static field. Screening is complete within a few Friedel oscillations ($\lesssim 10$ a.u.).

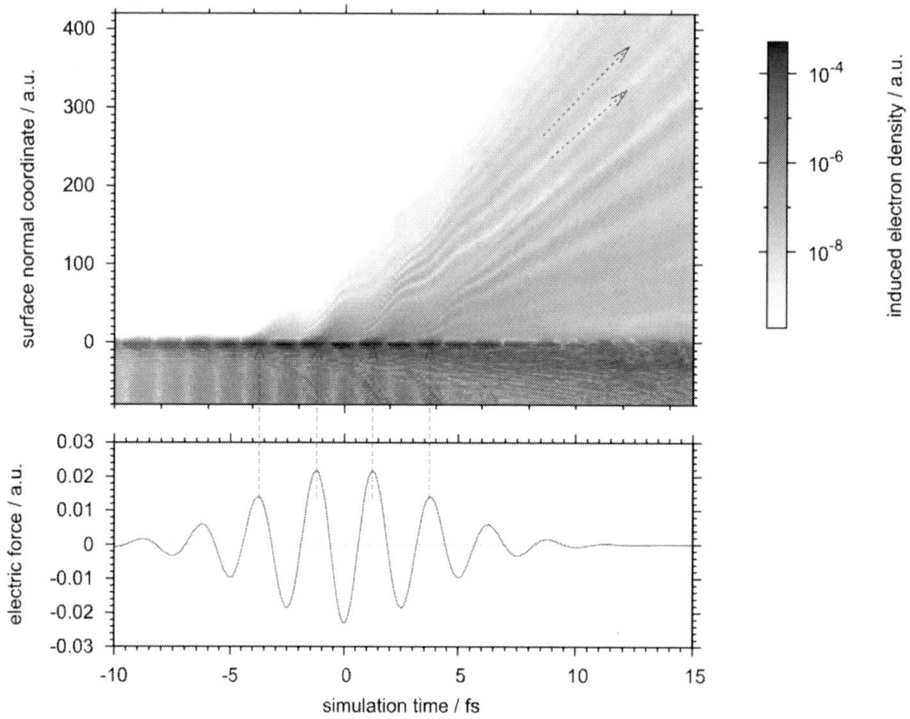

Figure 1. Sample simulation. Top panel: Absolute value of the time-dependent induced density $|n(z,t) - n(z,-\infty)|$ on a logarithmic scale (for comparison, the bulk background density is $n_0 = 1.88 \times 10^{-2}$ a.u.). Bottom panel: Time-dependent external electric field (red solid line). Ionization happens mostly at the maxima of the electric field (marked by vertical arrows). Electrons are subsequently driven back to the surface and rescatter. Interferences in position space with increasing slopes (velocities) are detected as equi-spaced multi-photon peaks in the energy spectra (marked by tilted arrows).

4. Dependence on exchange-correlation potential

As an illustration of the theoretical error bar and the robustness of our simulation to details of the model we show the dependence of our results on the exchange-correlation potential in fig. 2. Here, the enhanced laser field amplitude is $F_{\text{eff}} = 0.021$ a.u. and the remaining parameters are taken as in section 3. We compare two different parametrizations of the local density approximation (LDA), the original Wigner formula (green long dashed line) and the improved parametrization by Perdew and Zunger (PZ81 [14], red solid line), the asymptotically correct generalized gradient approximation (GGA) by van Leeuwen and Baerends (LB94 [18], purple dotted line) and the Hartree approximation, i.e. switching off V_{xc} (blue short dashed line). Remarkably, the main features of the spectra are well reproduced with all four XC potentials: the slope of the direct part, the height of the rescattered part relative to the direct part, and

Journal of Physics: Conference Series **399** (2012) 012010 doi:10.1088/1742-6596/399/1/012010

the position and slope of the cut-off are virtually indistinguishable. Slight deviations are found in phase-sensitive details of the spectra such as the visibility and position of the peaks, likely owing to details of the time-dependent mean field near the surface. The fact that the Hartree potential leads to reasonable results underscores that the time-dependent dynamics are strongly dominated by the external laser field and its time-dependent interaction with the electron density via the classical electrostatic interaction (Poisson equation).

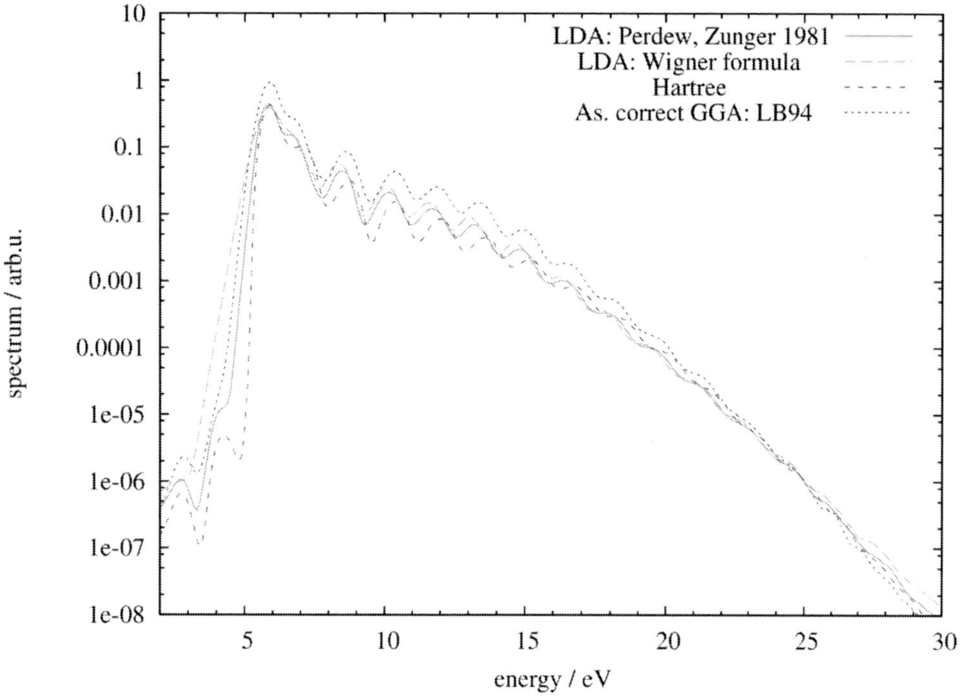

Figure 2. Calculated electron spectra for different exchange-correlation potentials. Local density approximation (LDA) using the Perdew-Zunger parametrization [14] (red solid line), LDA using the Wigner interpolation formula (green wide dashed line), exchange-correlation switched off (Hartree approximation, blue short dashed line), asymptotically correct generalized gradient approximation (GGA) by van Leeuwen and Barends [18] (purple dotted line).

5. Dependence on atomic core potential

Backscattering happens with a high probability when the impinging electron wave function experiences a steep potential gradient. In strong field studies with atomic systems, this potential gradient is provided by the Coulomb potential of the atomic nuclei. On the other hand, our jellium description presented above smears out the potentials of the atomic nuclei to a background potential, resulting in a smooth single-particle potential with a small reflection coefficient. It is therefore of interest to compare simulated spectra with and without the additional potential mimicking an atomic core (eq. 4).

Fig. 3 shows such a comparison of calculated spectra for an enhanced laser amplitude of $F_{\mathrm{eff}} = 0.021$ a.u. The low-energy parts of the spectra (first two peaks) look very similar but striking differences appear in the high energy parts. While a plateau structure is observed for a screened Coulomb potential (green wide dashed line) as well as a simple delta-like potential (blue short dashed line), a jellium potential alone (red solid line) does not show a plateau because the reflection coefficient is so low that it is covered by the direct part. We therefore infer that backscattering is indeed induced by atomic cores near the surface.

We note that in fig. 3 the essential features of the spectrum (decay, plateau, and cutoff) are qualitatively reproduced equally well by a screened Coulomb potential or by a delta function. The only difference lies in the relative height of the plateau which reflects the magnitude of the reflection coefficient for recolliding electrons. Details of the scattering potential, encoded in the scattering phase, cancel out in the interferences within the rescattered wave packets [4] and leave no signature in the electron spectra under the present conditions.

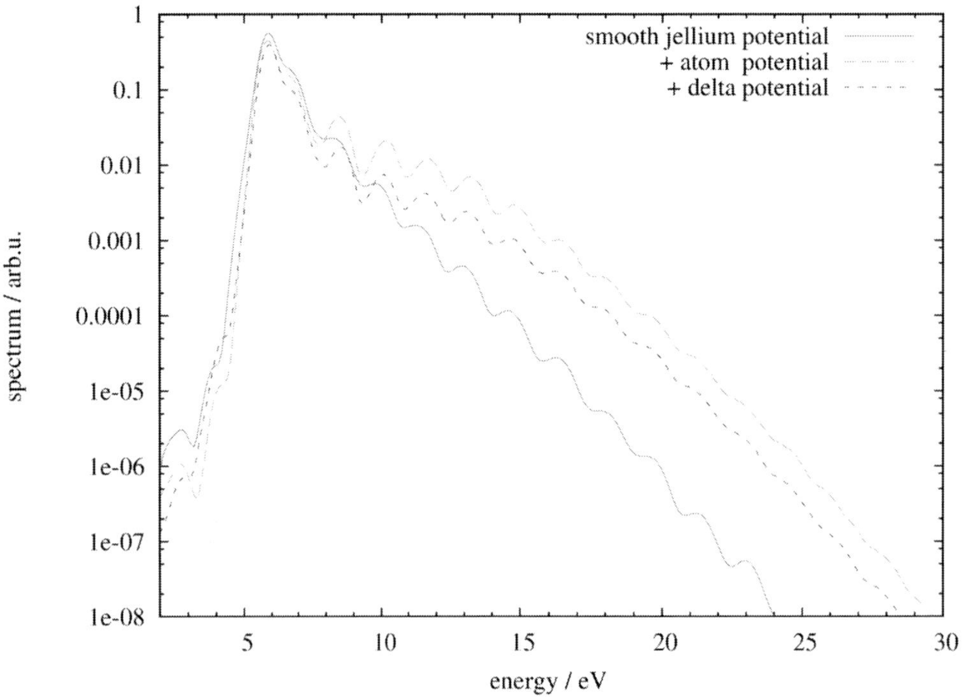

Figure 3. Calculated electron spectra for a jellium potential (red solid line), jellium with added screened smoothed Coulomb potential (green wide dashed line), and jellium with added delta potential (blue short dashed line). The added potentials boost the rescattering probability to levels necessary for the observation of the plateau. Apart from this effect, the different model potentials do not change the calculated spectra.

6. Conclusions

We have presented a theoretical study of electron emission from nano-scale metal tips by near-infrared few cycle laser pulses of moderate intensity. We have probed the sensitivity of the simulation to the employed approximation to the exchange-correlation potential demonstrating that exchange and correlation effects play only a small role. The dependence of the simulated electron spectra on the atomic core pseudo-potential has been investigated corroborating that rescattering is induced by the atomic cores of the atoms in the first layer.

Acknowledgments

This work was supported by the Austrian Science Foundation FWF under Proj. Nos. SFB-041 ViCoM and P21141-N16. G.W. thanks the International Max Planck Research School of

Advanced Photon Science for financial support. The authors would like to thank M. Schenk, M. Krüger and P. Hommelhoff for discussions.

References

[1] Schenk M, Krüger M and Hommelhoff P 2010 *Physical Review Letters* **105** 257601+
[2] Krüger M, Schenk M and Hommelhoff P 2011 *Nature* **475** 78–81 ISSN 0028-0836
[3] Wachter G, Lemell C, Burgdörfer J, Schenk M, Krüger M and Hommelhoff P 2012 *Physical Review B* **86** 035402+
[4] Krüger M, Schenk M, Hommelhoff P, Wachter G, Lemell C and Burgdörfer J 2012 *New Journal of Physics* **14** 085019+ ISSN 1367-2630
[5] Paulus G G, Becker W, Nicklich W and Walther H 1994 *Journal of Physics B: Atomic, Molecular and Optical Physics* **27** L703–L708 ISSN 0953-4075
[6] Liebsch A 1997 *Electronic excitations at metal surfaces* (New York: Plenum)
[7] Burke K and Gross E K U 1998 *Density Functionals: Theory and Applications* (Berlin: Springer-Verlag)
[8] Maitra N T, Burke K, Appel H, Gross E K U and van Leeuwen R 2002 Ten Topical Questions in Time Dependent Density Functional Theory *A celebration of the contributions of Robert Parr* ed Sen K D (Singapore: World Scientific) pp 1186–1225
[9] Lemell C, Tong X M, Krausz F and Burgdörfer J 2003 *Physical Review Letters* **90** 076403+
[10] Eguiluz A G, Campbell D A, Maradudin A A and Wallis R F 1984 *Physical Review B* **30** 5449–5459
[11] Burgdörfer J 1994 Atomic collisions with surfaces *Review of fundamental processes and applications of atoms and ions* ed Lin C D (World Scientific Pub Co Inc) pp 517+ ISBN 978-981-02-1537-8
[12] Yamamoto S, Saitou N and Fukuhara S 1978 *Surface Science* **71** 191–198 ISSN 00396028
[13] Chulkov E 1999 *Surface Science* **437** 330–352 ISSN 00396028
[14] Perdew J P and Zunger A 1981 *Physical Review B* **23** 5048–5079
[15] Pohl A, Reinhard P G and Suraud E 2000 *Physical Review Letters* **84** 5090–5093
[16] De Giovannini U, Varsano D, Marques M A L, Appel H, Gross E K U and Rubio A 2012 *Physical Review A* **85** 062515+
[17] Dinh P M, Romaniello P, Reinhard P G and Suraud E 2012 A critical analysis of the theoretical scheme to evaluate photoelectron spectra (*Preprint* `1206.3918`)
[18] van Leeuwen R and Baerends E J 1994 *Physical Review A* **49** 2421–2431

Modeling of plasma and plasma-surface interactions for medical, environmental and nano applications

A Bogaerts, R Aerts, R Snoeckx, W Somers, W Van Gaens, M Yusupov and E Neyts

Research group PLASMANT, University of Antwerp, Dept of Chemistry, Universiteitsplein 1, B-2610 Wilrijk, Belgium

Annemie.bogaerts@ua.ac.be

Abstract. In this paper, an overview is given of modeling investigations carried out in our research group for a better understanding of plasmas used for medical, environmental and nano applications. The focus is both on modeling the plasma chemistry and the plasma-surface interactions. The plasma chemistry provides the densities and fluxes of the important plasma species. This information can be used as input when modeling the plasma-surface interactions. The combination of plasma simulations and plasma – surface interaction simulations provides a more comprehensive understanding of the underlying processes for these applications.

1. Introduction

Gas discharge plasmas are used for a large number of applications. In recent years, there is a growing interest for medical, environmental and nano applications. Plasma medical applications include sterilization of, among others, heat-sensitive (surgery) equipment, deposition of biocompatible coatings, as well as therapeutic applications, such as wound healing, dental treatment, cancer treatment, microsurgery,... Details can be found in the many review papers and in several books (e.g., [1-3]). Environmental applications comprise air and water purification, volatile organic compound decomposition, as well as the conversion of greenhouse gases into value-added chemicals (e.g., [4-6]). Finally, nano applications include the fabrication of nanoscale devices by etching and deposition steps, as well as the growth of nanostructured materials, such as carbon nanotubes, graphene, nanowires,...(e.g., [7-9]).

To improve these applications, a good insight in the underlying physical and chemical processes, occurring in the plasma itself and in interaction with a substrate, is desirable. This can be obtained by experiments, but measurements inside the plasma are not always straightforward because of the small plasma dimensions. Moreover, detailed information, such as species densities, can only be achieved by dedicated and expensive (laser) diagnostics. An alternative is to obtain this information by computer modeling, i.e., by describing the behavior of the various plasma species inside the plasma, as well as their interactions with the surface. This is the focus of the present paper.

There exist different modeling approaches in literature for describing plasmas, such as analytical models (e.g., [10]), solving the Boltzmann equation (e.g., [11]), zero-dimensional (0D) chemical kinetics models (e.g., [12]), 1D or 2D fluid models (e.g., [13]), Monte Carlo (MC) and particle-in-cell – Monte Carlo collision (PIC-MCC) simulations (e.g., [14-15]), as well as hybrid models (e.g., [16]), which are a combination of the above models. Each of these models has its specific advantages and disadvantages. For describing a detailed plasma chemistry, which is the focus of our work, 0D chemical kinetics models or fluid models are the most convenient, as they can describe a large number of different plasma species without too much computational effort. Also hybrid MC-fluid models can be applied, when the plasma chemistry is treated in the fluid part. The examples that will be presented in this paper are obtained with either 0D chemical kinetics models or with a 2D hybrid model.

Modeling the plasma-surface interactions can be performed with detailed atomistic molecular dynamics (MD) simulations (e.g., [17-18]) or with MC or analytical surface kinetics models (e.g., [19]). However, in the latter case, the surface reaction coefficients are needed as input in the model. This is not required in MD simulations, which describe the behavior of all the atoms in the system in a self-consistent, deterministic way, without the need of *a priori* knowledge on the reaction mechanisms. The major disadvantage of MD simulations is the long calculation time, limiting both the attainable time and length scales. Moreover, an accurate interatomic interaction potential must be available for realistic simulations..

In the following sections, we will present some examples of modeling work for the plasma chemistry and plasma-surface interactions, carried out in our research group, for medical, environmental and nano applications, to illustrate the kind of information that can be obtained with this kind of modeling.

2. Medical applications

2.1. Modeling the plasma chemistry

A range of different plasma source designs have been developed for plasma medical applications, but the plasma jet type is quite often used. In our work, we consider a plasma jet formed by a needle electrode, as designed by Hofman and Bruggeman [20] and by Leys and coworkers [21]. In order to obtain a better understanding of the important plasma species in this plasma jet, we have developed a detailed plasma chemistry set for an Ar plasma, flowing into ambient (humid) air (i.e., N_2, O_2 and H_2O). The species taken into account in this model are listed in Table 1.

Table 1. Overview of the species included in the model for the Ar plasma jet flowing into ambient humid air.

Ground state species	Excited species	Charged species
Ar	$Ar(^4S)$, $Ar(^4P)$, Ar_2^* $(a\ ^3\Sigma^+_u)$	e^-, Ar^+, Ar_2^+
N_2, N	$N_{2\ vibrational}$, $N_{2\ rotational}$, $N_2(A\ ^3\Sigma^+_u)$, $N_2\ (a'\ ^1\Sigma^-_u)$, $N(^2D)$,	N_2^+, N_3^+, N_4^+, N^+
O_2, O_3, O	$O_{2\ vibrational}$, $O_{2\ rotational}$, $O_2\ (a\ ^1\Delta_g)$, $O_2\ (b\ ^1\Sigma^+_g)$, $O(^1D)$	O_2^+, O_4^+, O^+, O^-, O_2^-, O_3^-
NO, NO_2, N_2O, NO_3, N_2O_3, N_2O_4, N_2O_5		NO^+, NO_2^+, NO_2^-, NO_3^-
H_2, H, NH, HNO,	$H_{2\ vibrational}$, $H_{2\ rotational}$	H^+, H_2^+, H_3^+, H^-, ArH^+
HNO_2, HNO_3, HNO_4		
H_2O, H_2O_2, HO_2 , OH	$H_2O_{\ vibrational}$	H_2O^+, H_3O^+, $H_2O_2^-$, OH^+, OH^-

These species react with each other in a large number of collisions. In total, 264 electron impact collisions, 796 ion reactions and 591 neutral reactions are taken into account. More detailed information about this reaction set, as well as the corresponding rate coefficients or cross sections, can be found in [22].

The density evolution of these species, based on the production and loss processes as defined by these chemical reactions, is calculated by means of a 0D chemical kinetics model, developed by Kushner and coworkers, called Global_kin [12]. The rate coefficients of the ion and neutral reactions are assumed constant, dependent only on the gas temperature. The rate coefficients of the electron impact reactions are a function of the electron energy. They are calculated by a Boltzmann solver for a wide range of reduced electric field values, and subsequently tabulated as a function of electron energy. The electron energy is calculated with an energy balance equation, based on energy gain due to Joule heating and energy loss due to elastic and inelastic collisions.

Although Global_kin is a 0D model, it is used in our work to calculate the densities of the various plasma species as a function of position from the needle, as well as in the flowing afterglow of the plasma jet, by converting the spatial variation into a time variation, based on the flow velocity. Hence, the species densities in a certain volume element are calculated as a function of time, treating this volume element as a batch reactor, but mimicking the condition of a cylindrical plug flow reactor, where this volume element moves in one direction, along the needle axis and beyond, as if the species densities vary as a function of distance. For this purpose, a certain power profile is applied, with a maximum at the needle tip, and decreasing towards the flowing afterglow (see Figure 1 below). Moreover, the mixing of the flowing Ar plasma jet with the ambient air is described by assuming certain concentration profiles of the background gases, mimicking the diffusion process of the humid air gases into the argon flow. This means that the Ar gas density corresponds to atmospheric pressure inside the needle device, but drops to lower values in the ambient air region, and the opposite is true for N_2, O_2 and H_2O. The background gas densities, as well as the calculated electron density and electron temperature, are plotted as a function of position from the needle tip, both inside the plasma jet device and in the afterglow, in Figure 1.

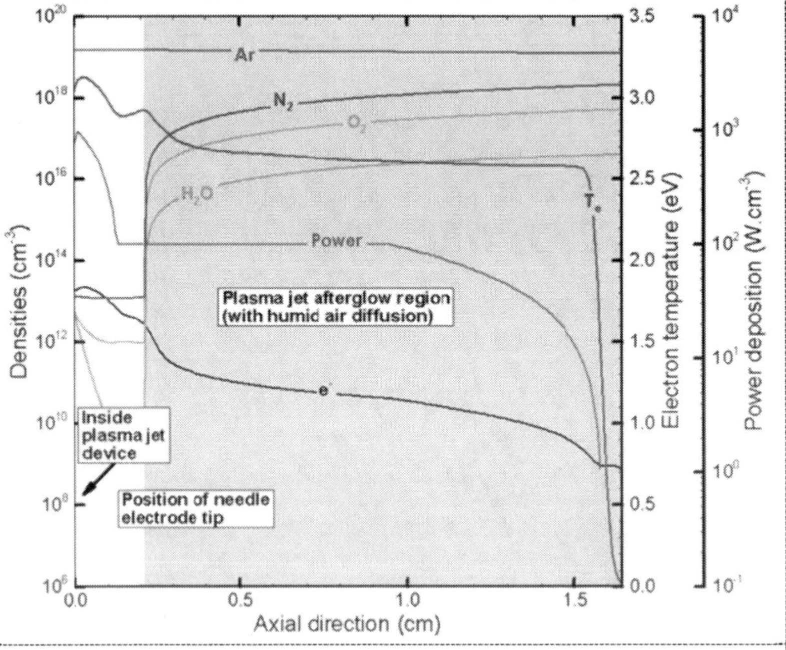

Figure 1. Densities of the background gases, as well as electron density and temperature, as a function of position from the needle tip, inside the plasma jet device and in the afterglow region.

The calculated densities of biomedically important species in the plasma jet afterglow region are illustrated in Figure 2. It is clear that O, O_3, NO, N, H, OH and H_2O_2 are present in large amounts. The O density, however, drops after about 1 cm from the needle tip, whereas the O_3 density, which is negligible near the needle tip, rises as a function of distance and becomes even higher than the O density. Furthermore, OH will convert into H_2O_2 after roughly 1.2 cm. Our calculations also predict that cluster formation is important in the ion chemistry. More details can be found in a forthcoming paper [22].

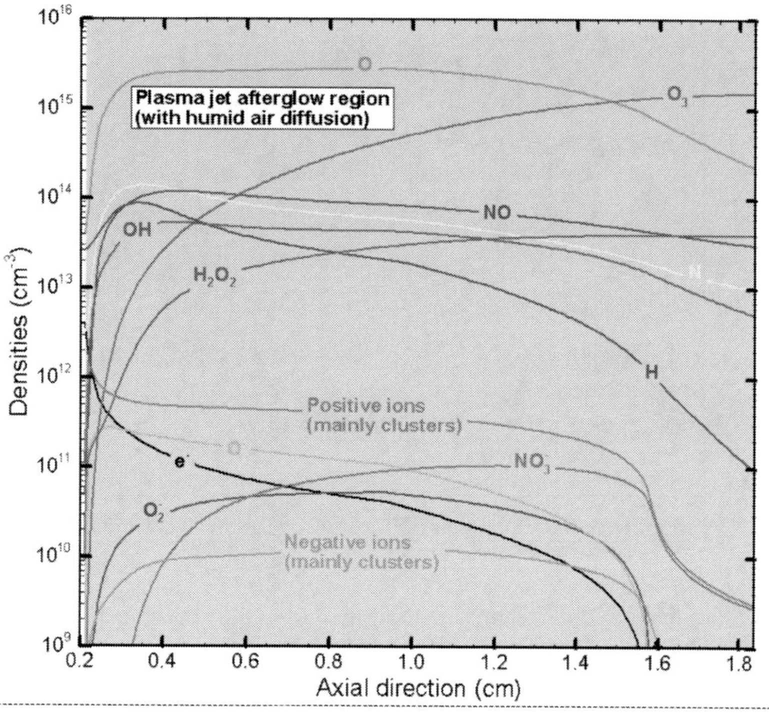

Figure 2. Densities of the biomedically important species, as a function of position from the needle tip, in the afterglow region.

2.2. Modeling the plasma-cell interactions

To elucidate the effect of plasma species for biomedical applications, it is of interest to study their detailed behavior when interacting with living cells. We use MD simulations for this purpose, and in a first attempt, we consider the interaction of plasma species with a bacterial cell wall, for sterilization purposes. More specifically, we describe the interaction with peptidoglycan (PG), which is the outer layer of the cell wall of gram-positive bacteria, such as Staphylococcus aureus.

The PG structure is already quite complicated, and consists of disaccharide layers, stem (tetrapeptide) and bridge (pentaglycine), as is shown schematically in Figure 3. More details about this structure can be found in [23].

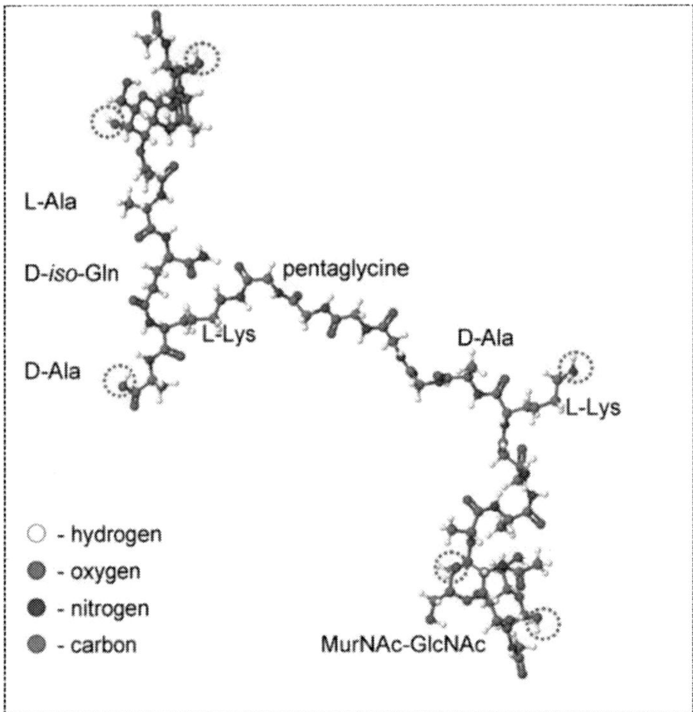

Figure 3. Structural unit of PG, as used for the MD simulations

The interaction of plasma species with this PG is described by atomic scale MD simulations. The behavior of all the atoms in the system, i.e., originating from the impacting plasma species, as well as from the PG structure, is followed as a function of time by solving Newton's laws. The forces acting on all the atoms are derived from the interatomic interaction potential. In our work, we assume the Reax force field (ReaxFF), developed by van Duin and coworkers [24], and the parameters describing the C-N-O-H interactions are adopted from [25].

The plasma species considered in this study, include O, OH, O_3 and O_2. They were found to be important in the plasma jet, as shown in Figure 2 above, but it is no guarantee that they are also the important species interacting with the bacterial cell walls. Indeed, the bacteria are typically covered by a liquid layer (biofilm) and it is well possible that the important plasma species react in this layer, with the formation of other species. This process will be investigated in our future work. However, as a first step, we consider here the interaction of O, OH, O_3 and O_2 species, to obtain some first insight in the possible role that plasma species can play when interacting with bacterial cell walls.

Our calculations predict that O_2 molecules do not result in bond breaking of the PG structure, but the other species investigated, i.e., O, OH and O_3, do result in the breaking of some important C-C, C-N and C-O bonds, and this can give rise to bacterial cell wall damage. We found that the C-O bonds in the disaccharide molecules are most easily broken, followed by the C-C bonds and the C-N bonds. The O atoms appear to be the most reactive, as is clear from Table 2. This table shows the fraction of C-N, C-O and C-C bond breaking events, upon impact of O, O_3 and OH species, when following the species for a sufficiently long time. As mentioned above, it is well possible that other plasma species will interact with the bacterial cell wall, instead of the species studied here. But this example is only the first step in our investigations, and it illustrates that MD simulations can be very valuable for obtaining atomic scale insight in the underlying mechanisms of plasma medicine. More detailed information, including insight in the bond breaking mechanisms, can be found in [23].

Table 2. Summary of the fraction of C-N, C-O and C-C bonds (in %) that break in the PG structure upon impact of O, O_3 and OH species, as obtained from MD simulations.

Incident plasma species	C-N bonds	C-O bonds	C-C bonds
O atoms	26	78	38
OH radicals	8	54	14
O_3 molecules	8	56	26

3. Environmental applications

3.1. Modeling the plasma chemistry

As mentioned in the introduction, plasmas are used for various environmental applications. The application we are most interested in, is the conversion of greenhouse gases (CO_2, CH_4) into value-added chemicals or new fuels. We performed simulations for a dielectric barrier discharge (DBD) reactor, consisting of two concentric cylinders, with a gap of a few mm, through which the gas is flowing. Again the 0D chemical kinetics code Global_kin [12] was applied. The advantage of this code is that the occurrence of filaments in the DBD plasma can be approximated, by considering a large number of power pulses as a function of time. This is explained in more detail in [26,27].

We investigated the splitting of CO_2 [26], as well as the dry reforming reaction between CO_2 and CH_4 [27]. The species taken into account in the CO_2 plasma are listed in Table 3. They include not only ground state species, but also a large number of vibrationally and electronically excited species, as is clear from this Table. These species react in 240 electron impact collisions, 176 ion and 85 neutral reactions. More details about this plasma chemistry set can be found in [26].

Table 3. Overview of the species included in the model for the CO_2 splitting.

Ground state species	Excited species	Charged species
CO_2, CO, C_2O	CO_2 v1 (010)	CO_2^+, CO_4^+, CO^+
C, C_2	CO_2 v2 ((100), (020))	$C_2O_2^+$, $C_2O_3^+$,
	CO_2 v3 (001)	
O_2, O_3, O	CO_2 v4 ((n00), (0n0))	$C_2O_4^+$
	CO_2 ($^1\Pi_g$)	
	CO_2 ($^1\Delta_u$)	C^+, C_2^+, CO_3^-,
	COv1 (sum of vibrations)	CO_4^-
	CO($A^3\Pi$)	
	CO($A^1\Pi$)	O_2^+, O^+, O_4^+
	Sum of (CO($A^3\Sigma$),CO($D^3\Delta$), CO($E^3\Sigma$), O($B^3\Sigma$))	O^-, O_2^-, O_3^-, O_4^-
	Sum of (CO($C^1\Sigma$),CO($E^1\Pi$), CO($B^1\Sigma$), CO($I^1\Sigma$), CO($D^1\Delta$)	Electrons
	O_2 v1 (n_v=1,2)	

It is stated by Fridman that the vibrationally excited CO_2 molecules can play a significant role in the plasma, and in the CO_2 splitting process, depending on the type of plasma [28]. Indeed, when the reduced electric field is in the order of 1-100 Td, and the electron temperature is around 1 eV, up to 97% of the energy can go into the vibrationally excited levels [28]. However, in the DBD discharge under study, the reduced electric field is typically above 200 Td, so it is expected that the vibrationally excited levels play a more moderate role. This was investigated in detail in [26]. Figure 4 shows the ground state density of CO_2, as well as the densities of the four vibrationally excited states, taken into account in the model, during one discharge pulse and its afterglow (a), as well as for five consecutive pulses with an interpulse time of 1 μs (b). It is clear from Figure 4(a) that the ground state density slightly drops when the discharge pulse starts, and the vibrationally excited states are formed. Upon pulse termination, the higher vibrationally excited levels will decay to the lower levels by V-V transitions, and this leads to a further increase of the lowest excited level (CO_2v1). However, after about 10 μs, this lowest excited level will also fall back to the ground state by V-T transitions, so that the ground state density rises again to more or less its initial value before the pulse. When five consecutive discharge pulses are followed (see Figure 4(b)), some accumulation effect can be observed, especially for the lower vibrationally excited densities, and after these five pulses, the density of the lowest vibrational level is only a factor of three lower than the ground state density.

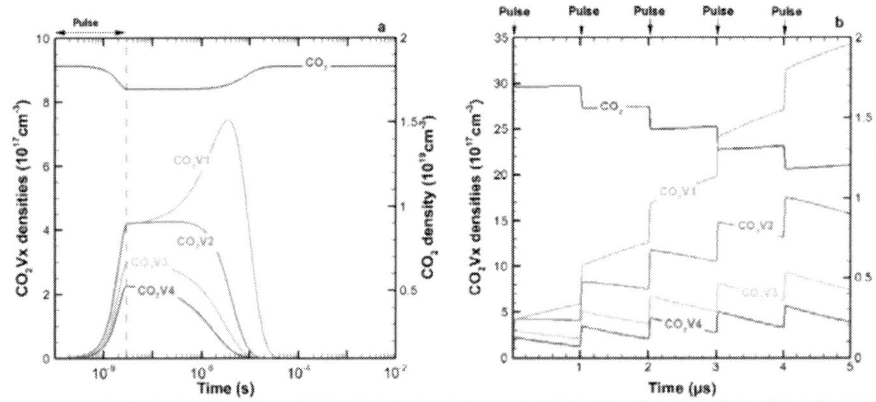

Figure 4. Calculated densities of the CO_2 ground state and four different vibrationally excited levels, in a DBD reactor, for one discharge pulse (filament) and its afterglow (a), as well as for five consecutive discharge pulses (filaments) with an interpulse time of 1 μs (b).

It should be noted that the consecutive pulses are applied in the model, to mimic the filamentary behavior of a DBD. This is a very approximative way, as we do not describe the filament formation process itself, but nevertheless it can give us some insight in the filamentary character of the DBD and its effect on gas conversion. We expect that the CO_2 gas molecules will pass several filaments when they flow through the DBD reactor, but the exact number of filaments they pass, and the typical time in between such filaments is actually not known. We have therefore investigated the effect of the interpulse time on the formation of the vibrationally excited levels, as well as on the fall back ratio of these levels, and the result is illustrated in Figure 5. It is clear from Figure 5(a) that for an interpulse time below 1 μs, a strong accumulation of the vibrationally excited levels is observed, as was also clear from Figure 4(b) above, and the fall back is found to be negligible (see Figure 5(b)). However, at an interpulse time above 10 μs, the accumulation effect is found to be negligible, and nearly all vibrationally excited levels will relax back the ground state. Therefore, the importance of the vibrationally excited levels will strongly depend on the actual interpulse time, or time between two filaments which the CO_2 molecules pass, in the DBD reactor.

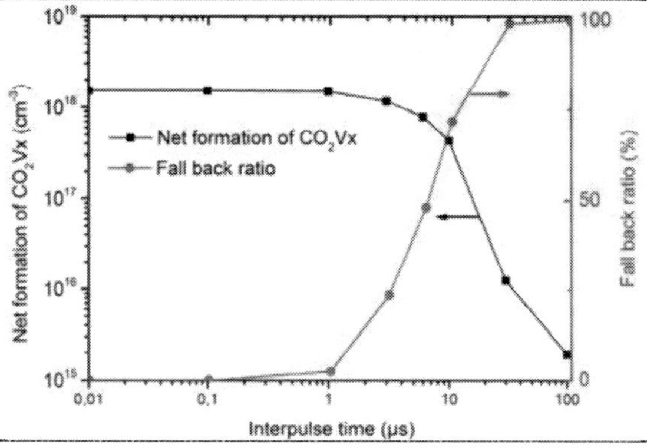

Figure 5. Calculated net formation of vibrationally excited CO_2 molecules (black curve; left axis), and fall back ratio to the ground state (red curve; right axis), as a function of the interpulse time, for consecutive discharge pulses (filaments).

The species considered in the model for the dry reforming process, i.e., in the CH_4/CO_2 gas mixture, are listed in Table 4. As is clear, several higher hydrocarbon species are included in the model, as well as possible interesting end products, such as syngas (CO/H_2), methanol (CH_3OH) and formaldehyde (CH_2O). These species react in 121 electron impact collisions, 87 ion reactions and 290 neutral reactions.

Table 4. Overview of the species included in the model for CH_4/CO_2 dry reforming.

Ground state molecules	Excited species	Charged species	Radicals
CH_4	CH_4*	CH_5^+, CH_4^+, CH_3^+, CH_2^+, CH^+	CH_3, CH_2, CH, C
C_2H_6, C_2H_4, C_2H_2	C_2H_6*, C_2H_4*, C_2H_2*	$C_2H_6^+$, $C_2H_5^+$, $C_2H_4^+$, $C_2H_3^+$, $C_2H_2^+$	C_2H_5, C_2H_3, C_2H
C_3H_8, C_3H_6, C_4H_2	C_3H_8*		C_3H_7
H_2	H_2*		H
O_2	O*	O_2^+, O^-	O
CO_2, CO	CO_2*, CO*	CO_2^+	
H_2O, H_2O_2	H_2O*	H_3O^+, OH^-	OH, HO_2
CH_2O, CH_3OH,		Electrons	CHO, CH_2OH, CH_3O
CH_3CHO, CH_2CO			C_2HO, CH_3CO
			CH_2CHO, $C_2H_5O_2$

Similar results as illustrated above were also obtained for this gas mixture, i.e., we studied the detailed behavior of species densities during one microdischarge pulse and its afterglow, as well as for five consecutive pulses. Moreover, we also investigated the long time scale behavior of the dry reforming process, by considering a large number of consecutive discharge pulses, for a total time of several seconds [27]. In this way, we could obtain information on the conversion of the CO_2 and CH_4 gas molecules, as well as on the selectivities of the reaction products, and on the energy cost and energy efficiency of the process, which is of course of primordial importance for this application. We cannot go into more detail here, due to space restrictions, but these results are presented in [27].

3.2. Modeling the plasma-catalyst interaction

Our plasma chemistry simulations reveal that the energy efficiency of the DBD reactor for greenhouse gas conversion is not yet competitive with existing classical reforming processes [27]. One way to improve the energy efficiency is by inserting a packing inside a DBD reactor, because this yields higher electric fields inside the voids between the packing pellets. Moreover, the pellets can be made of (or covered by) a catalytic material, which can improve the selectivity towards the desired end products. Currently, there is a lot of interest for so-called plasma catalysis, but again, the underlying mechanisms are not yet fully understood.

In an effort to obtain some better insight in the interaction mechanisms between plasma species and a catalyst surface, we performed MD simulations for the interaction of CH_3, CH_2 and CH species on a Ni surface, using again the ReaxFF force field, adopting parameters for the Ni-C-H system from [29]. Figure 6 illustrates, as an example, the molecules formed upon impact of CH_2 species on different Ni catalyst surfaces, i.e., Ni(111), which is the most stable Ni surface, and the most abundant surface facet in Ni catalysts [30], Ni(100), which has a higher surface energy and is also present in Ni catalysts [31], and two step-edged Ni(111) surfaces, since defect (step) sites may be formed at the catalyst surface [32]. It is clear that most CH_2 species are converted into CH_3 species, with a large fraction of non-adsorbed CH_3 species, and a minor fraction of adsorbed CH_3 species. Besides, also some CH_4, C_2H_4 and C_2H_6 molecules are formed, and even a small number of H_2 molecules, at least on the more reactive Ni(100) and stepped Ni(111) surfaces. More details can be found in [33]. These simulations are only a first step; in future work, we will investigate the effect of catalyst temperature and catalyst shape on the reactivity for different plasma species. Also other catalyst materials will be investigated in the future.

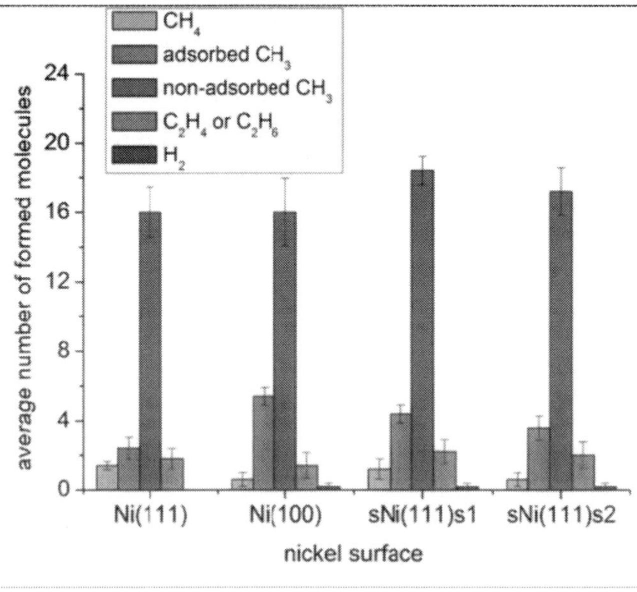

Figure 6. Molecules formed upon impact of CH_2 species on different Ni catalyst surfaces, as predicted by the MD simulations.

4. Nano applications

4.1. Modeling the plasma chemistry

CNT growth is one of the major topics under study in our group. CNTs are typically grown by catalytic chemical vapor deposition (CCVD), but we focus on plasma-based growth (PE-CVD). Most often, low pressure plasma's, such as inductively coupled plasma's (ICPs), are used for this purpose. At low pressure, a fluid (or 0D chemical kinetics) approach might be a too rough approximation, and therefore we made use of a hybrid MC – fluid model to describe the plasma chemistry. In the example shown here, we used the HPEM code, also developed by Kushner and coworkers [34].

CNTs can be grown in plasmas from different types of gas mixtures. Most commonly, a hydrocarbon gas (such as CH_4 or C_2H_2) is applied, in combination with a gas that can supply H-atoms (such as H_2 or NH_3) for etching the amorphous carbon from the CNT cap. We performed simulations for the different combinations, i.e., CH_4/H_2, C_2H_2/H_2, CH_4/NH_3 and C_2H_2/NH_3, to find out whether they yield a different plasma chemistry [35,36]. Again, a large number of different plasma species was taken into account, as is illustrated in Table 5 for the CH_4/NH_3 gas mixture. For more details about the chemistry sets, also for the other gas mixtures, we refer to [35].

Table 5. Overview of the species included in the model for the CH_4/NH_3 gas mixture.

Molecules	Radicals	Charged species
H_2	H	H^-, H_2^-, H_3^-
CH_4	C, CH, CH_2, CH_3	C^-, CH^-, CH_2^-, CH_3^-, CH_4^-, CH_5^-
C_2H_2, C_2H_4, C_2H_6 C_3H_8, C_4H_{10}	C_2, C_2H, C_2H_3, C_2H_5	C_2^-, C_2H^-, $C_2H_2^-$, $C_2H_3^-$, $C_2H_4^-$, $C_2H_5^-$, $C_2H_6^-$
NH_3	NH_2, NH	NH^-, NH_2^-, NH_3^-, NH_4^-
N_2H_2, N_2H_4	N_2H, N_2H_3	
N_2, N_2*	N, N*	N^-, N_2^-
HCN	H_2CN, CN	HCN^-
		Electrons

Our calculations predicted that the major species formed in the plasma are H, H_2, CH_4, C_2H_2, C_2H_4, N, N_2, NH_3, HCN [35,36]. The C concentration appeared to be very low, indicating that the C supply for CNT growth should originate from decomposition of hydrocarbons at the catalyst surface itself.

4.2. Modeling the CNT growth

Finally, also the CNT growth itself is modeled with the MD approach. In fact, only the growth of single-walled carbon nanotubes (SWCNTs) is currently simulated with MD, and most often only the cap formation, because of the long calculation times. We modelled the SWCNT cap formation from a Ni catalyst particle, consisting of 40 atoms, and made again use of the ReaxFF potential, employing the parameters adopted from [29].

As mentioned above, the C supply for the CNT growth originates from decomposition of hydrocarbons at the catalyst surface. However, as this decomposition might take place on a longer time scale than is accessible by MD, we assumed that the Ni catalyst particle, either floating [37] or surface bound [38], is bombarded directly by C atoms. We extended our MD simulations with a MC method, in order to capture the surface reorganization in between two carbon impacts. This allowed us to obtain, for the first time, CNT caps with definable chiralities [37,38]. We were able to simulate the growth of both a semi-conducting SWCNT (with (12,4) chirality) [37], as well as a metallic SWCNT (with (7,7) chirality) [38]. Moreover, our simulations have demonstrated that the chirality can still change during the cap formation [38].

The above simulations hold true for SWCNT growth from CCVD. However, in [39] we made some attempts to describe the PE-CVD growth, by including the effect of an electric field. It was found that electric field values in the order of 300-700 kV/cm at the CNT tip yielded vertical alignment of the SWCNT cap, as seen in Figure 7. Lower electric field values result in a random nucleation point, whereas at higher electric field values no nucleation at all occurred. This range of electric field values seems high for typical plasmas, but near the CNT cap the local electric field can be much higher than in the sheath due to curvature effects. This field effect can be explained by the difference in electronegativity between C and Ni atoms, which implies that the C-Ni bonds are slightly polar, and can therefore be affected by the electric field. There will be a competition between diffusion and migration of the C atoms. At low electric field values, diffusion dominates and nucleation is random. At high electric field values, migration is more important, and hence oriented nucleation occurs [39].

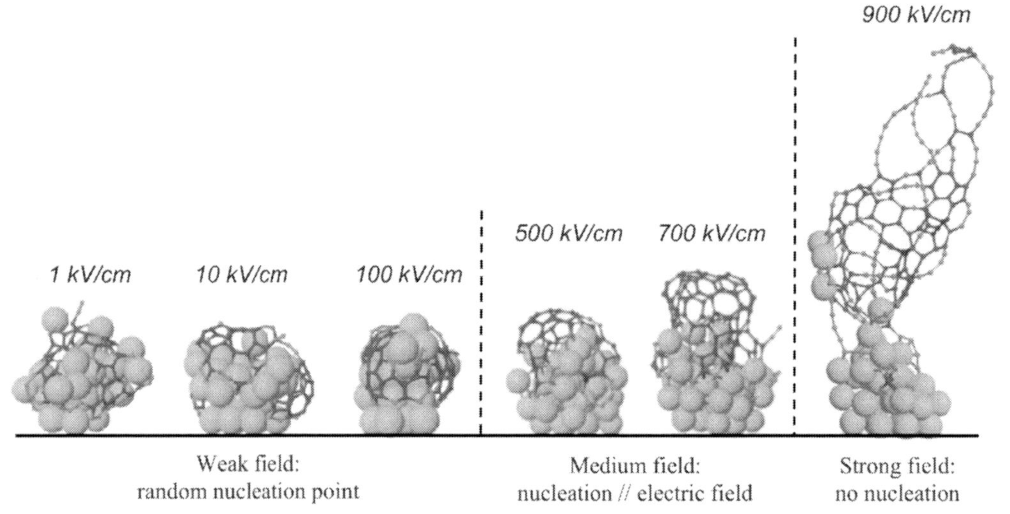

Figure 7. Effect of electric field on the nucleation of a SWNT. Reprinted with permission from [39], Copyright 2012 American Chemical Society

5. Conclusions

In this paper, we have shown some examples of modeling the plasma chemistry as well as plasma-surface interactions for plasma medicine, environmental and nano applications. For modeling the plasma chemistry, 0D chemical kinetics models or fluid models, or the fluid part of hybrid models are most suitable, as these model approaches can handle a large number of species without too much computational effort. These models give information on the densities and fluxes of the various species present in the plasma, as well as on the importance of the chemical reactions taking place in the plasma, yielding the production and loss of the various plasma species. Although a 0D model does not include spatial dimensions, it can still be used to give spatial information, as is illustrated here for a plasma jet (plasma needle). Moreover, a 0D model can approximate the filamentary behavior of a DBD reactor, by considering a large number of consecutive microdischarge pulses.

The plasma chemistry models provide information on the most important plasma species, which is of interest for plasma-surface interaction modeling. For this purpose we use MD simulations, which describe the behavior of the individual atoms in the system in a self-consistent way, without making approximations. We have shown examples in this paper from MD simulations for plasma medicine, i.e., the interaction of plasma species with bacterial cell walls, for plasma catalysis, i.e., the interaction of plasma species with Ni catalyst surfaces, and for nano applications, i.e., the growth of CNTs.

The plasma chemistry modeling and the atomic-scale MD simulations can give us a better insight in the plasma behavior and in its interaction with surfaces, respectively, which is of great importance for the applications.

References

[1] Dobrynin D, Fridman G, Friedman G, Fridman A 2009 *New J. Phys.*, **11** 115020
[2] Weltmann K D, von Woedtke T 2011 *IEEE Trans. Plasma Sci.* **39** 1015
[3] Laroussi M, Kong M G, Morfill G and Stolz W 2012 *Plasma Medicine: Applications of Low-Temperature Gas Plasmas in Medicine and Biology* (Cambridge University Press)
[4] Gutsol A, Rabinovich A, Fridman A 2011 *J. Phys. D: Appl. Phys.* **44** 274001
[5] Tu X, Gallon H J, Twigg M V, Gorry P A, Whitehead J C 2011 *J. Phys. D: Appl. Phys.* **44** 274007
[6] Fridman A, Chirokov A, Gutsol A 2005 *J. Phys. D: Appl. Phys.* **38** 1
[7] Vizireanu S, Stoica S D, Luculescu C, Nistor L C, Mitu B and Dinescu G 2010 Plasma Sources Sci. Technol. **19** 034016
[8] Ostrikov K, Cvelbar U and Murphy A B, 2011 *J. Phys. D: Appl. Phys.* **44** 174001
[9] Meyyappan M, 2011 *J. Phys. D: Appl. Phys.* **44** 174002
[10] Lee Y T, Lieberman M A, Lichtenberg A J, Bose F, Baltes H and Patrick R 1997 *J. Vac. Sci. Technol.* **15** 113
[11] Loffhagen D, Sigeneger F and Winkler R 2002 *J. Phys. D: Appl. Phys.* **35** 1768
[12] Dorai R, Kushner M and Hassouni K 2000 *J. Appl. Phys.* **88** 6060
[13] Passchier J D P, Goedheer W J 1993 *J. Appl. Phys.* **73** 1073
[14] Bogaerts A, van Straaten M and Gijbels R 1995 *Spectrochim. Acta Part B* **50** 179
[15] Bultinck E and Bogaerts A 2009 *New J. Phys.* **11** 103010
[16] Bogaerts A, Gijbels R and Goedheer W J 1995 *J. Appl. Phys.* **78** 2233
[17] Neyts E, Bogaerts A and van de Sanden M C M 2006 *J. Phys. D: Appl. Phys.* **39** 1948
[18] Graves D and Brault P 2009 *J. Phys. D: Appl. Phys.* **42** 194011
[19] Zhang D and Kushner M J 2000 *J. Appl. Phys.* **87** 1060
[20] Hofman S and Bruggeman P 2011 *IEEE Trans. Plasma Sci.* **39** 2332
[21] Li L, Nikiforov A, Xiong Q, Lu X, Taghizadeh L and Leys C 2012 *J. Phys. D: Appl. Phys.* **45** 125201
[22] Van Gaens W and Bogaerts A, paper in preparation
[23] Yusupov M, Neyts E C, Khalilov U, Snoeckx R, Van Duin A C T and Bogaerts A 2012 *New J. Phys.* (submitted).
[24] van Duin A C T, Dasgupta S, Lorant F and Goddard W A III 2001 *J. Phys. Chem. A* **105** 9396
[25] Rahaman O, van Duin A C T, Goddard W A III and Doren D J 2011 *J. Phys. Chem. B* **115** 249
[26] Aerts R, Martens T and Bogaerts A 2012 *J. Phys. Chem. C* (submitted).
[27] Snoeckx R, Aerts R and Bogaerts A, paper in preparation
[28] Fridman A 2008 *Plasma Chemistry* (New York: Cambridge University Press)
[29] Mueller J E, van Duin A C T, Goddard W A 2010 *J. Phys. Chem. C* **114** 4939
[30] Feng Q, Li T, Yue H, Qi K, Bai F and Jin J 2008 *Appl. Surf. Sci.* **254** 2262
[31] Huang Y Y, Shou Y C and Pan Y 2010 *Physica B* **405** 1335
[32] Bengaard H S, Norskov J K, Sehested J, Clausen B S, Nielsen L P, Molenbroek A M and Rostrup-Nielsen J R 2002 *J. Catal.* **209** 365
[33] Somers W, Neyts E C, van Duin A C T and Bogaerts A 2012 *J. Phys. Chem. C* (submitted).
[34] Kushner M J 2003 *J. Appl. Phys.* **94** 1436
[35] Mao M and Bogaerts A 2010 *J. Phys. D: Appl. Phys.* **43** 205201
[36] Mao M and Bogaerts A 2010 *J. Phys. D: Appl. Phys.* **43** 315203
[37] Neyts E C, Shibuta Y, van Duin A C T and Bogaerts A 2010 *ACS Nano* **4** 6665
[38] Neyts E C, van Duin A C T and Bogaerts A 2011 *J. Am. Chem. Soc.* **133** 17225
[39] Neyts E C, van Duin A C T and Bogaerts A 2012 *J. Am. Chem. Soc.* **134**, 1256

Experimental transition probability measurements in pulsed lamps: critical points

Juan A. Aparicio[1], María T. Belmonte[1], Ramón J. Peláez[1,2], Stevica Djurović[3] and Santiago Mar[1]

[1]Departamento de Física Teórica, Atómica y Óptica, Universidad de Valladolid, P. Belén 7, 47011 Valladolid, Spain

[2]Laser processing Group, Instituto de Óptica, CSIC, C/ Serrano 121, 28006 Madrid, Spain

[3]Faculty of Sciences, Department of Physics, Trg Dositeja Obradovića 4, 21000 Novi Sad, Serbia

E-mail: apa@opt.uva.es

Abstract. Pulsed linear lamps may be very useful experimental devices to obtain transition probability data. They can provide valuable information of the spectral emission of spectral lines in very different conditions of electron density and temperature. However, they require a very careful previous analysis, which include many experimental aspects e.g., spatial and temporal emission homogeneity, good characterization of equilibrium status or a very precise knowledge of the equipment employed to make measurements: spectrometer transmittance, self-absorption control or lamp windows dirtying. Most of these aspects will be considered along this work, their particular relevance as well as the way they have been controlled in the Plasma Physics Laboratory at the University of Valladolid (Spain).

1. Introduction

The importance of transition probabilities (A_{ki}-values) is well established. In many scientific, industrial or astrophysical fields, transition probabilities are used to determine plasma temperature and, from this, element abundances and other physical parameters of interest can also be obtained. Pulsed lamps may be very promising sources of high quality/low uncertainty transition probability data. Among its intrinsic advantages, it is remarkable the possibility of making measurements at very different plasma conditions. However, several critical points should be very carefully considered when designing and using these light sources for this scientific purpose.

The calculation of the geometry involved in the process emission-detection of the radiation coming from a spectral line makes commonly the determination of its absolute transition probability an unfeasible task. In this case, only relative transition probabilities are available. However, when only one relative scale is tried and lines coming from different upper levels are involved, the characterization of the equilibrium state in the plasma becomes a new significant problem. In this case, the existence of excitation equilibrium along the measured plasma volume should be carefully guaranteed. The Boltzmann-plot technique is a useful tool to check the exponential population of the excited states of the radiation emitters. An absolute scale may be again attainable if a reliable set of

Published under licence by IOP Publishing Ltd

transition probabilities exists previously in the literature and is used to perform the Boltzmann-plot and therefore determine the population temperature.

The experimental difficulties are not negligible either. If the uncertainty associated to the determined transition probability value is tried to be low, it is necessary a high homogeneity in the emitting plasma volume whose radiation is detected. This requires, among others, the absence of cold boundary layers which may emit the same spectral lines at very different intensities than those emitted in the hottest regions. Furthermore, although the homogeneity might be checked in a reliable way, this is not enough. Self-absorption may become a serious problem whose control is of vital importance. When the quantification of this effect is possible and is not too severe, different experimental and numerical procedures exist in the literature which may help us to reconstruct the original emitted profile and to make possible the comparison of line intensities.

From the experimental point of view, other cautions should also be considered and its effects quantitatively evaluated. Among them, we could remark the need of a precise calibration of the spectrometric system and the possible influence of electrode sputtering on spectral transmittance of lamp windows.

In this work, we will face most of the critical points that should be considered for a precise determination of the A_{ki}-values as well as the strategies employed in the Plasma Laboratory at the University of Valladolid.

2. Experimental considerations

From now on, we will be concerned about plasma spectroscopy emission experiments where the plasma is generated by discharging a capacitor bank, charged up to 7000-10000 V, through a cylindrical Pyrex tube. The required characteristics of the electrical discharge employed in our laboratory are the following: a) pulse width in the range 100-300 μs, b) peak current in the order of kA with an energy near kJ to produce dense plasmas, c) smooth current variations with time, which prevent the plasma shock waves and so, of instabilities. d) about a 40 μs plateau, that gives measuring facilities and, finally, a high reproducibility in the pulses.

The excitation unit has been designed, constructed and tested in our laboratory. Further details can be found elsewhere [1]. The capacitors are connected to one another and to the discharge system (a Spark-Gap is used) by self-inductions which reduce the temporal derivative of the intensity of the current. The self-induction values are selected in order to obtain a current pulse with a stable 40 μs plateau at the first discharge instants, followed by a downfall close to critical damping, which actually shows a small plateau between the instants 90-110 μs (see Figure 2 in [2]). In order to guarantee the best plasma repetitiveness the excitation unit has a continuous current source of low intensity (several mA) that allows working with a preionised gas in the lamp.

In its basic design, the plasma source consists of a cylindrical tube of Pyrex glass, 155 mm in length with an inner diameter of 18 mm. The figure 3 in [2] shows a section of the lamp in which all the relevant data have been specified. The glass tube is projected farther than the electrodes, made of aluminum, to reduce sputtering. The windows, which enclose the lamp, are separated 10 mm from the electrodes by a metal ring, electrically isolated by means of viton o-rings from the electrodes. In this configuration the distance between windows is 175 mm. This basic design allows several thousands of discharges to be performed without significant damage on the window surfaces. The electrical connections to the electrodes are made using 4 rigid rods on each electrode, parallel to the lamp axis and symmetrically distributed around it, to avoid plasma inhomogeneities during the discharge. The lamp is connected to the vacuum and gas systems by standard connections placed on the electrodes.

The gas or mixture of gases are flowing continuously through the tube at pressures of several hundreds or thousands Pa. An example of the typical experimental arrangement has been previously published [3]. In this arrangement the lamp is placed in one of the arms of a Twyman-Green interferometer. Two laser beams (543.0 nm and 632.8 nm) pass through the lamp in the axial direction in order to obtain the temporal evolution of plasma refractivity and from this, the electron density (N_e) with an uncertainty lower than 10%. Typical highest values in N_e are around $2x10^{23}$ m^{-3}. Spectroscopic and interferometric end-on measurements are usually performed simultaneously during the plasma life, both at 2 mm off the lamp axis in a symmetrical configuration. The spectroscopic beam is limited by two diaphragms 2-3 mm in diameter, and focused by a concave mirror of 150 mm focal length onto the entrance slit (35-70 μm width) of a Jobin-Yvon monochromator (1.5 m focal length, 2400 lines mm^{-1} holographic grating). The spectra are currently recorded with either an optical multichannel analyser (OMA) of 512 channels (EG&G1455R-512HQ) or an intensified charge couple device ICCD (4QuickE Stanford Computer Optics). Measurements are usually performed in first order of diffraction with an exposure time around 1-5 μs. Excitation temperature is determined by a Boltzmann-plot of the species of interest and is usually in the range 15000–35000 K. The error in temperature measurements is estimated to be around 10%.

3. Spatial and temporal homogeneity of the discharge

One of the key points which should be considered when measuring transition probabilities from the line emission of spectral lines is that the plasma parameters should be reasonably homogeneous in the measured plasma volume during the exposure times of the detection system.

Concerning the temporal homogeneity, only the instants where temperature and electron density show the lowest gradients should be selected. In relation to the spatial homogeneity, the situation is much more complex and very often only indirect information can be obtained from the experiment after a very laborious analysis.

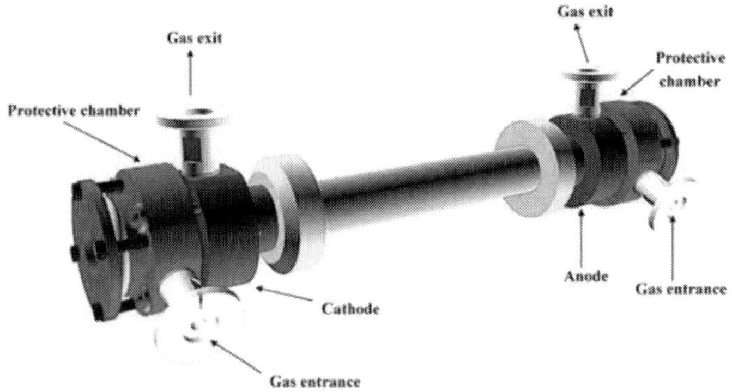

Figure 1. Scheme of the lamp employed for Stark parameters and transition probability measurements of Si II and Si III spectral lines. Protective chambers 30 mm long were employed in both extremes of the lamp.

An example of the analysis of the inhomogeneities along the plasma column was performed for a specific design of lamp built for measuring Stark parameters and transition probabilities of Si II and Si III [4,5]. The plasma was generated on a mixture of silane (SiH_4) and helium. In Fig 1 this design is shown. To the basic design described above, protective chambers were added to both extremes of the lamp. Furthermore, entrance and exit of gas in the lamp is made from these two chambers. This technique reduces very efficiently the lamp windows dirtying by sputtering and the deposition of sand on the extremes of the lamp by creating a laminar flow of gas just on the inner side of windows surfaces. Chambers of different lengths were tested by comparing the electron density evolution curves obtained from interferometry and from Stark broadening of some He I lines. Results are shown in Fig. 2. In the determination of the interferometric electron density curve, it has been assumed that the plasma column length is equal to the lamp length.

As concluded from Fig. 2, the very good agreement obtained between both determinations of N_e when no chambers are included (basic design) or with chambers 30 mm length in the specific design, informs us that the plasma column fills completely the lamp under these two designs and thus, if cold boundary layers exist, they do not affect significantly to the electron density determination. For longer chambers, the agreement is much worse and it might be reasonably to conclude that the plasma column length changes during the discharge.

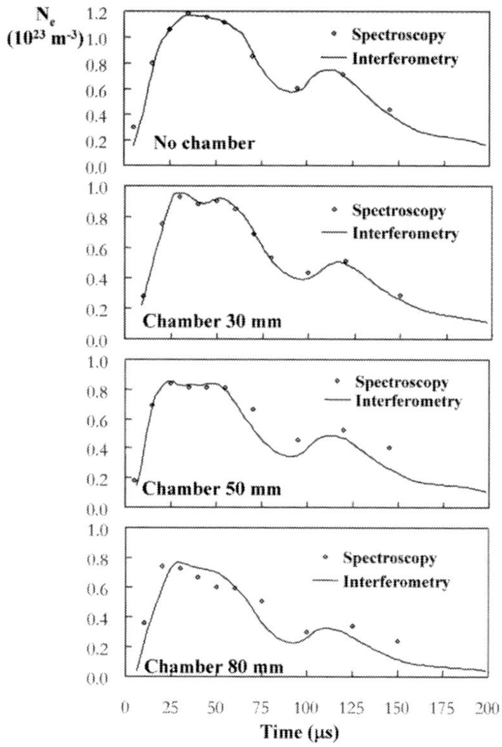

Figure 2. Comparison of the interferometric and spectroscopic determinations of N_e for three different chamber lengths and for our *basic design* (no chamber).

Additional evidences of the plasma homogeneity can be obtained with some more experimental work. In previous works [2], the results obtained by axially measuring the interferometric electron density evolution curve at different distances from the lamp axis were published. A similar analysis was performed for the He I excitation temperature evolution curve obtained from Boltzmann-plot. From these experimental results we conclude that, after some tens of microseconds, there exists a highly homogenous and symmetric core within the cylindrical plasma column. This would allow, for instance, to work with light beams of 2-3 mm diameter in the plasma column for spectroscopic and interferometric standard measurements without important loss of spatial resolution. This symmetry also allows to perform interferometric and spectroscopic measurements in separated but symmetrical plasma column relative to the lamp axis. Furthermore, the temporal distribution of the main plasma parameters with their soft evolution allows good spectroscopic registers within the time scale of pulsed plasmas, and along a wide range of different values.

An additional study which would inform us about the spatial resolution of the discharge is the analysis of the symmetry of theoretically symmetric profiles, as, for instance, some hydrogenic lines. This analysis was performed by us for the He II P_α line [6]. Results revealed levels of profile symmetry never published before in the literature.

4. Plasma equilibrium mode

The level of knowledge of the plasma equilibrium state depends, among others, on the procedure employed for obtaining the A_{ki}-values. If we desire to obtain the transition probabilities of lines coming from the same upper energy level in an emission spectroscopy experiment, no equilibrium specifications are required. Only the comparison between intensities of the involved lines will provide the transition probabilities in a relative scale, which can be converted in an absolute one if one of them is previously known.

If lines coming from different upper energy levels are pursued in a common scale, the situation becomes more complex. In this case, the minimum equilibrium requirement corresponds to the partial local thermodynamic equilibrium (pLTE). Under these conditions, the upper energy levels are assumed to be populated according to the Boltzmann law. The classical Boltzmann-plot is a good way to check this type of equilibrium. This procedure requires that some transition probabilities are previously known. The presence of a linear behaviour in these plots allows us to define an excitation temperature T_{exc}, parameter from which the A_{ki}-values can be obtained by simply considering its measured intensity. A possible departure from this linear behaviour would point out to the existence of non balanced atomic interaction processes inside the plasma [7]. However, it is important to remark that, both the analysis of the plot linearity and the final quality of the obtained transition probabilities are conditioned by the average uncertainty of the A_{ki}-values taken from the literature. Furthermore, the existence of a linear behaviour in the plot only guarantees the pLTE state for the range of upper energy levels involved, not for higher or lower ones. Transition probabilities obtained, from this T_{exc}-value, for lines coming from upper energy levels out of this range should be very carefully considered.

In Fig. 3 some of these plots are performed for different instants of the plasma life in a mixture of xenon and helium, as well as the obtained T_{exc}. Only Xe II lines are considered. As shown in this figure, a very good linear dependence is obtained for the instants in which this analysis has been made.

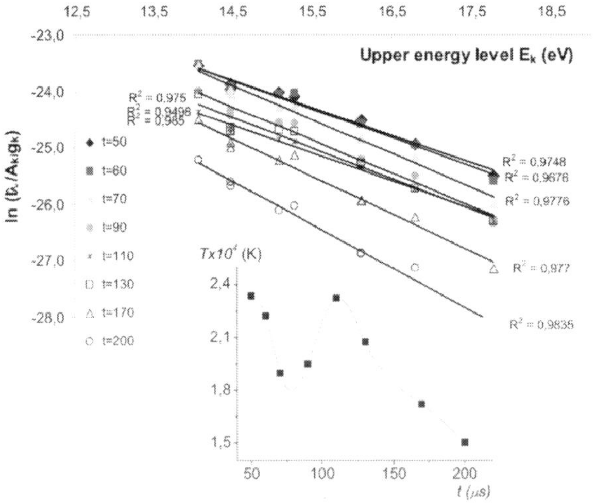

Figure 3. Boltzmann-plots and T_{exc}-values obtained from Xe II lines. Considered times of the plasma life are expressed in microseconds and the correlation factor of the linear fit is included as an index of the level

Although other equilibrium models are possible, they usually require more theoretical assumptions not always easy to be checked experimentally. These assumptions finally derive in greater uncertainties and mistrust on the final transition probabilities.

5. Control of self-absorption

It is frequent that, at the plasma conditions from which the pursued transition probabilities are available, some profiles may show self absorption. It is important to control this parameter if we desire to properly measure the intensity emitted by the line. In order to make this control, it is usual to place a mirror behind the lamp. This mirror allows the radiation leaving lamp in the opposite direction to the detection system, to reenter the lamp and adds to the radiation directly emitted towards the spectrometer. In a certain instant of the plasma life and for each wavelength, the absence or presence of selfabsorption in the measured profile can be checked by simply comparing the light registered from each line when the self absorption mirror (M3 in Fig. 1 of [3]) is covered and when this mirror is not covered. The ratio of the registered intensity in these two spectra, gives us the effective reflection coefficient of the mirror which should be constant if selfabsorption does not exist (see Fig. 4). If self-absorption appears in some part of the spectrum, this reflection coefficient is strongly diminished as happens in Fig. 4 for the Ar II line 493.3 nm. Different reconstruction techniques of the original profiles are available [8]. All them require a high enough spectral transmittance in the lamp window closer to the self absorption mirror.

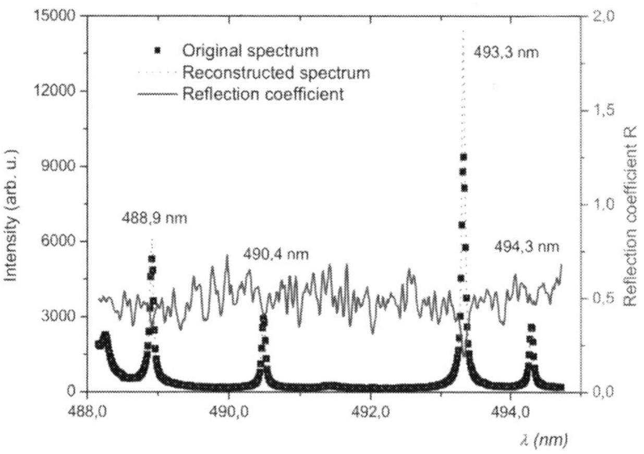

Figure 4. Original (solid squares) and reconstructed (dotted line) spectra containing some Ar II lines. Reflection coefficient is also plotted by a solid line.

However, the sputtering coming from the electrodes after each discharge produces an embedding of electrode materials in the lamp windows inner surfaces. This produces two annoying effects which should be considered all along the process of transition probability determination. Fist, the sputtering strongly reduces the spectral transmittance of the lamp window closer to the self-absorption mirror. This makes more difficult the self-absorption check and even may make impossible the reconstruction process of selfabsorpted spectra. Second, even although selfabsorption may not exist, the loss of spectral transmittance of the window lamp closer to the spectrometer should be considered to obtain the really emitted intensity of spectral lines by the measured plasma column. In Fig. 5 several curves of spectral transmittance of the lamp windows are shown. The upper one corresponds to a clean window and the following ones below it correspond to the spectral transmittances obtained with an increasing number of discharges. As can be clearly seen, the loss of transparency is dramatic, particularly at the lower wavelengths.

As can be seen in this figure, there are wavelengths ranges where transparency increases in spite of the increasing number of discharges. This is due to the interference effect of a multilayer deposition on a surface, whose spectral transmittance depends on its thickness and wavelength.

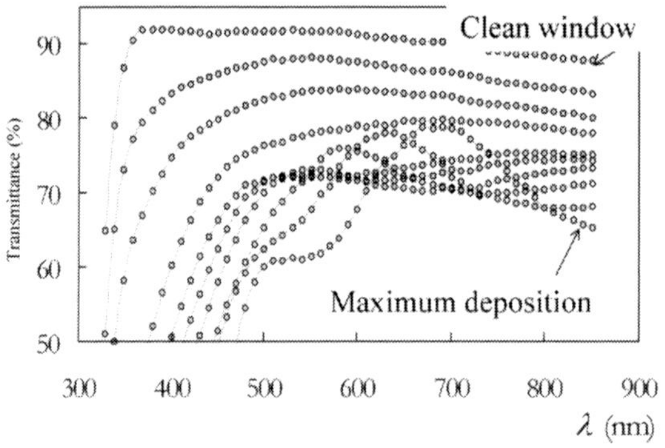

Figure 5. Spectral transmittance for different number of discharges performed.

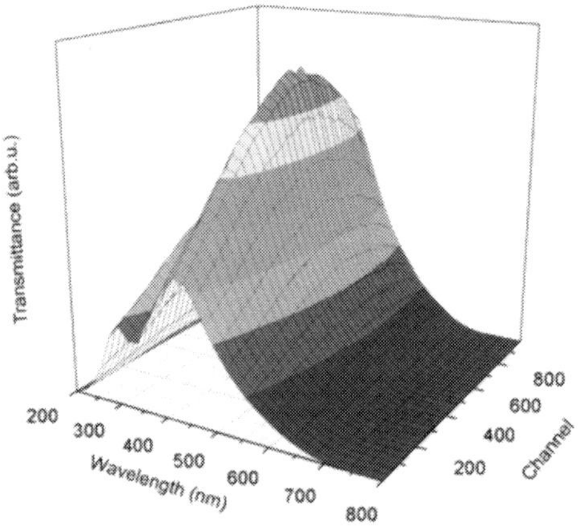

Figure 6. Example of spectral transmittance function of a spectrometer as a function of wavelength and channel detector.

Concerning the spectral transmittance of the detection system, the most significant aspect to be considered is the effect of the spectrometer on the final detected counts per line. Both the monochromator and the detector have a spectral response strongly dependent on wavelength. Furthermore, each channel of an unidimensional or a bidimensional detector array may have different spectral responsivity, which complicates the correction of the detected spectra of the spectral transmittance function.

In Fig. 6 the bidimensional spectral transmittance function (in arbitrary units) of a spectrometer formed by a Jobin-Yvon monochromator, first order of diffraction) and an ICCD (4QuickE Stanford Computer Optics) is shown. This function is obtained by considering the obtained spectra from an incandescent lamp emitting as a blackbody at a colour temperature of 3041 °K from which a previous spectral calibration is obtained. As shown in this figure, the spectrometer employed in this experiment is enhanced for the ultraviolet radiation detection. As mentioned before, the wavelength is the dominant effect on the spectral transmittance function but however, the influence of the detector channels where a spectra line is registered is very important on the final number of counts registered. This is a typical effect of the optics involved in the ICCDs.

6. Conclusions

Measurement of transition probabilities in pulsed discharge lamps is a complex experimental task. After the experience compiled during the last 30 years in the Plasma Laboratory at the University of Valladolid, we have learnt that, although theoretical assumptions concerning equilibrium are usually verifiable and not too strong (pLTE), the plasma source experimental set-up, its proper featuring and the control of all the aspects involved in the line radiation measurement are difficult challenges. Very good spatial and temporal homogeneity features are required and much previous experimental work is to be done until a plasma discharge can be considered as valid for this kind of measurements. Other factors, such as selfabsorption, spectral transmittance or the control of windows dirtying should also be very carefully considered if a low uncertainty is aimed in the final A_{ki}-values.

Acknowledgements

We thank S. González for his work on the experimental device, Dr J. A. Aparicio wants to express his personal acknowledgement to the ONCE for help. S. Djurović thanks to Ministry of Education, Science and Technological Development of Republic of Serbia for support in Project 171014.

References

[1] González, I., Mar, S. and Cardeñoso, V.: 1986, Atti Fond. Giorgio Ronchi, 41, 501.
[2] del Val, J. A., Mar, S., Gigosos, M. A., de la Rosa, I., Pérez, C., González, V. R.: 1998, Jpn. J. Appl. Phys., 37, 4177.
[3] Peláez, R. J., Djurović, S., Ćirišan, M., Rodríguez, F., Aparicio, J. A., Mar, S.: 2008, Astrophys. J., 687, 1423.
[4] González, V. R., Aparicio, J. A., del Val, J. A. and Mar, S.: 2000, Astron. Astrop-hys., 363, 1177.
[5] González, V. R., Aparicio, J. A., del Val, J. A. and Mar, S.: 2002, J. Phys. B: At. Mol. Opt. Phys., 35 3557.
[6] Peláez, R. J., Pérez, C., González, V. R., Rodríguez, F., Aparicio J. A. and Mar, S.: 2005, J. Phys. B: At. Mol. Opt. Phys., 38, 2505.
[7] van der Mullen, J. A. M.: 1990, Phys. Rep., 191, 109.
[8] Aparicio, J. A.:1996. PhD Thesis, Universidad de Valladolid.

Journal of Physics: Conference Series **399** (2012) 012013 doi:10.1088/1742-6596/399/1/012013

Cathode sheath and hydrogen Balmer lines modelling in a micro-hollow gas discharge

Dj Spasojević

University of Belgrade Faculty of Physics, 11001 Belgrade, P.O.Box 368, Serbia

djordjes@ff.bg.ac.rs

Abstract. We present a model of the cathode sheath (CS) processes responsible for the broadening of the hydrogen Balmer beta line recorded from a micro-hollow gas discharge (MHGD) and used for simultaneous diagnostics of plasma and CS parameters. The MHGD was generated in a microhole (diameter 100 μm at narrow side and 130 μm at wider side) of a gold-alumina-gold sandwich in the pressure ranges: (100–900) mbar in argon with traces of hydrogen, and (100–400) mbar in pure hydrogen. The electron number density is determined from the plasma broadened line width of the central part of Balmer beta profile, while the average value of electric field strength in the CS and the CS thickness are determined from the extended line wings induced by the dc Stark effect.

1. Introduction

Microplasma is μm to mm size plasma operated in dc to GHz regime at higher (atmospheric) pressure. Due to small size and low power consumption, microplasma devices have various applications in material processing, biomedical treatment, and production of novel materials. Integrated into other technologies, microplasmas are also relevant for nanoparticle synthesis, thin film deposition, and production of printers, displays, semiconductor circuits, etc – see [1,2].

Micro-hollow gas discharges (MHGD) are miniaturized versions of classical hollow cathode gas discharges in noble gases or gas mixtures. They are realized in sandwich geometry (two metallic layers separated by insulator) with a central laser drilled hole (dia 100-300 μm) in which weakly ionized microplasma is ignited when a dc voltage is applied. Since its discovery (Schoenbach *et al* [3]), MHGD attracted a significant research interest and found many applications in industry and medicine. In this paper we report our results [4,5,6] of modelling of cathode sheath processes and of formation of profile of hydrogen Balmer beta line used for MHGD diagnostics.

2. Experimental

In figure 1 we show the schematic diagram of our experimental setup. The discharge source was a laboratory made MHGD device placed inside a vacuum tight chamber. The discharge was generated in a microhole (dia 100 μm at narrow and 130 μm at wider side) laser drilled in the center of a 250 μm thick alumina square piece (12 mm x 12 mm) with circular (dia 10 mm) gold electrodes.

The experiments were carried out with argon-hydrogen mixture (Ar+0.9% H_2 by volume) in the pressure range (100-900) mbar, and with pure hydrogen (purity 99.999%) in the pressure range (100-400) mbar. The pressure and continuous gas flow of (0.1-1.0) L/min were regulated by two vacuum valves and one two-stage mechanical vacuum pump. The discharge was maintained by a current stabilized power supply (0-2 kV, 0-100 mA) current limited by a 50 kΩ series ballast resistor.

The light from the discharge was focused with unity magnification by an achromatic lens (focal length 75.8 mm) onto the entrance slit of a Czerny-Turner type spectrometer f/4.7 (0.67 m focal length with 1800 g/mm refection grating; the reciprocal dispersion of 0.83 nm/mm in first diffraction order). The overall instrumental profile was very close to Gaussian with 0.029 nm FWHM. Signals from air-cooled-CCD detector (2048x512 square pixels, size 12 μm) were A/D converted and collected by PC.

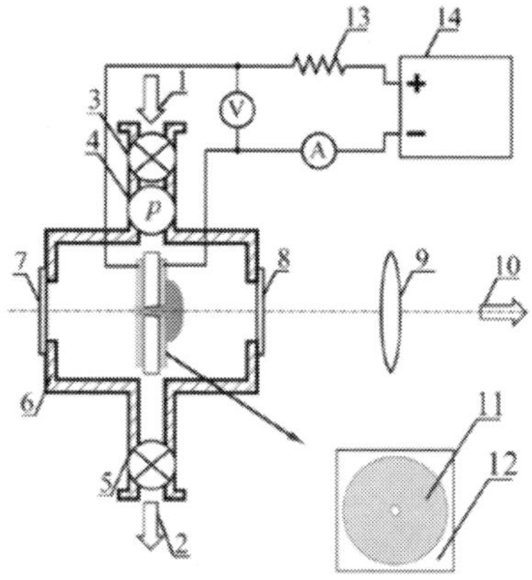

Figure 1. Schematic diagram of the experimental setup: (1) Gas inlet, (2) gas outlet to vacuum pump, (3) gas inlet valve, (4) vacuum gauge and manometer, (5) vacuum valve, (6) vacuum chamber, (7) rear window, (8) front window, (9) achromatic lens, (10) to spectrometer, (11) Au electrodes, (12) alumina, (13) ballast resistor R=50 kΩ, (14) dc power supply (2kV, 100mA).

3. Experimental hydrogen Balmer beta line profiles from hydrogen seeded argon MHGD

In figure 2 we present the hydrogen Balmer beta (H_β) line profiles recorded end-on from the cathode side of argon MHGD seeded by hydrogen for the optical emission spectroscopy diagnostics. Although all hydrogen Balmer lines are electric field sensitive (via linear Stark effect) and easily detectable, the H_β line was selected due to proper intensity and well separation from other spectral lines – see [5].

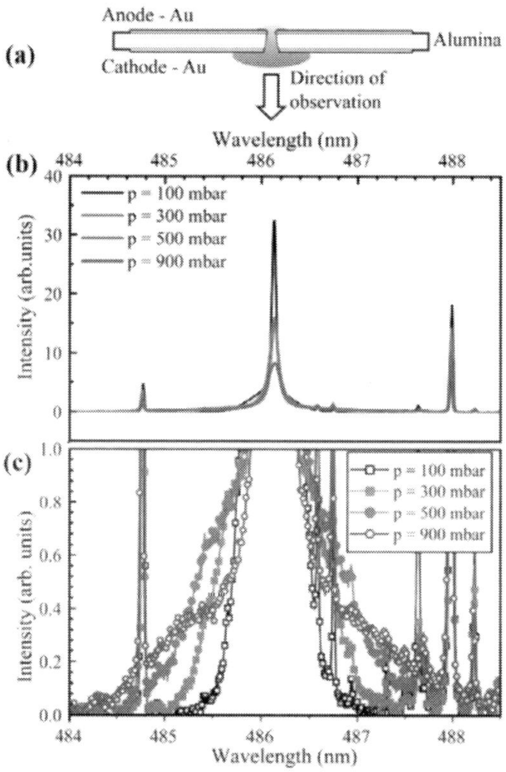

Figure 2. The H_β line profiles recorded end-on from the cathode side of hydrogen seeded argon MHGD, operated in the pressure range 100 – 900 mbar. Panel a) – schematic depiction of the observation side. Panel b) – profiles of the H_β line recorded at four different values of discharge pressure. Panel c) – pedestals of the foregoing profiles, zoomed for better visibility of anomalously broadened wings.

The recorded H_β line profiles have two components: 1) narrow central part, and 2) anomalously broadened wings. The appearance of the central part is well know – it is dominantly formed by the radiation of thermalized H atoms excited by electrons in negative glow (NG) and plasma producing a Voigt-type contribution $G_V(\lambda)$ of the overall line profile. On the other hand, the spread of wings suggest two things: i) the wings are comprised of radiation emitted by fast H atoms with kinetic energies $E_k \approx eU_D$, where U_D is the discharge voltage and e is the elementary charge, and ii) these atoms radiate in a strong electric field E of the cathode sheath (CS) and the radiation profile is therefore broadened by the linear dc Stark effect. The aim of our theoretical model is to explain formation of fast H atoms, describe their distribution, and predict the shape of radiation wings.

4. Theoretical model

4.1. *Formation of fast H atoms*

We consider that the fast H atoms originate from positive H ions. These ions are mostly created in NG and thermalized in collisions with matrix gas. Those of them that diffuse into the tiny CS of discharge are subsequently accelerated towards the cathode by a strong cathode sheath electric field E gaining kinetic energy E_k that is comparable to the whole available energy eU_D.

For H^+ ions, the mean free path λ is greater than the CS thickness; for instance, at 500 mbar, the CS thickness is ~ 25 µm, and λ (estimated at 1000 K with the aid of data from [7]) is ~ 55 µm. So, H^+ ions move almost without collisions through the CS, forming a beam of ions directed normally towards the cathode; in the crudest approximation, the contribution of H_2^+ and H_3^+ ions can be neglected because their λ's are 3 µm and 8 µm under the same discharge conditions.

The profile of the H^+ ion beam at the cathode is taken in the form

$$B(v_z) = I_B \exp\left[-\frac{m_H}{2\Delta E_B}\left(v_z + \sqrt{2E_B/m_H}\right)^2\right], \tag{1}$$

where I_B and E_B are the beam intensity and mean energy, respectively, $\Delta E_B \approx 0.1$ eV is the beam energy width inherited from the cold H atoms upon creation and thermalization of H^+ ions in NG, while m_H is the mass and v_z is the z-component of velocity of H^+ ion; note here that the z-axis is oriented away from the cathode, so $\langle v_z \rangle_B = -\sqrt{2E_B/m_H}$ is the mean z-component of velocity of H^+ ions in the beam.

At the cathode H^+ ions are neutralized. The majority of them are backscattered as fast H atoms, while the rest of them are captured and released later as cold atoms which don't contribute to formation of radiation wings. Note also that other processes, like charge-exchange collisions, contribute negligibly to formation of fast H atoms because they tale place in the CS which is very thin. According to our simulations, performed with the aid of publicly available SRIM /TRIM program [8], the distributions of energies and directions of backscattered atoms are independent for normal incidence of (monoenergetic) H^+ ions. We have found that the probability distribution of backscattering angle θ can be approximated by

$$dP_{Ang}^{(BS)}(\theta) = \sin(2\theta)d\theta, \tag{2}$$

and the probability distribution of backscattering energy by

$$dP_{En}^{(BS)}(\varepsilon) = A(\varepsilon_b - \varepsilon)e^{k\varepsilon}d\varepsilon \tag{3}$$

for $\varepsilon \le \varepsilon_b$ and 0 otherwise; here: A is the normalization constant, $\varepsilon \equiv E_{BS}/E_i$ is the relative backscattering energy (i.e. quotient of backscattering energy E_{BS} and incident energy E_i), while the threshold ε_b and exponent k are the material constants for the cathode material in question. In the course of our simulations we have found that $\varepsilon_b = 0.99$ and $k = 3.5$ for gold. We have also found that our final results are rather insensitive to details of assumed energy distribution. Thus, the variation of ε_b from 0.8 to 0.99 and k from 2.5 and 6.5 change the values of the best fit parameters less than 1%. The same conclusion applies to the form of the beam profile, i.e. instead of (1) we could use any other sufficiently narrow beam profile of the same energy E_B, and the only real advantage of profile (1) is that it enables estimation of gas temperature $T_g \approx T_B$ via beam temperature $T_B = \Delta E_B / k_B$.

4.2. *Radiation of the backscattered H atoms*

On their subsequent motion though matrix gas, the backscattered H atoms occasionally radiate after being excited. The major part (p_{ISO}) of them are excited in collisions with the most abundant heavy particles – Ar atoms in the case of MHGD in Ar, and H_2 molecules in the case of discharge in pure hydrogen. In the first approximation, these collisions can be treated as collision of rigid spheres because both colliding particles are neutral.

Collision of rigid spheres is isotropic in the centre of mass frame of reference. Due to mass ratio of 1/40, the collision of H atom and Ar atom is practically isotropic in the laboratory frame of reference too. The consequence is that the directional motion of H atom is on average wiped out after single collision, so the resulting end-on profile in Ar discharge is (almost) symmetric. On the other hand, the mass ratio for collision of H atom and H_2 molecule is ½, so the colliding H atoms retain on average 1/3 of their velocity. Nevertheless the radiation is again almost (but less) symmetric because H atoms experience on average more than one collision before excitation. Note also that a small $1-p_{ISO}$ part of the backscattered H atoms are excited by electrons and their radiation is blue-shifted.

4.3. *The role of dc Stark broadening*

The Doppler profile of radiation is broadened by the linear dc Stark effect in a strong CS electric field E. The broadened profile is the convolution $[S_E * D](\lambda)$ of the radiation Doppler profile $D(\lambda)$ and the Stark shift operator

$$S_E(\lambda) = \sum_j c_j \delta(\lambda - s_j E), \tag{4}$$

where s_j and c_j are the Stark manifold and intensity coefficients for the line in question.

The magnitude of Stark broadening is linearly dependent on the magnitude of electric field E which vary throughout the CS. For the sake of simplicity, we have assumed that the effective distribution of electric field is

$$E(z) = E_0 (1 - z / z_g) \tag{5}$$

where E_0 is the electric field at the cathode surface, z is the distance from the cathode, and z_g is the CS thickness. Hence, the voltage in the CS is $U_{CS} = E_0 z_g / 2$, and the average electric field (a more robust quantity independent on details of electric field distribution in CS) is $E_a = U_{CS} / z_g = E_0 / 2$.

In end-on observations we record the emissions from all perceptible parts of discharge. Due to an approximate plan-parallel symmetry of the CS region, the Doppler profile varies only with distance z from the cathode, $D(\vec{r}, \lambda) = D_z(\lambda)$, so the end-on profile of wings is

$$G_W(\lambda) = \int dz \left[S_{E(z)} * D_z \right](\lambda). \tag{6}$$

In this model we took that $D_z(\lambda)$ is

$$D_z(\lambda) = e^{-az} D_0(\lambda), \tag{7}$$

where $D_0(\lambda)$ is the Doppler profile at the cathode, and a is a coefficient of exponential attenuation of profile's magnitude with distance z. Successive stages of wings formation according to our model are shown in figure 3.

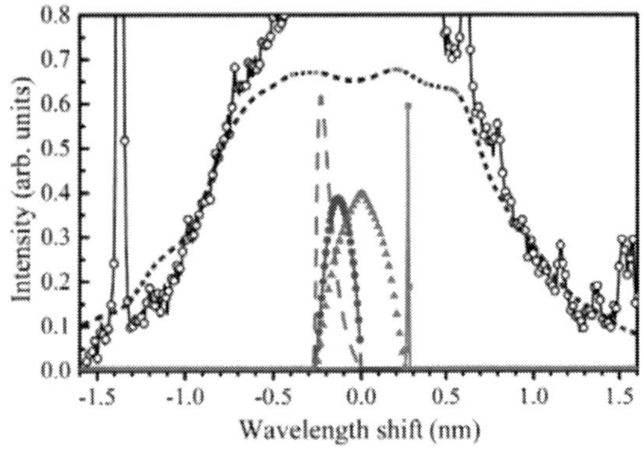

Figure 3. Successive stages of wings formation for MHGD at $p = 500\,\text{mbar}$. Experimental points are shown by empty circles connected by straight lines. The narrowest curve (line-connected solid squares) represents a beam of 140eV H^+ ions with normal incidence at the cathode. The backscattered beam is represented by the dashed curve prior to the angular scattering and by solid circles after the angular scattering. The effect of isotropic scattering is shown by triangles. Finally, the profile of wings is shown by the dotted curve.

4.4. *The H_β line model function*

The final form of our model function for experimental line profiles is

$$G(\lambda) = \Im^*\left[G_W(\lambda) + G_V(\lambda)\right] + b \tag{8}$$

where $G_W(\lambda)$ is the model function for anomalously broadened wings, and $G_V(\lambda)$ is the Voigt profile pertaining to the central part of line profile. Here, the overall theoretical profile

$$G_{th}(\lambda) = G_W(\lambda) + G_V(\lambda), \tag{9}$$

i.e. the profile that would be recorded by perfect instruments, is convolved with the instrumental profile $\Im$, taken in the form of a (zero-centred) Gaussian with 0.029 nm FWHM in accordance with calibration of our experimental system. Finally, the constant b is introduced to take into account possible experimental offset of the data base line.

The model function $G_W(\lambda)$ for the shape of line wings has seven free parameters. The three parameters: intensity I_B, mean energy E_B, and energy-width ΔE_B, describe the beam of H^+ ions at the cathode surface (1). The next two parameters, the voltage U_{CS} and thickness z_g, are characteristics of cathode sheath. The last two parameters are the attenuation coefficient a of the Doppler profile (7), and p_{ISO} - the part of the photons emitted by isotropically scattered H atoms. These two parameters describe fast H atoms, formed by the neutralization and backscattering of H^+ ions at the cathode. In addition, the Voigt profile $G_V(\lambda)$ is characterized by its intensity I_V, Lorentz width w_L, and Doppler width w_D.

5. Applications

5.1. *Hydrogen seeded MHGD in argon*

In four panels of figure 4 we present the experimental H_β line profiles together with their best fit curves (8); implicitly involved Stark parameters s_j and c_j for the H_β line are the same as in [9]. The experimental profiles are recorded in the pressure range 100-900 mbar. The discharge parameters and the values of the best fit parameters are given in table 1 together with calculated values for other relevant parameters. Parameter variation on discharge pressure is presented in figure 5.

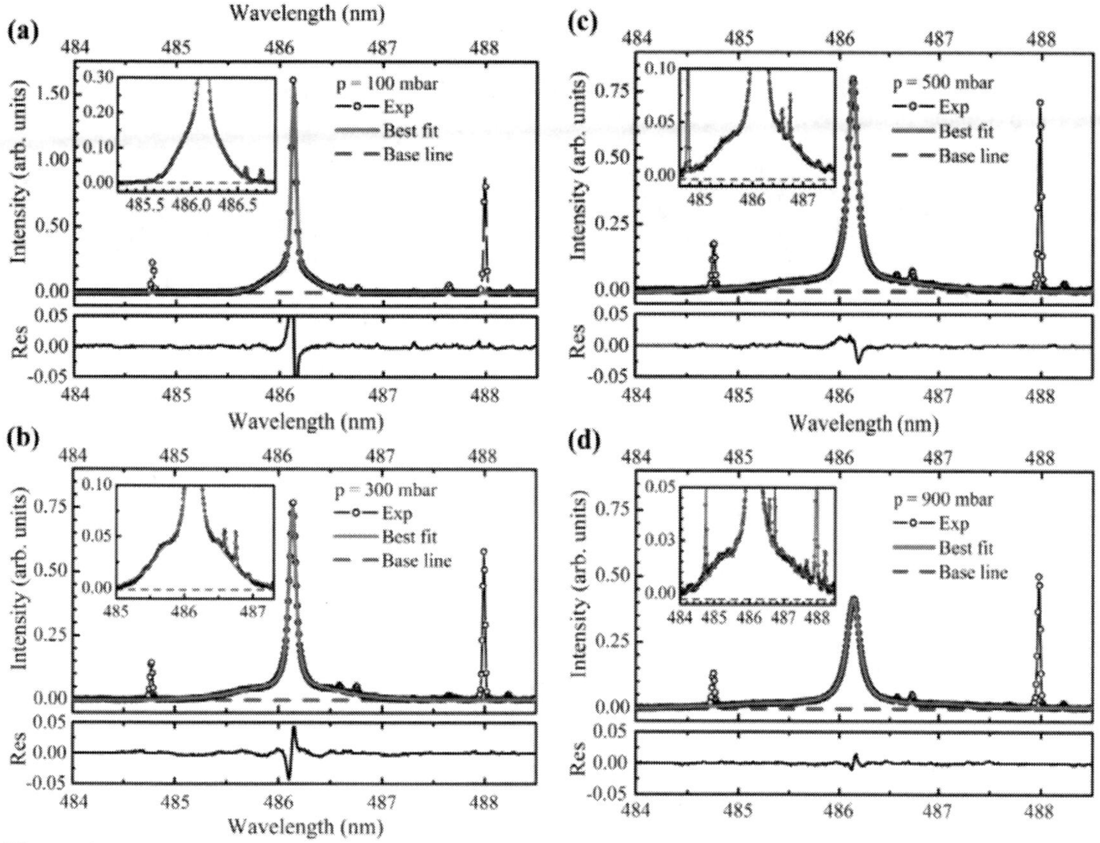

Figure 4. Experimental H_β profiles (symbols) , their best fit curves (solid lines), and the base line level (dashed line). Zoomed wings are presented in insets, while the residuals are shown below each graph.

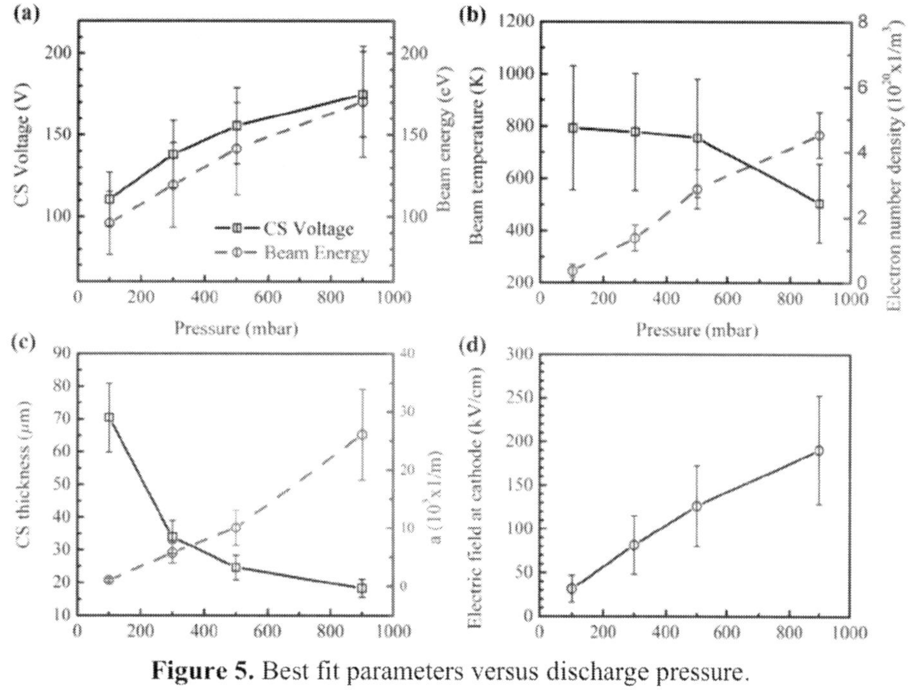

Figure 5. Best fit parameters versus discharge pressure.

Table 1. Best fit parameters and calculated parameters (E_a, E_0, w_V, w_S, and N_e); percentage errors are given in parentheses.

	p=100 mbar (U_D=183 V, I_D=5.0 mA)	p=300 mbar (U_D=190 V, I_D=5.5 mA)	p=500 mbar (U_D=190 V, I_D=8.0 mA)	p=900 mbar (U_D=190 V, I_D=8.0 mA)
E_B (eV)	95 (20%)	120 (20%)	140 (20%)	170 (20%)
T_B (K)	790 (30%)	780 (30%)	750 (30%)	500 (30%)
U_{CS} (V)	110 (15%)	140 (15%)	155 (15%)	175 (15%)
z_g (μm)	70 (15%)	34 (15%)	25 (15%)	18 (15%)
a (1/m)	1000 (30%)	5700 (30%)	10000 (30%)	26000 (30%)
p_{ISO}	0.73 (20%)	0.80 (20%)	0.75 (20%)	0.90 (20%)
w_L (nm)	0.027 (10%)	0.067 (10%)	0.10 (10%)	0.15 (10%)
w_D (nm)	0.0025	0.0018	0.0010	0.0012
E_a (kV/cm)	16 (30%)	41 (30%)	62 (30%)	95 (30%)
E_0 (kV/cm)	32 (30%)	82 (30%)	125 (30%)	190 (30%)
w_{VdW} (nm)	-	-	0.017	0.028
w_S (nm)	0.22	0.053	0.087	0.119
N_e (10^{20} m^{-3})	0.4 (50%)	1.4 (30%)	2.9 (20%)	4.5 (15%)

The electron number density N_e is calculated from parameters of observed Voigt profile $\Im * G_V(\lambda)$ pertaining to the central part of the H_β line. This profile is convolution of instrumental profile $\Im$ and the true Voigt profile $G_V(\lambda)$, i.e. the Voigt profile which would be observed with an ideal spectrometer. The true Voigt profile is

$$G_V(\lambda) = G_L(\lambda) * G_D(\lambda), \tag{10}$$

where $G_L(\lambda)$ is the resulting Lorentz profile with FWHM w_L, and $G_D(\lambda)$ is the Doppler profile due to velocity distribution of H atoms that emit the H_β radiation; the uncertainty of $G_D(\lambda)$'s FWHM w_D is very large, so the values of w_D in table 1 are given only for illustration purposes.

The resulting Lorentzian is

$$G_L(\lambda) = G_S(\lambda) * G_{VdW}(\lambda), \tag{11}$$

where $G_S(\lambda)$ is the Stark Lorentzian (induced by Stark broadening in microscopic electric field), and $G_{VdW}(\lambda)$ is Van der Waals (VdW) Lorentzian (induced by collisions of Ar atoms with H atoms). Therefore, $w_L = w_S + w_{VdW}$, where w_S is FWHM of $G_S(\lambda)$ and w_{VdW} is FWHM of $G_{VdW}(\lambda)$.

The estimation of w_{VdW} is carried out as in [10] with remark that due to great uncertainty of w_D the beam temperature T_B was used instead of gas (Doppler) temperature T_g, calculated from w_D. Finally, N_e is evaluated from the Stark width $w_S = w_L - w_{VdW}$ by empirical formula (2a) from reference [11].

5.2. MHGD in pure hydrogen

In figure 6 we show the H_β experimental profiles with superimposed molecular H_2 lines recorded in pure hydrogen MHGD at 100 mbar, 200 mbar, and 400 mbar. The experimental setup was the same as for the MHGD in argon. The profiles were fitted by the modified model function (8) which reads

$$G(\lambda) = \Im * \left[G_{th}(\lambda) + G_{H_2}(\lambda) \right] + b. \tag{12}$$

Here, $G_{th}(\lambda)=G_W(\lambda)+G_V(\lambda)$ is previously defined theoretical H_β profile (9) and $G_{H_2}(\lambda)$ is a Voigt-type model function for the molecular hydrogen line 0.04 nm red-shifted relative to H_β center. Being poorly resolved from the central Voigt, the molecular line obstructs fitting and greatly increases the uncertainty of the best fit parameters.

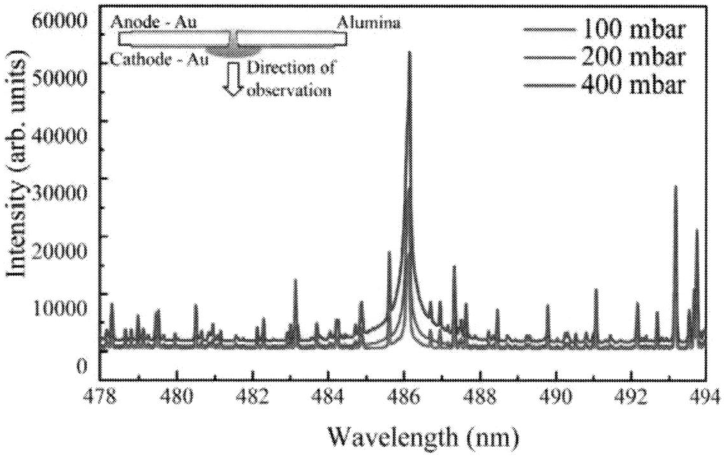

Figure 6. Hydrogen MHGD end-on spectra recorded around the H_β line. Several molecular H_2 lines, superimposed on the H_β profile, are visible.
Inset: schematic depiction of MHGD and direction of observation.

In figure 7 we show the best fits of the foregoing experimental profiles, and in table 2 we give the corresponding values of the best fit parameters together with the values of calculated parameters.

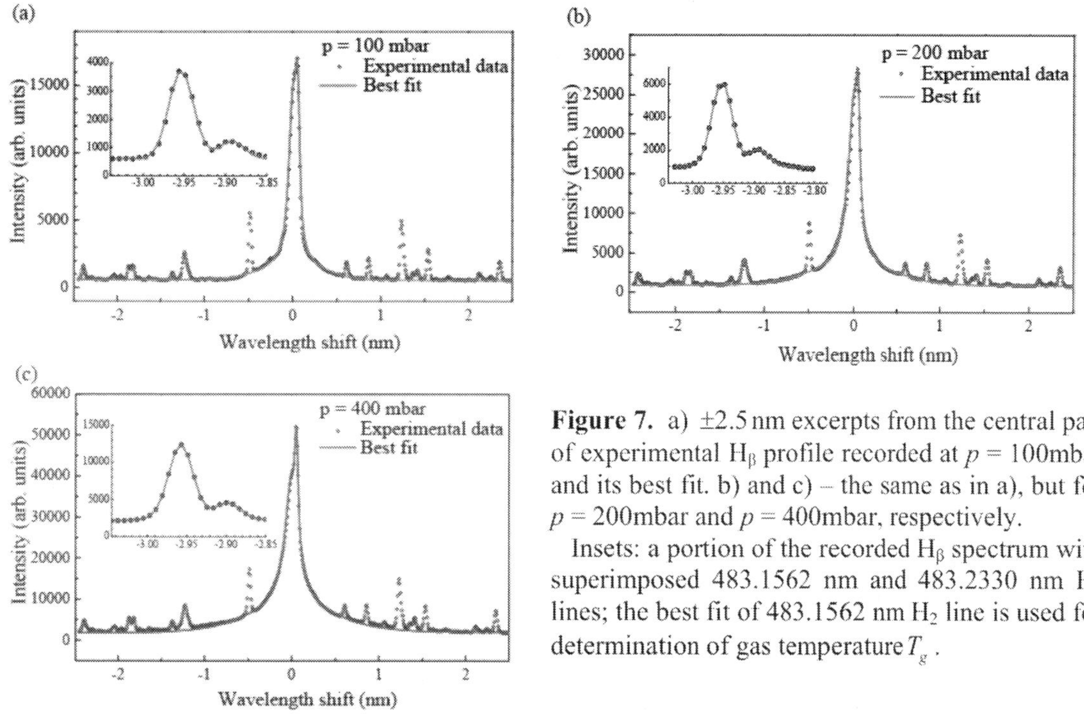

Figure 7. a) ±2.5 nm excerpts from the central part of experimental H_β profile recorded at p = 100mbar and its best fit. b) and c) – the same as in a), but for p = 200mbar and p = 400mbar, respectively.
Insets: a portion of the recorded H_β spectrum with superimposed 483.1562 nm and 483.2330 nm H_2 lines; the best fit of 483.1562 nm H_2 line is used for determination of gas temperature T_g .

Table 2. The most relevant best fit parameters and calculated values for E_a, E_0 and Tg, quoted with their percentage errors.

	p=100 mbar (U_D=300 V, I_D=4.5 mA)	p=200 mbar (U_D=250 V, I_D=7.0 mA)	p=100 mbar (U_D=400 V, I_D=12 mA)
E_B (eV)	300 (20%)	240 (20%)	300 (20%)
w_B (pm)	34 (30%)	11 (30%)	6.6 (30%)
U_{CS} (V)	300 (15%)	250 (15%)	300 (15%)
z_g (μm)	120 (15%)	58 (15%)	65 (15%)
w_L (pm)	99 (20%)	144 (20%)	195 (20%)
w_G (pm)	5.1 (60%)	5.2 (60%)	4.5 (60%)
E_a (kV/cm)	25 (30%)	43 (30%)	46 (30%)
E_0 (kV/cm)	50 (30%)	86 (30%)	92 (30%)
T_g (K)	735 (0%)	785 (0%)	870 (0%)

5.3. Comparison with Kushner's simulation data

In figure 8, we compare our data with the Kushner's simulation data [12] obtained in similar geometry and in the same pressure range, but with constant discharge current of 2 mA and variable discharge voltage. The discharge voltage and corresponding CS voltage vary with pressure roughly the same, whereas the data for electron number density N_e almost agree within the experimental error bars.

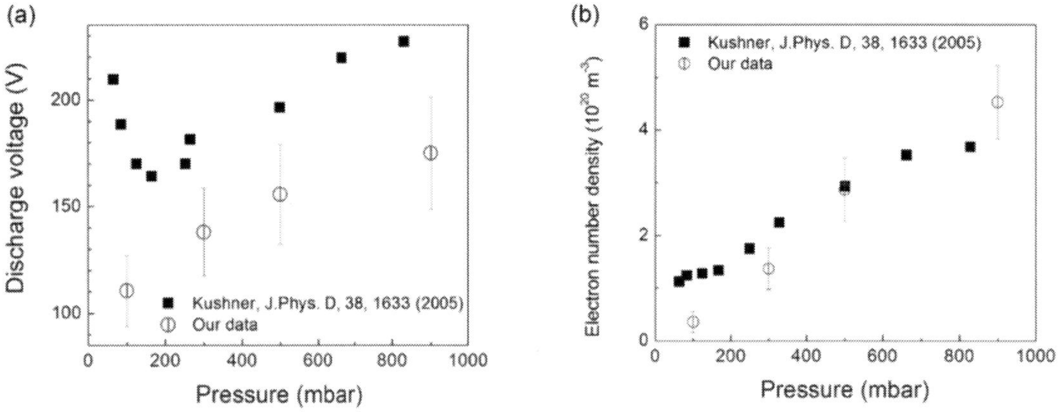

Figure 8. Comparison of our data with Kushner's simulation data [12].
(a) Discharge voltage (simulations) and the corresponding CS voltage (our MHGD).
(b) Electron number densities.

6. Discussion

In this work we have applied our model to the case of H_β line for two reasons: firstly, compared to other hydrogen Balmer lines, separation of H_β from neighbouring spectral lines is better, and secondly, the H_β line is neither too strong nor too weak; otherwise, the self-absorption has to be taken into account and/or the signal to noise ratio is poor. Nevertheless, the use of other lines is possible.

Next, although we didn't test our model outside the current scope we expect that it should be valid for other MHGD's provided that requisite changes in values of structural parameters (like Stark manifold and intensity coefficients, threshold ε_b and exponent k for different cathode material, etc) are done. For this reason we expect that different MHGD's manifest many similarities under like discharge conditions and MHGD design. On the other hand, some differences are to be expected as well. Thus, in the case of hydrogen seeded MHGD in argon and MHGD in pure hydrogen the following differences have to be noticed and discussed

- CS thickness in pure hydrogen is two times larger than in argon MHGD, most likely due to smaller momentum transfer cross-section of H^+ ions in H_2.
- In Ar MHGD the majority of H^+ ions are the ionized H atoms produced by destruction of H_2 molecules in collisions with Ar metastables; therefore, the beam temperature T_B is approximately equal to the gas (i.e. argon) temperature T_g. In pure hydrogen, H^+ ions from dissociative ionization of H_2 molecules are only partially thermalized and a higher beam temperature T_B is expected; for this reason, the gas temperature T_g is estimated from the width of the H_2 483.1562nm line.
- The N_e calculations in pure H_2 are not reliable due to self-absorption, giving faulty larger N_e values.

7. Conclusion

In conclusion, we have proposed a model of cathode sheath (CS) processes and formation of hydrogen Balmer lines' profiles in micro-hollow gas discharges (MHGD). The model was applied to the hydrogen Balmer beta profiles (recorded in hydrogen seeded argon and in pure hydrogen MHGD), and used for simultaneous diagnostics of plasma and CS parameters. The average value of CS electric field and CS thickness are determined from the extended line wings, while the electron number density is found from the width of the central part of line profile. Our findings are in a good agreement with the results obtained by numerical simulations [12], whereas the agreement with the results of other similar experiments [13,14,15,16] is less satisfactory and requires explanation.

Acknowledgments

This work is supported by the Ministry of Science and Technological Development of the Republic of Serbia under Project No. 171027.

References

[1] Becker K H, Schoenbach K H and Eden J G 2006 *J. Phys.D: Appl. Phys.* **39** R55
[2] Tachibana K 2012 Microplasmas in The 2012 Plasma Roadmap, *J. Phys.D: Appl. Phys.* **45** 253001
[3] Schoenbach K H, Verhappen R, Tessnow T, Peterkin F E and Byszewski W W 1996 *Appl. Phys. Lett.* **68** 13
[4] Spasojević Dj, Cvejić M, Šišović N M and Konjević N 2010 *Appl. Phys. Lett.* **96** 241501
[5] Cvejić M, Spasojević Dj, Šišović N M and Konjević N 2011 *J. Appl. Phys.* **110** 033305
[6] Spasojević Dj, Cvejić M, Šišović N M and Konjević N 2012 *J. Appl. Phys.* **111** 096103
[7] Tabata T and Shirai T 2000 *Atomic Data and Nuclear Tables* **75** 1; Phelps A V 1990 *J. Phys. Chem. Ref. Data* **19** 653
[8] Ziegler J F, Ziegler M D, and Biersack "SRIM-2008" (www.SRIM.org)
[9] Videnović I R, Konjević N and Kuraica M M 1996 *Spectrochim. Acta Part B* **51** 1707
[10] Djurović S and Konjević N 2009 *Plasma Sources Sci. Technol.* **18** 035011
[11] Ivković M, Jovićević S and Konjević N 2004 *Spectrochim. Acta Part B* **59** 591
[12] Kushner M J 2005 *J. Phys. D: Appl. Phys* **38** 1633
[13] Penache C, Miclea M, Bräuning-Demian A, Hohn O, Schössler S, Jahnke T, Niemax K and Schmidt-Böcking H 2002 *Plasma Sources Sci. Technol.* **11** 476
[14] Moselhy M, Petzenhauser I, Frank K and Schoenbach K H 2003 *J. Phys. D: Appl. Phys* **36** 2922
[15] Lazzaroni C, Chabert P, Rousseau A and Sadeghi N 2010 *Eur. Phys. J. D* **60** 555
[16] Sismanoglu B N, Grigorov K G, Caetano R, Rezende M V O and Hoyer Y D 2010 *Eur. Phys. J. D* **60** 505

Physico-Chemistry of Planetary Atmospheric Entry Plasmas

Arnaud Bultel, Julien Annaloro and Vincent Morel

CORIA, UMR CNRS 6614, Université de Rouen, BP 12, Avenue de l'Université, 76801 Saint-Etienne du Rouvray Cedex, France

E-mail: arnaud.bultel@coria.fr

Abstract. This contribution deals with the description of the physico-chemistry in the reactive flow produced near the surface of a body entering a planetary atmosphere in hypersonic regime. A shock layer is formed in which a boundary layer is developed where energy is released to the body surface. N_2 is chosen as a test-case of Earth atmospheric reentry to illustrate the main characteristics of the flow. The vibrational and electronic specific Collisional-Radiative model CoRaM-N_2 is described and implemented in a one-dimensional Euler flow solver. The well-known FIRE II flight experiment conditions are used to illustrate the behaviour of the different ground and excited states of atomic and molecular species. The results show that the three successive phases occur: vibrational excitation, dissociation and ionisation. Radiation plays a minor role. According to the upstream thermodynamic conditions, the boundary layer edge can be in local thermodynamic equilibrium or not. The different strategies using thermal protection systems are discussed to reduce the damaging of the entering body.

1. Introduction

Three weeks before the SPIG 2012 conference in Zrenjanin, the news reported an example of perfect control of a planetary atmospheric entry phase. Indeed, NASA succeeded in landing the rover Curiosity on the ground of Mars. After a long journey of 8 months and more than 550 millions of kilometres, the rover entered the atmosphere mainly composed by CO_2 and N_2 molecules at a speed close to 6 km s^{-1}. The ground pressure level on Mars is of 650 Pa which is 160 times lower than on Earth. Density of the atmosphere is close to 10^{-2} kg m^{-3} on the ground level and decreases according to a law of exponential type with altitude. At z = 100 km, density is 10^{-7} kg m^{-3}. These values correspond to weak order of magnitude with respect to those observed on Earth. Nevertheless, Martian atmosphere is dense enough to be strongly slowed down near the probe which leads to a strong compression. The resulting increase of the pressure induces a force exerted on the probe the intensity of which is strong enough to slow down the probe in turn during its free fall. Afterward, using supersonic parachutes and a sky crane, the probe landed with a very small velocity close to 1 m s^{-1}.

In terms of energy, the probe lost all its kinetic energy over the phase of entry. This lost energy is converted into thermal energy during the compression process at the probe surface. The characteristic time scales of this compression are smaller than the ones needed to reach equilibrium. The flow near the probe surface therefore departs from equilibrium.

The main characteristics of the flow near the entering body will be described in the present contribution. The discussion will be focused on thermal and chemical non equilibrium. In

Published under licence by IOP Publishing Ltd

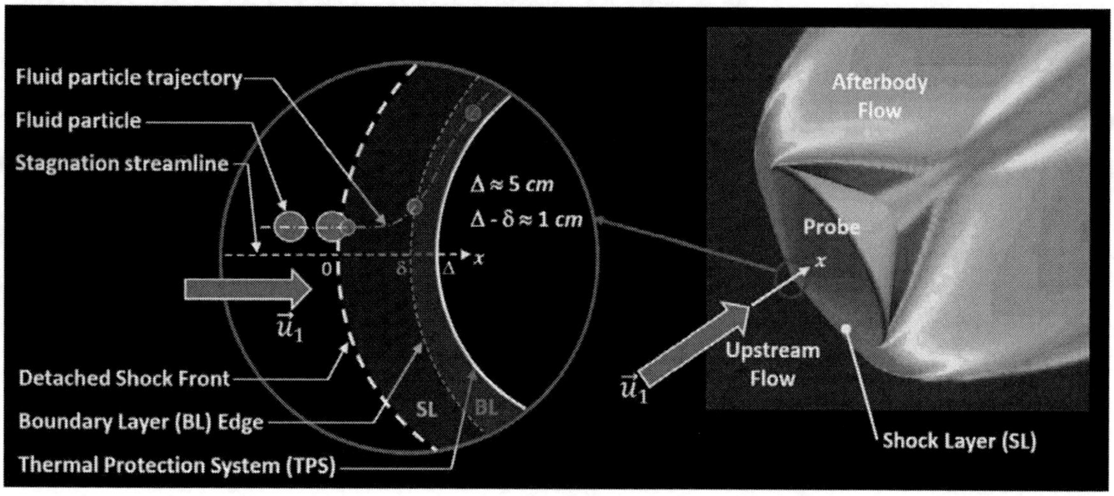

Figure 1. Schematic structure of the flow around a probe entering a planetary atmosphere. Focus on the stagnation streamline corresponding to the strongest gradients zone. A Shock Layer (SL) is formed by the sudden decrease of the flow velocity near the surface. A typical fluid particle undergoes a strong compression crossing the shock front limiting the SL, and a strong heating. Flowing close to the probe surface, this fluid particle releases part of its energy to the wall which is protected by a Thermal Protection System (TPS) forming a heatshield.

particular, state-to-state simulations based on the elaboration of Collisional-Radiative (CR) models will be detailed.

2. Post-shock front relaxation

The strong decrease of the flow velocity u_1 near the wall induces the formation of a Shock Layer (SL) where density is largely higher than in the upstream flow. This SL is limited by the probe surface and a detached shock front. Its typical thickness Δ is of several centimetres. Figure 1 displays a schematic view of the flow structure around the entering body. Incident fluid particles cross the detached shock front and undergo the strong compression. Despite a rough decrease, their velocity u_2 is then high enough to prevent Local Thermodynamic Equilibrium (LTE). Let us consider the first Damköhler number

$$Da_1 = \frac{\Delta}{u_2\,\tau_{CR}} \tag{1}$$

where τ_{CR} is the characteristic time scale to reach equilibrium through collisional and radiative elementary processes. Indeed, its order of magnitude is of unity which means that the flow is in local chemical non equilibrium. The kinetic energy converted into thermal energy leads to ionisation phenomena. A gas $\rightarrow$ plasma transition is therefore observed. The characteristic time scale required to obtain Maxwell-Boltzmann equilibrium for each type of particles (heavy particles and electrons) is short enough to have second Damköhler number $Da_2 = \Delta/(u_2\,\tau_{MB})$ much higher than unity. The concept of temperature can then be used to characterize the translation mode of each type of particle (T_A for the heavy particles and T_e for electrons). Unfortunately, the characteristic time τ_{e-A} to obtain thermal equilibrium between T_A and T_e is significant. The third Damköhler number $Da_3 = \Delta/(u_2\,\tau_{e-A})$ is lower than unity. The flow then departs from thermal equilibrium.

Table 1. Species and states involved in the nitrogen Collisional-Radiative model CoRaM-N_2 described in section 2. Elementary processes between these states and species are listed in table 2.

Species	States
N_2	$X^1\Sigma_g^+(v=0\rightarrow v_{max}=67)$, $A^3\Sigma_u^+$, $B^3\Pi_g$, $W^3\Delta_u$, $B'^3\Sigma_u^-$, $a'^1\Sigma_u^-$, $a^1\Pi_g$, $w^1\Delta_u$, $G^3\Delta_g$, $C^3\Pi_u$, $E^3\Sigma_g^+$
N_2^+	$X^2\Sigma_g^+$, $A^2\Pi_u$, $B^2\Sigma_u^+$, $a^4\Sigma_u^+$, $D^2\Pi_g$, $c^2\Sigma_u^+$
N	${}^4S^o_{3/2}$, ${}^2D^o=({}^2D^o_{5/2}+{}^2D^o_{3/2})$, ${}^2P^o=({}^2P^o_{1/2}+{}^2P^o_{3/2})$, ${}^4P_{1/2}$,... (63 states)
N^+	${}^3P_0$, ${}^3P_1$, ${}^3P_2$, ${}^1D_2$, ${}^1S_0$, ${}^5S^o_2$, ${}^3D^o_3$, ${}^3D^o_2$, ${}^3D^o_1$
e^-	-

The plasma flow is therefore in thermal and chemical non equilibrium. Owing to the specific energy of the flow around several 10^7 J kg^{-1}, the excited states population density is strong which leads to significant radiative losses. In order to simulate with relevance the SL dynamics, a state-to-state approach is mandatory. This approach consists in considering each species on its excited states as an independent species. Denoting x the normal coordinate to the wall and y_{X_i} the mass fraction of this species X on its excited state i, the balance equation can be written as

$$\frac{dy_{X_i}}{dx}=\frac{m_{X_i}\left([\dot{X}_i]_C+[\dot{X}_i]_R\right)}{\rho u} \tag{2}$$

where m_{X_i} is the mass of the species, ρ the density of the flow, u its velocity and $[\dot{X}_i]_C+[\dot{X}_i]_R$ the collisional and radiative source terms.

Earth and Mars atmospheres have a common compound. Indeed, N_2 molecules are found in both atmospheres in different abundance. The mole fraction of these molecules is 0.78 for the Earth's atmosphere and 0.025 for the Martian atmosphere. Titan, the largest satellite of Saturn, has an atmosphere the nitrogen molecules mole fraction of which is 0.98. N_2 is therefore particularly interesting for entry situations. This explains why N_2 is considered as a benchmark and is often chosen for simulations or experiments in ground test facilities. In typical entry situations, N_2 molecules can lead to the formation on ground and excited states of N_2, N_2^+, N and N^+. Table 1 summarizes these species and states. We have developed a Collisional-Radiative model for them. This CR model is named CoRaM-N_2. The involved elementary processes are listed in table 2.

This CR model accounts for vibrational processes, electronic excitation, excitation transfer, dissociation, and ionisation either by electron impact or heavy particle impact. Charge exchange and dissociative recombination are also taken into account. Each backward rate coefficient is derived from the forward rate coefficient using the detailed balance principle. Radiative transitions are also involved in the Collisional-Radiative model. Escape factors are implemented in order to simulate the possible self-absorption of the plasma.

In 1965, NASA launched the FIRE (Flight Investigation of Reentry Environment) II probe in the framework of the Apollo program. This probe experienced a reentry at hypersonic velocity (mach number $\mathcal{M}_1>30$) in order to provide information on heat flux on the probe surface [19]. The results of this experiment are still intensively studied and the related conditions are retained here to illustrate the behaviour of the species inside the shock layer (elapsed time from the launch $t=1640\sim1641$ s, altitude $z\simeq58$ km).

Assuming a discontinuity at the shock front (located in $x=0$) treated with the classical

Table 2. Elementary processes of the nitrogen Collisional-Radiative model CoRaM-N_2 involving the species and states listed in table 1.

Type	Elementary processes	References
Vibrational processes	$N_2(X,v) + e^- \rightarrow N_2(X,w) + e^-$	[3]
	$N_2(X,v) + e^- \rightarrow 2\ N(^4S^o_{3/2}) + e^-$	[3]
	$N_2(X,v) + (N_2 \text{ or } N) \rightarrow N_2(X,w) + (N_2 \text{ or } N)$	[4, 5]
	$N_2(X,v)+N(^4S^o_{3/2}) \rightarrow 3\ N(^4S^o_{3/2})$	[4, 5]
	$N_2(X,v_{max})+N_2 \rightarrow 2\ N(^4S^o_{3/2}) + N_2$	[4, 5]
	$N_2(X,v_1)+N_2(X,v_2) \rightarrow N_2(X,w_1)+N_2(X,w_2)$	[4]
Electronic excitation	$N_2(i) + e^- \rightarrow N_2(j) + e^-$	[6]
	$N_2(i) + (N_2 \text{ or } N) \rightarrow N_2(j) + (N_2 \text{ or } N)$	[6, 7, 8]
	$N_2^+(i) + e^- \rightarrow N_2^+(j) + e^-$	[9]
	$N(i) + e^- \rightarrow N(j) + e^-$	[1, 10]
	$N(i) + (N_2 \text{ or } N) \rightarrow N(j) + (N_2 \text{ or } N)$	[6, 7, 8, 11]
	$N^+(i) + e^- \rightarrow N^+(j) + e^-$	[1]
	$N^+(i) + (N_2 \text{ or } N) \rightarrow N^+(j) + (N_2 \text{ or } N)$	[6, 7, 8]
Excitation transfer	$N_2(A) + N_2(A) \rightarrow N_2(X) + N_2(B)$	[6]
	$N_2(A) + N_2(A) \rightarrow N_2(X) + N_2(C)$	[12]
	$N_2(A) + N_2(B) \rightarrow N_2(X) + N_2(C)$	[11]
	$N_2(A) + N(^4S^o_{3/2}) \rightarrow N_2(X) + N(^2P^o)$	[6]
	$N_2(B) + N(^4S^o_{3/2}) \rightarrow N_2(X) + N(^2P^o)$	[11]
	$N_2(C) + N(^4S^o_{3/2}) \rightarrow N_2(X) + N(^2P^o)$	[11]
Dissociation	$N_2(i \neq X) + e^- \rightarrow N(j) + N(k) + e^-$	[9]
	$N_2^+(i) + e^- \rightarrow N(j)+ N^+(k) + e^-$	[9]
Ionisation	$N_2(i) + e^- \rightarrow N_2^+(j)+ 2\ e^-$	[9]
	$N_2(i) + (N_2 \text{ or } N) \rightarrow N_2^+(j) + e^- + (N_2 \text{ or } N)$	[7, 8]
	$N(i) + e^- \rightarrow N^+(j)+ 2\ e^-$	[1, 10, 13]
	$N(i) + (N_2 \text{ or } N) \rightarrow N^+(j) + e^- + (N_2 \text{ or } N)$	[7, 8]
Charge exchange	$N_2(X) + N^+(^3P_0) \rightarrow N_2^+(X) + N(^4S^o_{3/2} \text{ or } ^2P^o)$	[12]
	$N_2(X) + N^+(^3P_0) \rightarrow N_2^+(A) + N(4S^o_{3/2})$	[12]
Dissociative recombination	$N_2^+(X) + e^- \rightarrow N(^4S^o_{3/2}) + N(^2D^o \text{ or } ^2P^o)$	[14]
	$N_2^+(X) + e^- \rightarrow N(^2D^o) + N(^2D^o)$	[14]
Radiation	$N_2(B^3\Pi_g) \rightarrow N_2(A^3\Sigma_u^+) + h\nu$ (1st positive)	[10, 15]
	$N_2(C^3\Pi_u) \rightarrow N_2(B^3\Pi_g) + h\nu$ (2nd positive)	[10, 15]
	$N_2^+(B^2\Sigma_u^+) \rightarrow N_2^+(X^2\Sigma_g^+) + h\nu$ (1st negative)	[10, 15]
	$N(i) \rightarrow N(j < i) + h\nu$	[2]
	$N^+(i) \rightarrow N^+(j < i) + h\nu$	[2]

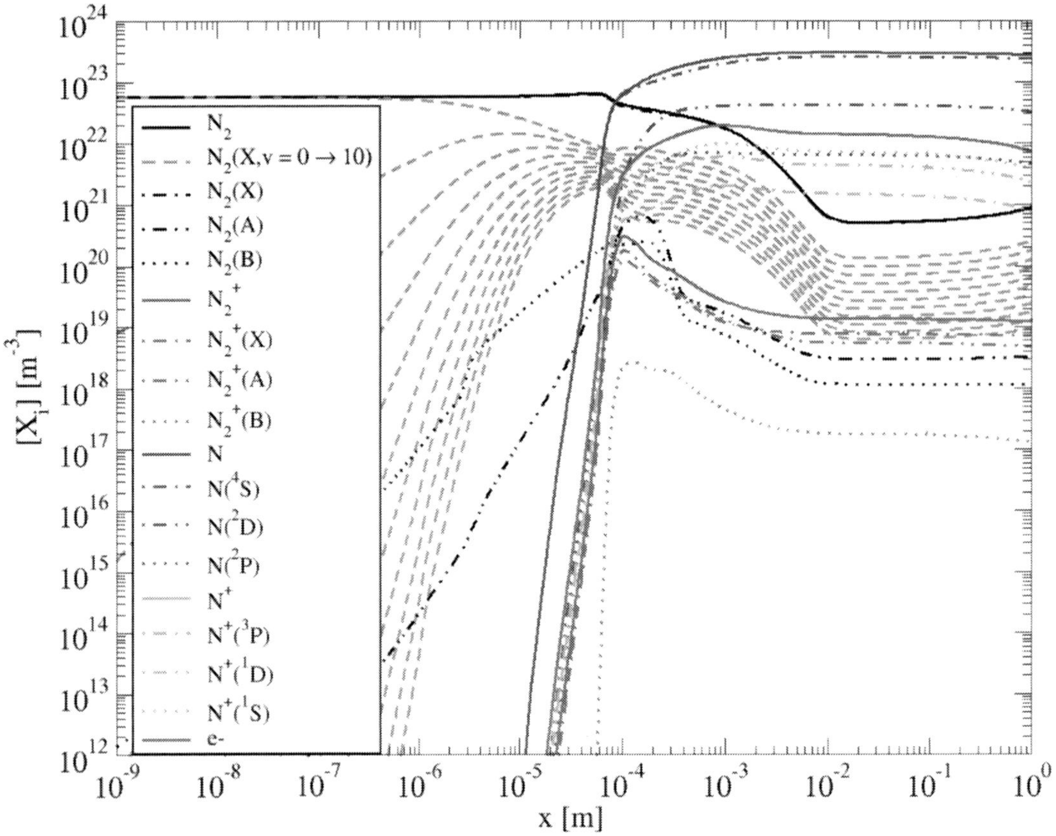

Figure 2. Species number density profiles along the stagnation streamline in the FIRE II flight conditions [19]. Rankine-Hugoniot equations are used crossing the shock front (all inner modes of energy storage frozen, except the translation increased from 250 K to 50 000 K approximately by energy conservation). We can note that the plasma goes successively through the following phases. First, the molecules dissociates. Then, the atoms are ionised. After $x \simeq 10^{-2}$ m, the relaxation of the plasma is observed. The dissociation and ionisation degrees reach 0.98 and 0.1, respectively.

Rankine-Hugoniot equations, figure 2 displays the spatial profiles of the species densities along the x-axis on the stagnation streamline for the upstream flow conditions $\mathcal{M}_1 = 33$, $p_1 = 33$ Pa, and $T_1 = 250$ K. Equation (2) is solved in coupling with the momentum and energy balance equations according to a one-dimensional (1D) Euler flow solver. A 2T (two temperatures) approach is considered for which translation of electrons and heavy particles are separately treated owing to the Da_2 and Da_3 Damkhöhler numbers order of magnitude discussed previously.

Just behind the shock front, a rapid vibrational excitation is observed. The concomitant excitation of the electronic excited states of N_2 takes place. The vibrational excitation allows the dissociation of N_2 either under N_2 molecules impact at the beginning or under N impact when the dissociation degree reached high enough values. Ionisation of the gas closely follows dissociation. The ionisation degree becomes quickly of the order of 0.1. At $x = 10^{-4}$ m from the shock front, the dissociation and ionisation phases are achieved. The remaining relaxation is much more quiet. To better understand these different phases, a thorough examination of the

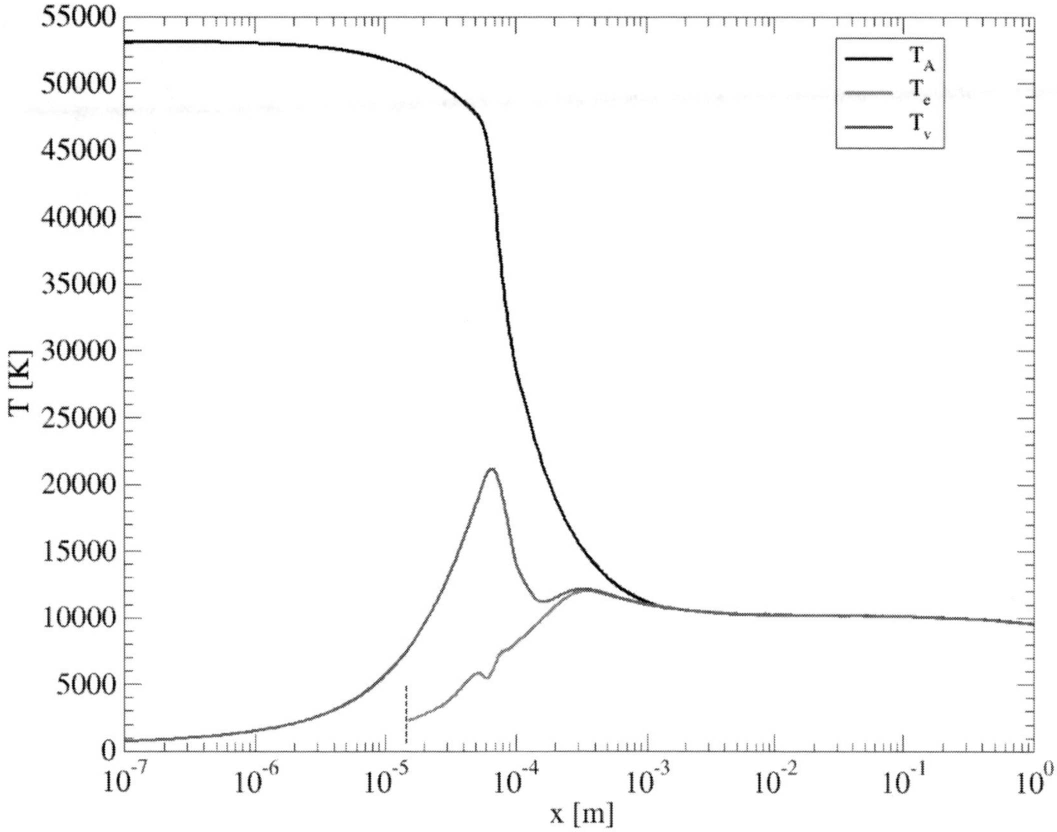

Figure 3. T_A, T_e and T_v profiles in the conditions of figure 2. Electron temperature is plot starting from $\simeq 10^{-5}$ m for which electron density is high enough and the concept of temperature relevant. For $x > 10^{-3}$ m, thermal equilibrium is observed.

temperatures of the flow is performed.

We define the vibrational temperature T_v as the excitation temperature of the first five vibrational levels of $N_2(X)$ calculated with

$$T_v = -\frac{1}{k_B \left[\frac{d}{dE_{X,v}}\left(\ln[N_2(X,v)]\right)\right]_{lsl}} \tag{3}$$

where lsl means that the derivative is the slope of the least square line and $E_{X,v}$ is the vibrational energy. Figure 3 displays the profile of T_v compared with that of T_e and T_A. Just behind the shock front, T_A has strongly increased owing to the Rankine-Hugoniot assumptions while the shock does not affect electrons owing to their weak mass. Indeed, the regime is subsonic for electrons and their temperature after the shock front remains at the same level as that in upstream flow. Nevertheless, the electron density is negligibly small so that the concept of electron temperature is questionable. Therefore, figure 3 reports T_e for $x \gtrsim 10^{-5}$ m for which electron density has sufficiently increased.

Since the vibration is frozen at the shock front, T_v equals T_1. The vibrational levels are quickly excited thanks to the vibration-translation elementary processes. T_A decreases while T_v increases. Energy is taken from translation and given to excitation, ionisation and mainly to

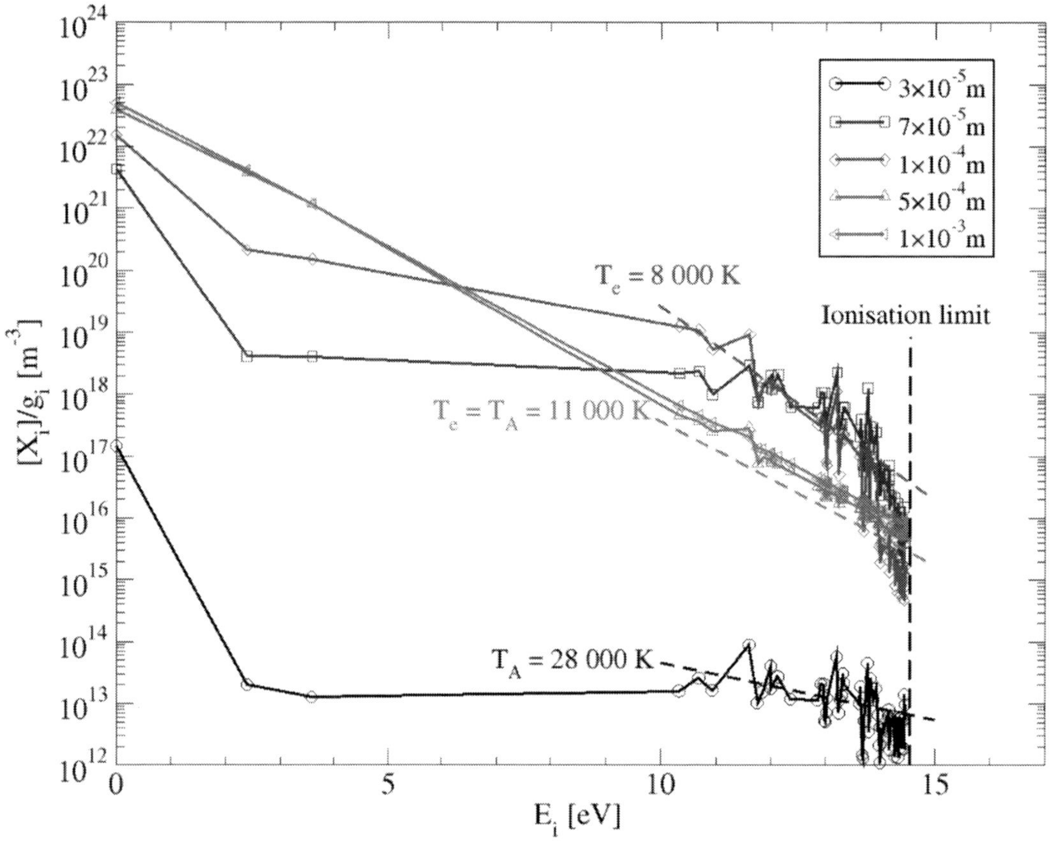

Figure 4. Boltzmann plot of N atoms at different x-values from the shock front. A rapid relaxation is observed toward an excitation temperature close to $T_A = T_e$ at thermal equilibrium.

dissociation. When T_v reaches a maximum ($x \simeq 10^{-4}$ m), the T_A decrease rate is maximum. Afterward, the relaxation goes on until the coupling between all temperatures ($x \simeq 10^{-3}$ m) which means that thermal equilibrium is then reached.

The plasma relaxation can be further examined using Boltzmann plots displayed on Figure 4 relative to the N atoms excited states. The distribution is first far from equilibrium with strong underpopulation with respect to T_A. Indeed, at $x = 3 \times 10^{-5}$ m, the main collision partners of N atoms are N_2 and N. If equilibrium were fulfilled, the excited states should be more populated according to a ratio close to 40. When approaching $x \simeq 10^{-3}$ m, the coupling takes place by collision with all types of particles and stabilizes with an excitation temperature close to $T_e = T_A = 8\ 000$ K. Although many transitions are optically thin, the departure from the Boltzmann equilibrium is negligibly small for $x > 10^{-3}$ m.

Such a study can also be performed for the molecular excited states. Their relaxation is slower than for atoms. Between 10^{-3} m and 10^{-2} m, the excitation equilibrium is not reached. The complete relaxation of the plasma in LTE is in fact achieved for $x \simeq 10^{-2}$ m. The influence of radiation is negligibly small. Nevertheless, we can observe the influence of these radiative losses. Indeed, they cause the slow decrease of the temperature of the plasma.

Of course, Earth or Mars atmospheres are not composed by N_2 molecules only. The case of post-shock flows in N_2-O_2 mixtures related to entry into the Earth's atmosphere has been treated recently in a simplified way [20, 21]. Conversely to the model of the present section, the state-to-state approach has been developed for electronic excited states of molecules and atoms only. The N_2 and O_2 vibrational states have not been considered as independent species. They have been lumped together and the concept of global dissociation or recombination similar to the global ionisation thoroughly described in [22] has been used. The more complete model CoRaM-Air accounting for the vibrational states of N_2, O_2, NO (obtained by Zeldovich neutral exchange processes) and electronic excited states of Ar has been elaborated [23]. The equivalent model CoRaM-Mars has been elaborated for Martian atmosphere entries. At the moment, only relaxation calculations at constant pressure and temperature can be performed. Their coupling with energy and momentum balances for post-shock calculations is in progress.

3. Plasma-wall interaction

The previous study focused on the post-shock relaxation put forward the high levels of temperature and the thermodynamic equilibrium of the plasma near the entering body surface. The pressure equals 4×10^4 Pa at $x \simeq \delta$ from the shock front. The collision frequency is then high enough and the velocity sufficiently small to ensure LTE. Of course, these conditions are particular. If the calculations are performed for higher altitude where the Mach number $\mathcal{M}_1$ is higher and temperature T_1 and pressure p_1 lower, the non equilibrium zone is thicker and the boundary layer edge is not in LTE. But the levels of temperatures are higher. In any case, the entering body therefore interacts with a high temperature plasma and can be strongly heated.

The heating process has different contributions. The first contribution results from the conduction of all energy storage modes (translation, rotation, vibration and electronic excitation). These conduction phenomena take place owing to the difference between the plasma temperature and the surface temperature which is necessarily limited by its melting or sublimation point. This conductive part induces the formation of boundary layers characterized by gradients of translation temperatures and ground and excited states number densities. The second contribution results from radiation emitted everywhere in the plasma. Indeed, a part of this emission can reach and partially be absorbed by the entering body surface. The third contribution is due to dissociated species which can be adsorbed on the surface and recombine by catalysis. The dissociation energy then released to the surface enhances the heat flux observed at the wall.

The previous contributions generally are of the same order of magnitude. No one can then be ignored in the energy balance at the wall. They lead to significant values of the parietal heat flux, typically between 1 and 10 MW m^{-2}. Such heat fluxes induce a strong heating therefore a damaging of the entering body. To reduce the damage as much as possible, two strategies have been developed. First, refractory materials treated against oxidation can be used as TPS. They reduce the part of the heat flux resulting from the recombination of oxygen in O_2 molecules catalyzed by the surface. These materials are in fact used as heatshield in the case the heat flux corresponds to weak values. They are reusable and were in the past developed for the American Space Shuttle and the Russian Bor and Buran vehicles [24]. The second strategy consists in using the damage resulting from the heat flux for the protection of the surface. Ablative materials are used. During heating, the pyrolysis of the material is obtained which leads to the endothermic formation of hydrocarbons gases. Their mixing with other species and heat transfer inside the boundary layer then reduce the net heat flux. Since the material ablates, its behaviour in known conditions has therefore to be particularly well characterized in order to optimize the heatshield sizing. The material used for the entry of the Curiosity rover in the Martian atmosphere (Phenolic Impregnated Carbon Ablator, PICA) belongs to this second type of TPS [25].

4. Conclusion

In this contribution, the physico-chemistry of flows produced by bodies entering into a planetary atmosphere has been thoroughly examined. A shock layer is produced in the vicinity of the entering body's surface owing to the compression of the incident hypersonic flow. This compression leads to the dissociation and the ionisation of the incident gas molecules. Then, the plasma thus produced releases part of its energy to the surface with significant heat fluxes. The entering body is protected using either reusable or ablative material heatshield.

The description of the behaviour of the nitrogen shock layer flow resulted from the implementation of our electronic and vibrational specific Collisional-Radiative model CoRaM-N_2 in a 1D Euler flow solver accounting for the momentum and energy balance equations. The shock front is treated as a discontinuity in the framework of the Rankine-Hugoniot assumptions. Results put forward the successive phases allowing the conversion of the initially cold gas into plasma. N_2 is vibrationally excited, dissociate and the atoms are then ionised.

Based on CoRaM-N_2, two other vibrational and electronic specific CR models (CoRaM-Air and CoRaM-Mars) are in elaboration in the purpose of their implementation in 1D Euler flow solvers. They will be used to simulate the physico-chemistry inside shock layers formed during atmospheric entries into Earth and Mars atmospheres. Although many entries were successful, a lot of work remains to be performed to improve the sizing of heatshield. In the future, missions to Mars (ExoMars in 2016-2018 by ESA and Russia), to Titan (Titan-Saturn-System-Mission in 2020 by NASA and ESA) as well as round-trip using the Soyouz vehicle for astronauts coming back from the International Space Station, will require an ever deeper understanding of this type of physical situation.

References

[1] Drawin H W 1966 *Collision and Transport Cross Sections* EUR-CEA-FC 383
[2] http://www.nist.gov/pml/data/asd.cfm
[3] Bourdon A and Vervisch P 1997 *Phys. Rev.* E **55** 4634
[4] Armenise I, Capitelli M, Colonna G and Gorse C 1996 *J. Thermophys. Heat Transfer* **10** 397
[5] Esposito F, Armenise I and Capitelli M 2006 *Chem. Phys.* **331** 1
[6] Capitelli M, Ferreira C M, Gordiets B F and Osipov A I 2000 *Plasma Kinetics in Atmospheric Gases* (Springer-Verlag, Berlin)
[7] Park C 1988 *J. Thermophysics* **2** 8
[8] Lotz W 1968 *Z. Phys.* **216** 241
[9] Teulet P, Sarrette J-P and Gomes A-M 1999 *J. Quant. Spectrosc. Radiat. Transfer* **62** 549
[10] Bultel A, Chéron B G, Bourdon A, Motapon O and Schneider I F 2006 *Phys. Plasmas* **13** 043502
[11] Starik A M, Titova N S and Arsentiev I V 2010 *Plasma Sources Sci. Technol.* **19** 015007
[12] Kossyi I A, Kostinsky A Yu, Matveyev A A and Silakov V P 1992 *Plasma Sources Sci. Technol.* **1** 207
[13] Kim Y K and Desclaux J P 2002 *Phys. Rev.* A **66** 012708
[14] Peterson J R, Le Padellec A, Danared H, Dunn G H, Larsson M, Larson A, Peverall R, Stromholm C, Rosen S, af Ugglas M and van der Zande W J 1998 *J. Chem. Phys.* **108** 1978
[15] Laux C O and Kruger C H 1992 *J. Quant. Spectrosc. Radiat. Transfer* **48** 9
[16] Holstein T 1947 *Phys. Rev.* **72** 1212
[17] Cauchon D L 1972 *Radiative Heating Results from the FIRE II Flight Experiment at a Reentry Velocity of 11.4 Kilometers per Second* NASA TM-X-1402
[18] Matsuda A, Fujita K, Shunichi S and Abe T 2004 *J. Thermophys. Heat Transfer* **18** 342
[19] Cornette E S 1966 *Forebody Temperatures and Calorimeter Heating Rates Measured during Project FIRE II Reentry at 11.35 Kilometers per Second* NASA TM-X-1305
[20] Panesi M, Magin T, Bourdon A, Bultel A and Chazot O 2009 *J. Thermophys. Heat Transfer* **23** 236
[21] Panesi M, Magin T, Bourdon A, Bultel A and Chazot O 2011 *J. Thermophys. Heat Transfer* **25** 361
[22] Annaloro J, Morel V, Bultel A and Omaly P 2012 *Phys. Plasmas* **19** 073515
[23] Bultel A, Annaloro J, Morel V and Omaly P 2012 *ESCAMPIG 2012* 325
[24] Kolesnikov A 1999 *Combined Measurements and Computations of High Enthalpy and Plasma Flows for Determination of TPM Surface Catalycity* VKI Lecture Series RTO-EN-AVT-8
[25] Beerman A F, Lewis M J, Santos J A and White T R 2010 *10th AIAA/ASME Joint Thermophysics and Heat Transfer Conference* 2010-4313 1

Diagnostics and biomedical applications of radiofrequency plasmas

Saša Lazović[1,2]

[1]Institute of Physics, University of Belgrade, Pregrevica 118, 11080 Belgrade, Serbia

[2]Jozef Stefan Institute, Jamova cesta 39, 1000 Ljubljana, Slovenia

lazovic@ipb.ac.rs

Abstract. In this paper we present spatial profiles of ion and atomic oxygen concentrations in a large scale cylindrical 13.56 MHz capacitively coupled plasma low pressure reactor suitable for indirect biomedical applications (like treatment of textile to increase antibacterial properties) and direct (treatment of seeds of rare and protected species). Such reactor can easily be used for the sterilization of medical instruments by removing bacteria, spores, prions and fungi as well. We also discuss electrical properties of the system based on the signals obtained by the derivative probes and show the light emission profiles close to the sample platform. In the case of seeds treatment, the desired effect is to plasma etch the outer shell of the seed which will lead to the easier nutrition and therefore increase of the germination. In the case of textile treatment the functionalization is done by bounding atomic oxygen to the surface. It appears that antibacterial properties of the textile are increased by incorporating nanoparticles to the fibres which can successfully be done after the plasma treatment. From these two examples it is obvious that the balance of ion and atomic oxygen concentrations as well as proper choice of ion energy and power delivered to the plasma direct the nature of the plasma treatment.

1. Introduction

Non-thermal plasmas are widely used in many industrial processes due to high and controllable chemical reactivity at low gas temperatures [1, 2]. This enables treatment of sensitive materials like cells, polymers, textile and seeds [3, 4, 5, 6, 7, 8]. It is much easier to achieve non-equilibrium conditions at low pressure compared to atmospheric pressure (due to lower ionization growth) and uniformity over the large volumes [9]. Of course, this comes at the cost of using a pumping system and not being able to apply plasma directly to patients.

The specifics of material that we wanted to treat with plasma led us to develop a large scale, low-cost reactor of simple design. Sensitive materials like textile and seeds were treated in air plasma [10, 11]. After the treatment of textile, the surface became hydrophilic leading to increased wettability and therefore easier dying of the samples. Plasma treatment also makes it possible to incorporate nanoparticles like that of silver or TiO_2 which increases antibacterial properties of the textile [3, 4]. Plasma treated textile can be used in hospitals, surgery, military, food production facilities to lower the chance of bacterial infections. In the case of seeds, plasma improves the nutrition of the seeds by etching the outer shell. In nature it is done mechanically, usually by the birds and it also can be done chemically. Nevertheless, plasma shows good results in increasing the seeds germination which is especially important for the rare and protected species. Similar effects can be achieved with the atmospheric pressure plasma [12] but the advantage of a low pressure plasma reactor is the uniform covering of large treatment area. Damaging or even destruction of the seeds is inevitable if the plasma is too aggressive so the proper optimization of the discharge parameters in necessary.

In this paper we will discuss plasma properties of the asymmetric cylindrical reactor and discuss characteristic features introduced by using such geometry. Ion and atomic oxygen concentrations were radially measured using Langmuir and catalytic probes. Development of plasma light emission was recorded using ICCD camera. Derivative probes that we have used to measure the power delivered to the plasma can be used for all sources in the MHz range regardless of the operating pressure (low or atmospheric). Finally, plasma influence to the germination of *Gentiana asclepiadea* L. seeds will be shown.

2. Experimental setup

The discharge is developed in a cylindrical vessel made of stainless steel, 2.5 m long and 1.17 m in diameter. The powered electrode made of aluminium (1.5 m long) is placed axially in the chamber. The reactor is powered at 13.56 MHz using a Dressler Cesar 1310 together with Variomatch matching network. The chamber wall is grounded. Low pressure is maintained using a mechanical pump.

The samples are placed beneath the powered electrode on a platform which can be moved closer or away from the electrode. Derivative probes are placed as close as possible to the powered electrode as shown at Figure 1. The signals form the probes are collected using an oscilloscope and a PC and then processed [13, 14].

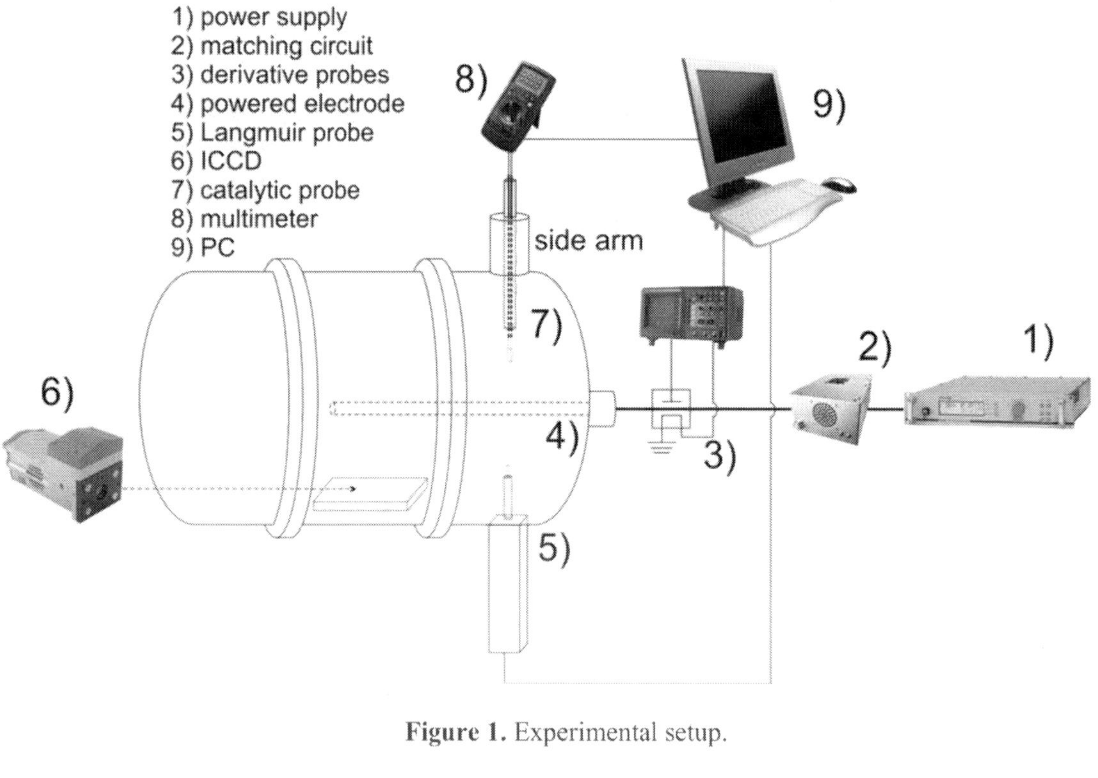

Figure 1. Experimental setup.

Radial profiles of ion and atomic oxygen concentrations are recorded using Langmuir probe (Hiden ESPION) and a nickel catalytic probe [15, 16, 17] respectively. ICCD camera (Andor iStar DH720-18U-03) placed end on to the sample platform was used to record the light emission in the vicinity at the location where sample are to be placed.

3. Plasma diagnostics

3.1. Electrical properties

Signals from the derivative probes were recorded for the range of pressures from 100 to 800 mTorr and the results are presented at Figure 2. The working gas was air.

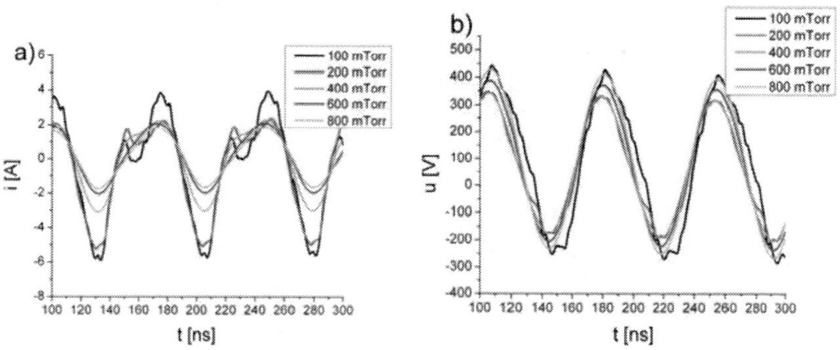

Figure 2. Current a) and voltage b) signals obtained by derivative probes in air at 100, 200, 400, 600 and 800 mTorr. RF generator power was 200 W.

We can see that the shape of both current and voltage are influenced by the working pressure. With the increase of the pressure, the amplitudes of current signals are decreasing and starting to be sine-like. At low pressures the voltage signals are less distorted compared to the current signals. Their shape is changing from the triangle-like to the sine-like when increasing the pressure.

The changes of current and voltage with pressure lead to the change in the power calculated as the integral of their product as shown at Figure 3. We can see that the power transmitted to the plasma is actually higher than 75% in all cases. At 600 mTorr this percent is even higher meaning that the power transfer efficiency is slightly higher compared to 100 mTorr.

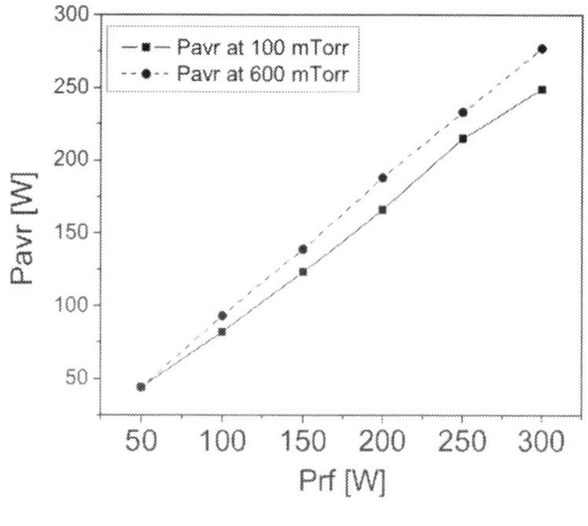

Figure 3. Average power delivered to the plasma at 100 and 600 mTorr.

Measuring of the power delivered to the plasma represents a good parameter for characterizing the treatment procedures and insures its reproducibility. Derivative probes are capable of measuring small powers (less than 1 W) achieving good accuracy of 0.1 W.

3.2. Light emission profiles

Light emission profiles were recorder using ICCD camera placed end-on to the sample platform underneath the powered electrode (see the top picture at Figure 4.). The generator power was increased from 20 to 400 W. The maximum intensity is moving towards the platform, more intensive light emission starts to develop across the surface of the platform where samples are to be placed.

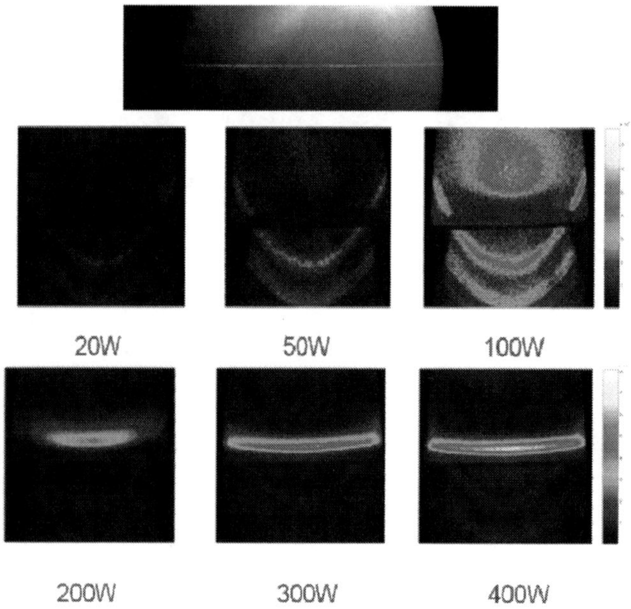

Figure 4. Light emission profile recorded end-on to the sample platform. The gas was air at 100 mTorr.

For smaller powers, after the plasma ignition, light emission intensity is maximal around the powered electrode. With the power increase emission intensity is also increasing around the powered electrode, but at the same time another region of increased intensity can be observed between the powered electrode and sample platform (Fig. 4. - 100W). With further increase in power, the maximum of emission moves towards the sample platform and for powers higher than 200 W the emission is centered on the platform. The light emission is uniform across the large area of the platform for high powers.

3.3. Ion and atomic oxygen concentrations

As mentioned before in the text, the nature of the treatment is strongly dependent on the choice of plasma parameters where concentrations of ions and reactive atomic oxygen represent one of the key players for many of the biomedical applications. Radial profiles were investigated using Langmuir and catalytic probes for the discharge in air by placing the probes on several distances from the powered electrode.

Figure 5. shows the results of ion concentrations which are of the order of 10^{15} m^{-3} to 10^{16} m^{-3}. The measurements are made close to the grounded chamber wall (50.5 cm from the powered electrode), at 20.5 cm from the powered electrode and in-between at 35.5 cm. With the increase of rms voltage (by increasing the power given by the generator) the concentrations increase but the changes are much less pronounced compared to the changes introduced by moving the probes closer to the electrode.

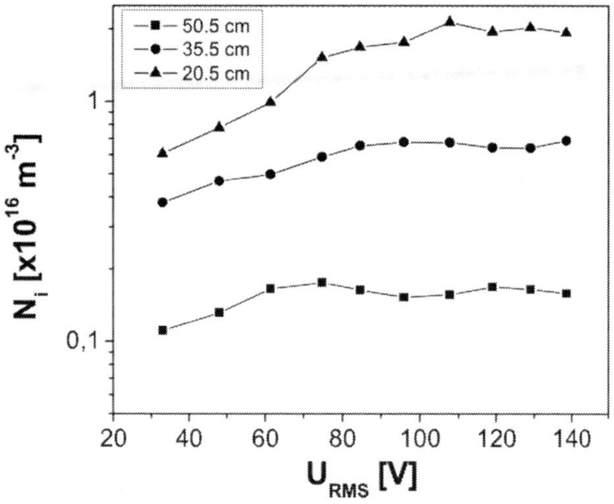

Figure 5. Ion concentrations as a function of rms voltage recorded radially for three distances from the powered electrode (air at 100 mTorr).

This leads to the conclusion that a specific value of ion concentration can be achieved either by increasing the power or by moving closer to the electrode. The results also show that is more practical to obtain higher concentrations by adjusting the distance of the treated sample than to increase the power.

The last conclusions can also be made for the atomic oxygen concentrations. These concentrations are of the order of 10^{19} m^{-3} which is less than in the ICP reactors by 1 to 2 orders of magnitude [18, 19].

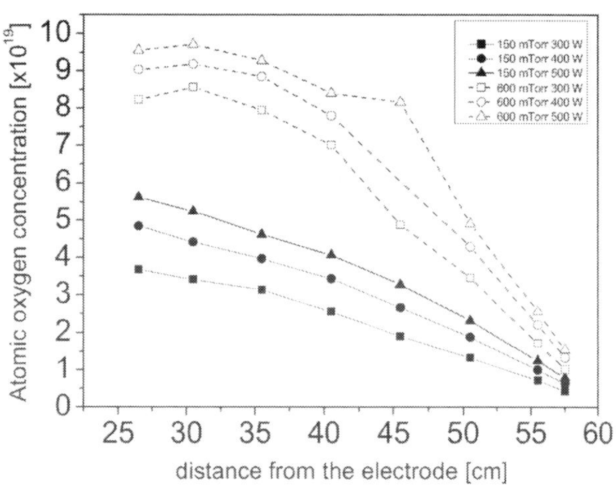

Figure 6. Radial profiles of atomic oxygen concentrations recorded in air at 150 and 500 mTorr.

We can see that the concentrations are decreasing when moving towards the stainless steel wall which also recombines atomic oxygen and contributes to its loss term. At higher pressure the concentrations are higher as expected.

3.4. Biomedical application

Here we present the effects of the plasma to the germination of the *Gentiana asclepiadea* L. (see Figure 7.). The seeds were treated in air plasma at 200 mTorr for the generator power of 100 W. Depending on the treatment time the germination can be increased but if the treatment is too long, the seeds are damaged and the germination is decreased (20 min exposure).

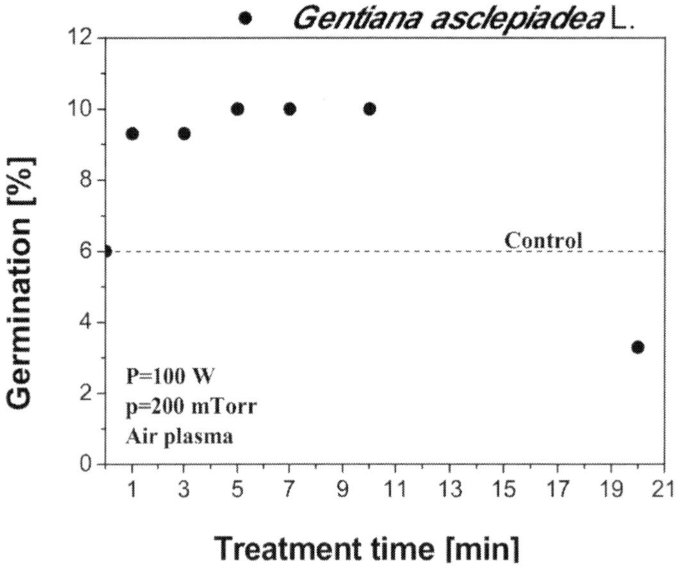

Figure 7. Effects of the plasma to the germination of the *Gentiana asclepiadea* L. seeds..

4. Conclusions

We have shown the properties of the low pressure cold plasma capable of indirect (textile treatment) and direct (seeds treatment) biomedical applications. Antibacterial properties of textile and increased germination of seeds are accomplished due to the functionalization of fiber surface by oxygen atoms and then incorporating the nanoparticles and plasma etching of the seed outer shell, respectively [3, 4, 6]. The advantages of cylindrical geometry of the reactor are shown by presenting the ion and atomic oxygen concentrations at different distances from the powered electrode making it possible to adjust their ratio to the certain amount and achieve different effects. The ion energy is also strongly dependent on the position where the sample is placed.

The uniformity of plasma light emission at the location where samples are to be placed is demonstrated for higher powers. The power delivered to the plasma is measured with good accuracy providing proper treatment characterization and reproducibility. The low pressure cold plasma is not an alternative to atmospheric plasma sources when it comes to biomedical applications but is the preferred tool for all materials that may be treated under low pressure including organic materials and even some biological samples such as seeds. Full understanding of the processes and phenomena occurring in such plasmas as well as optimization of new applications may be achieved by using swarm based data [20, 21], plasma models [21,22] and possibly known scaling [23] to connect to other types of discharges.

Acknowledgments
This work was performed as a part of requirement for a PhD thesis which was completed under the supervision of Dr. Nevena Puač and Prof. Zoran Lj. Petrović. I would also like to express gratitude to my co-supervisor Dr. Gordana Malović, Kosta Spasić and colleagues from the Biological institute "Siniša Stanković". This research has been supported by the Ministry of Education, Republic of Serbia, project numbers III41011 and ON171037.

References
[1] Makabe T and Petrovic Z Lj 2006 Plasma Electronics (Taylor and Francis:New York); Lieberman M A and Lichtenberg A J 2005 Principles of Plasma Discharge and Materials Processing, (Wiley:Hoboken)
[2] Roth J R 1995 Industrial Plasma Engineering (Institute of Physics:Bristol, UK); Kelly-Wintenberg K, Hodge A, Montie T C, Deleanu L, Sherman D, Roth J R, Tsai P and WadsworthL 1999 *J. Vac. Sci. Technol.* A **17** p 1539
[3] Ilic V, Saponjic Z, Vodnik V, Lazovic S, Dimitrijevic S, Jovancic P, Nedeljkovic J M and Radetic M 2010 *Ind Eng Chem Res* **49** p 7287
[4] Mihailovic D, Saponjic Z, Radoicic M, Lazovic S, Baily C J, Jovancic P, Nedeljkovic J and Radetic M 2011 *Cellulose* **18** p 811
[5] Lazović S, Puač N, Miletić M, Pavlica D, Jovanović M, Bugarski D, Mojsilović S, Maletić D, Malović G, Milenković P and Petrović Z 2010 *New Journal of Physics* **12** p 083037
[6] Zivkovic S, Puac N, Giba Z, Grubisic D and Petrovic Z L 2004 *Seed Sci Technol* **32** p 693
[7] Junkar I, Vesel A, Cvelbar U, Mozetič M, and Strnad S 2009 *Vacuum*, **84**(1) p 83
[8] Vesel A, Junkar I, Cvelbar U, Kovač J, and Mozetič M 2008 *Surf. Interface Anal.*,**40**(11) p 1444
[9] Petrović Z Lj, Puač N, Lazović S, Maletić D, Spasić K and Malović G 2012 *J. of Phys.:Conf. Ser.* **356** p 012001
[10] Radetić M 2012 *Journal of Materials Science* DOI 10.1007/s10853-012-6677-7
[11] Gorensek M, Gorjanc M, Bukosek V, Kovac J, Petrovic Z and Puac N 2010 *Textile Research Journal* **80** p 1633
[12] Puač N 2008 *Journal of Physics: Conference Series* **133** p 012007
[13] Puač N, Petrović Z Lj, Malović G, Đorđević A, Živković S, Giba Z and Grubišić D 2006 *J. Phys. D: Appl. Phys.* **39** p 3514
[14] Puač N, Petrović Z Lj, Živković S, Giba Z, Grubišić D and Đorđević A 2005 Plasma Processes and Polymers ed. R d'Agostino, P Favia, C Oehr, M R Wertheimer (Wiley-VCH:Berlin) p 193
[15] Drenik A, Vesel A and Mozetic M 2005 *Inform. Midem* **35** p 85.
[16] Mozetic M, Vesel A, Cvelbar U and Ricard A 2006 *Plasma Chem. and Plasma Proc.* **26** (2) p 103
[17] Vrlinic T, Mille C, Debarnot D and Poncin-Epaillard F 2009 *Vacuum* **83** p 792
[18] Primc G, Zaplotnik R, Vesel A and Mozetic M 2011 *AIP Advances* **1** p 022129
[19] Zaplotnik R, Vesel A and Mozetic M 2011 *EPL (Europhysics Letters)* **95** p 55001
[20] Dujko S, Ebert U, White R D and Petrović Z Lj 2011 *Japanese Journal of Applied Physics* **50** p 08JC01
[21] Petrović Z Lj, Dujko S, Marić D, Malović G, Nikitović Ž, Šašić O, Jovanović J, Stojanović V and Radmilović-Radenović M 2009 *J. Phys. D: Appl. Phys.* **42** p 194002
[22] Savić M, Radmilović-Radjenović M, Šuvakov M, Marjanović S, Marić D and Petrović Z Lj 2011 *IEEE Trans. Plasma Sci.* **39** (11) p 2556
[23] Petrović Z Lj, Škoro N, Marić D, Mahony C M O, Maguire P D, Radmilović-Rađenović M and Malović G 2008 *J. Phys. D: Appl. Phys.* **41** p 194002

26th Summer School and International Symposium on the Physics of Ionized Gases (SPIG 2012) IOP Publishing
Journal of Physics: Conference Series **399** (2012) 012016 doi:10.1088/1742-6596/399/1/012016

Near-threshold photoelectron angular distributions from two-photon resonant ionisation of He and Ne atoms

A Mihelič[1], M Žitnik[1,2], P O'Keeffe[3], P Bolognesi[3], A Moise[4], R Richter[4], L Avaldi[3]

[1] Jožef Stefan Institute, Jamova cesta 39, Ljubljana, Slovenia
[2] Faculty of mathematics and physics, University of Ljubljana, Jadranska ulica 19, Ljubljana, Slovenia
[3] CNR-IMIP, Area della Ricerca di Roma 1, Monterotondo Scalo, Italy
[4] Sincrotrone Trieste SCpA, Area Science Park, Basovizza (Trieste), Italy

E-mail: andrej.mihelic@ijs.si

Abstract. We present an overview of two recent experiments in which photoelectron angular distributions from light-polarised He and Ne atoms are studied, with the emphasis on theoretical description of photoionisation. Polarised He (Ne) atoms are prepared in the 1s4p, 1s5p, or 1s6p ($2p^5 3d$) states with linearly polarised synchrotron light and ionised with a tunable infrared laser. The asymmetry parameters extracted from the measured velocity map imaging spectra agree well with the calculated asymmetry parameters.

1. Introduction

Photoionisation of polarised atoms has proven to be a useful technique for probing details regarding photoionisation dynamics. This point was noted by Klar and Kleinpoppen [1] who pointed out that measurements of photoelecron angular distributions (PADs) from polarised atomic targets can provide "complete" information on the photoionisation process (the so called complete experiment). Of particular interest to us here are PADs from noble gas atoms prepared in polarised excited states for low kinetic energies of the photoelectrons (see [2] for a recent review). A way of preparing atoms in polarised atomic states is by photoexcitation with linearly polarised light. These polarised excited atoms are then ionised by absorbing the second photon – either from the same or from a different radiation source.

Here we review two recent experiments [3, 4], in which photoelectron angular distributions from light-polarised He and Ne atoms are measured, and present a theoretical framework for their description. We study angular distributions of photoelectrons ejected in two-photon ionisation (figure 1) of ground-state helium and neon atoms by simultaneous irradiation with synchrotron photons (linearly polarised, polarisation $\hat{e}_1$, energy ω_1) and Ti:Sapphire laser photons (polarisation $\hat{e}_2 = \hat{e}_1$, energy ω_2):

$$|g\rangle \xrightarrow{\omega_1} |\nu\rangle \xrightarrow{\omega_2} |\boldsymbol{k};\gamma\rangle, \tag{1}$$

where $|g\rangle$ denotes the ground state, $|\nu\rangle$ the chosen intermediate state, and $|\boldsymbol{k};\gamma\rangle$ the final continuum state. With $\boldsymbol{k}$ we denote the wave vector of the photoelectron and with γ additional quantum numbers of the final state. The energy of the synchrotron photons is chosen to lie on resonance with the intermediate state ($\omega_1 = E_\nu - E_g$).

In the first experiment, we examine PADs from the ground state helium atoms

$$\mathrm{He\,1s^2} \xrightarrow{\omega_1} \mathrm{He\,1s}n\mathrm{p} \xrightarrow{\omega_2} \mathrm{He^+\,1s + e^-}, \tag{2}$$

where the singlet 1snp intermediate states with $4 \le n \le 6$ are considered. We use LS coupling scheme for the description of bound and continuum states.

In the second experiment, we examine photoelectron angular distributions from the ground state neon atoms:

$$\mathrm{Ne\,2p^6} \xrightarrow{\omega_1} \mathrm{Ne\,2p^5\,(^2P_{1/2})\,3d\,[3/2]_1} \xrightarrow{\omega_2} \mathrm{Ne^+\,(^2P_{1/2}) + e^-}, \tag{3a}$$

$$\mathrm{Ne\,2p^6} \xrightarrow{\omega_1} \mathrm{Ne\,2p^5\,(^2P_{3/2})\,3d\,[3/2]_1} \xrightarrow{\omega_2} \mathrm{Ne^+\,(^2P_{3/2}) + e^-}, \tag{3b}$$

where jK coupling notation [5] is used for the intermediate $2p^5 3d$ states with the angular momentum of the core (2p) electrons equal to $J_c = 1/2$ and $J_c = 3/2$ and the total angular momentum $J = 1$, as described by (3a) and (3b), respectively.

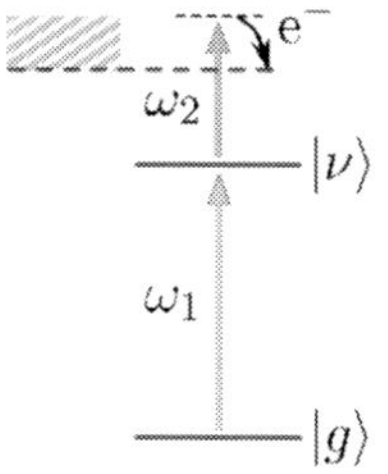

Figure 1. Resonant two-photon ionisation of an atom in the ground state $|g\rangle$. The intermediate state is denoted by $|\nu\rangle$, and the energy of the first and the second photon with ω_1 and ω_2, respectively.

It has been assumed in (2), (3a), and (3b) that the dipole approximation is valid. Higher-order transition amplitudes (M1, E2, etc.) are expected to be negligible in comparison with the dipole amplitude in the first and in the second step for both He and Ne. Since, in addition, $\hat{\mathbf{e}}_1 = \hat{\mathbf{e}}_2$ holds, the PADs are axially symmetric, and the differential cross section has the form [3]

$$\frac{\mathrm{d}\sigma^{(2)}}{\mathrm{d}\Omega} = \frac{\sigma^{(2)}_{\mathrm{int}}}{4\pi}\Big(1 + \beta_2 P_2(\cos\theta) + \beta_4 P_4(\cos\theta)\Big), \tag{4}$$

where $\cos\theta = \hat{\boldsymbol{k}} \cdot \hat{\mathbf{e}}_1 = \hat{\boldsymbol{k}} \cdot \hat{\mathbf{e}}_2$, P_q denotes Lagrange polynomials of order q, $\sigma^{(2)}_{\mathrm{int}}$ is the angle integrated generalised two-photon ionisation cross section, and β_2 and β_4 are the asymmetry parameters.

Both experiments were performed at the Gas Phase Photoemission beamline at Elettra synchrotron, Italy [6]. The details of the experimental setup are described in [3, 7, 8] and the references therein. In short, the monochromatised synchrotron light is focused into the interaction region of a velocity map imaging (VMI) spectrometer, where it is aligned with a counter-propagating Ti:Sapphire laser beam tunable in the wavelength range of 670–1000 nm. For Ne, the laser can be used in a continuous-wave (CW) mode since the lifetime of the intermediate states is relatively long, and in the case of He, mode locking is used (pulse duration of 15 ps, 83.3 MHz repetition rate). The acquired images are inverted using the pBasex method [9], which is used to reconstruct the original 3-dimensional distributions from the recorded 2-dimensional projections.

2. Photoelectron angular distributions

2.1. Helium

The measured electron yield is proportional to the generalised two-photon ionisation cross section (e.g., see [3]). Since the first step of the two-photon excitation is resonant, the generalised ionisation cross section can be written as:

$$\frac{\mathrm{d}\sigma^{(2)}}{\mathrm{d}\Omega} \propto \sum_{\mu,\mu_s,m_s} |\langle \boldsymbol{k}; \lambda, \mu, \mu_s, m_s | D(\hat{\mathbf{e}}_2) | \nu \rangle|^2, \tag{5}$$

where λ, μ, and μ_s are the angular momentum of the core electron, its projection on the z axis, and the projection of its spin, respectively, and m_s is the spin projection of the continuum electron. The dipole transition operator which describes the absorption of the infrared (IR) photon is denoted by $D(\hat{\mathbf{e}}_2)$.

In LS coupling, the intermediate states are described well within the single-configuration approximation even for the lowest-lying member ($n = 2$) of the 1snp series [10]. We may therefore use the uncoupling formula [5] to uncouple the "inner" electron part and the "outer" electron part of the wave function. This results in a relatively simple parametrisation for β_2 and β_4:

$$\beta_2 = \frac{40 - 140X\cos\Delta}{28 + 35X^2}, \tag{6a}$$

$$\beta_4 = \frac{72}{28 + 35X^2}, \tag{6b}$$

where $X(\epsilon) = \mathfrak{d}_0(\epsilon)/\mathfrak{d}_2(\epsilon)$ is the ratio of single-electron dipole matrix elements pertaining to the 1sϵs ($\ell = 0$) and 1sϵd ($\ell = 2$) partial waves of the continuum wave function accessible by two-photon absorption, and $\Delta(\epsilon) = \sigma_0(\epsilon) + \pi\kappa_0(\epsilon) - \sigma_2(\epsilon) - \pi\kappa_2(\epsilon)$ is the phase difference between these waves. We have used $\epsilon = k^2/2$ for the kinetic energy of the photoelectron, $\pi\kappa_0$ and $\pi\kappa_2$ for the non-Coulomb (short-range potential) phase shifts of the two partial waves, σ_0 and σ_2 for the Coulomb phase shifts, and

$$\mathfrak{d}_\ell(\epsilon) = \int_0^\infty \mathrm{d}r\, r^2\, R_{n\mathrm{p}}(r)\, r\, R_{\epsilon\ell}(r), \quad \ell = 0, 2, \tag{7}$$

for the radial parts of the single-electron dipole matrix elements. It should be noted that unlike X, phase difference Δ is independent of the choice of the intermediate state. The notation used in (6a) and (6b) is the same as in [11].

Real configuration-interaction coupled Coulomb-Sturmian basis functions [12] have been used to calculate the bound and continuum states and the relevant dipole matrix elements. The values of $X(\epsilon)$ just above the ionisation threshold have been checked to agree with the values obtained from extrapolated density of oscillator strength of bound 1sn's and 1sn'd states. The calculated ratios for $n = 2$ and $n = 3$ agree well with ratios extracted from the partial single-photon ionisation cross sections given in [13] for $\epsilon \geq 0.5$ eV (figure 2). Near the threshold, the extracted values deviate slightly from our results, but the ratios obtained from the cross sections given in [14] for $2 \leq n \leq 5$ – using the same approach as in [13] – which focus on the near-threshold region ($\epsilon \lesssim 0.11$ eV) may be seen to match the present calculations. Our measured and calculated values of X are shown in figure 3, and are seen to agree reasonably well for the 1s4p, 1s5p, and 1s6p states. Furthermore, the measurements of Haber *et al* [11] are in good agreement with the theory for the 1s3p state (filled triangles), whereas there is a minor discrepancy for the lower of the two points for the 1s4p state (filled squares), for which the calculated value lies outside the experimental error bars.

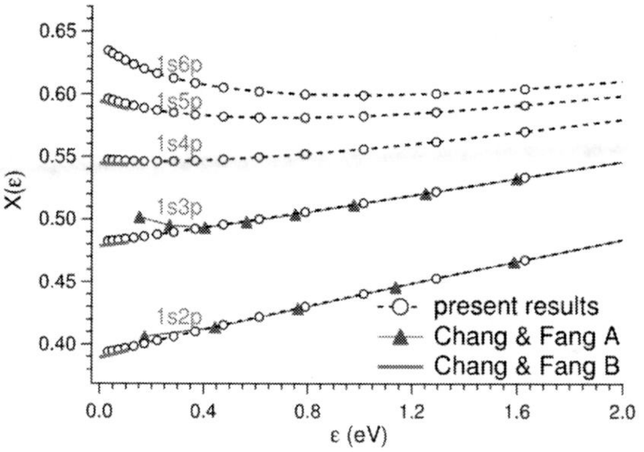

Figure 2. Calculated ratio $X(\epsilon)$ for the 1snp intermediate states (empty circles, dashed lines). The results obtained from [13] (Chang & Fang A) are plotted with filled triangles and the results from [14] (Chang & Fang B) with solid red lines.

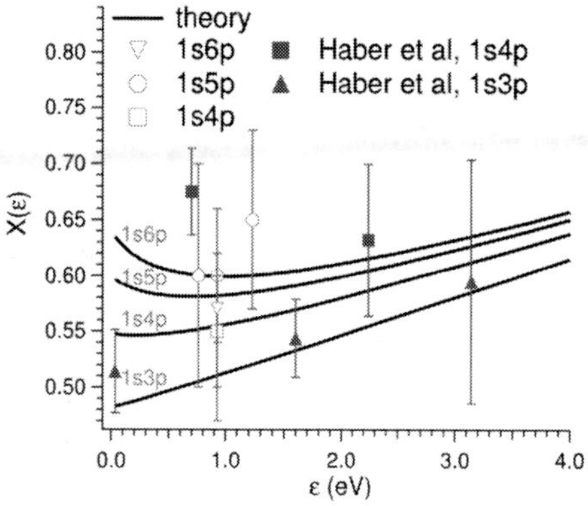

Figure 3. Calculated (solid line) and measured (empty symbols) ratio $X(\epsilon)$ for the 1s4p, 1s5p, and 1s6p intermediate states. The experimental values from [11] (1s3p and 1s4p) are plotted with filled symbols.

The phase differences between the 1sϵs and 1sϵd waves have been obtained by extrapolating the energy dependence of the quantum defects κ_0 and κ_2 of the 1sn's and 1sn'd bound states across the threshold. The difference $\pi(\kappa_0 - \kappa_2)$ has been found to be negligible in comparison with the difference between the Coulomb shifts $\sigma_0 - \sigma_2$ in the investigated energy region. The

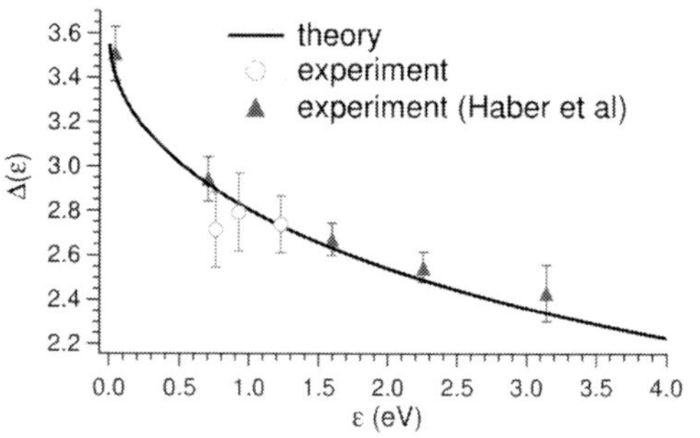

Figure 4. Calculated (solid line) and measured (empty circles) phase difference $\Delta(\epsilon)$. The results from [11] are plotted with filled triangles.

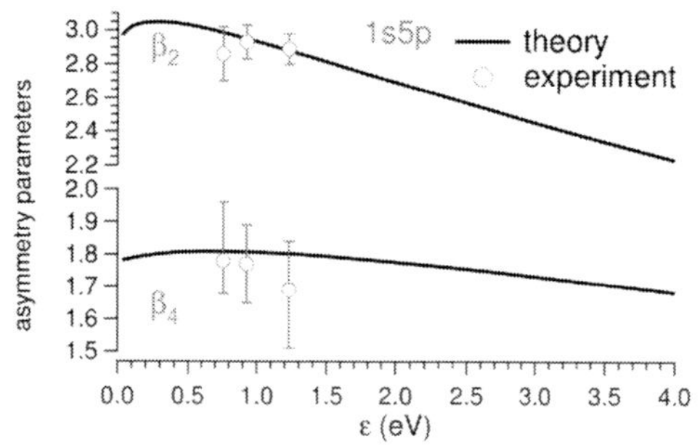

Figure 5. Calculated (solid line) and measured (circles) asymmetry parameters for the 1s5p intermediate state.

dependence of the phase difference Δ between the outgoing 1sϵs and 1sϵd waves on the kinetic energy of the photoelectron is shown in figure 4. The calculations are in good agreement with our experimental results and with the measurements presented in [11].

The asymmetry parameters close to the threshold do not depend strongly on the principal quantum number n of the intermediate states considered here. While β_2 increases slightly with n above the threshold ($\epsilon \lesssim 2$ eV), β_4 decreases with n in this energy region. An example of a typical energy dependence of the asymmetry parameters is shown in figure 5 for the 1s5p intermediate state. Similar plots have been obtained also for the 1s4p and 1s6p states.

2.2. Neon

As for helium, the measured electron yields for two-photon ionisation of ground-state neon atoms are proportional to [3]:

$$\frac{d\sigma^{(2)}}{d\Omega} \propto \sum_{M_c, m_s} |\langle \boldsymbol{k}; J_c, M_c, m_s | D(\hat{\boldsymbol{e}}_2) | \nu \rangle|^2, \tag{8}$$

where J_c is the angular momentum of the 2p electrons, M_c its projection on the z axis, and m_s the spin projection of the continuum electron. In the two sets of the measured PADs, the energy of the synchrotron photons is chosen to lie on resonance with the $J_c = 1/2$ $2p^5 3d$ $[3/2]_1$ state (3a) or the $J_c = 3/2$ $2p^5 3d$ $[3/2]_1$ state (3b). Calculations show that the intermediate states are described well within a single configuration approximation if jK coupling is used: $\boldsymbol{K} = \boldsymbol{J}_c + \boldsymbol{\ell}$, $\boldsymbol{J} = \boldsymbol{K} + \boldsymbol{s}$, where $\boldsymbol{J}$, $\boldsymbol{\ell}$, and $\boldsymbol{s}$ are the total angular momentum, orbital angular momentum of the continuum electron, and its spin, respectively.

The energy region of the final $2p^5$ continuum states consists of two sub-regions. The first sub-region, which lies between the $2p_{3/2}$ and $2p_{1/2}$ thresholds and is referred to as the autoionisation region, contains resonances which consist of $J_c = 1/2$ discrete components (closed channels or configurations) mixed with $J_c = 3/2$ continuum components (open channels). In the second sub-region, which lies above the $2p_{1/2}$ threshold, all the $2p^5$ channels are open. In our experiment, the energy of the IR photon is tuned to avoid double-resonance excitation in the first sub-region. This is possible since unlike the $2p^5\epsilon d$ resonances accessible with single-photon ionisation [15], the accessible $2p^5\epsilon p$ and $2p^5\epsilon f$ resonances are narrow [7]. This fact allows for a simpler theoretical treatment in which the coupling between the $J_c = 1/2$ and $J_c = 3/2$ channels can be neglected. This has already been assumed in (8), where it has been taken that the final state has a well defined momentum J_c.

For convenience, we also use jK coupling to describe the final continuum states. The accessible final state with $J_c = 3/2$ is written as a linear combination of the ϵp $[5/2]_2$, ϵp $[3/2]_2$, ϵp $[1/2]_0$, ϵf $[3/2]_2$, and ϵf $[5/2]_2$ channel wave functions (the notation used is $\epsilon\ell$ $[K]_J$, where $2p^5$ has been omitted). These channel functions will be denoted with indices $\mathfrak{A}$, $\mathfrak{B}$, $\mathfrak{C}$, $\mathfrak{F}$, and $\mathfrak{G}$, respectively. The $J_c = 1/2$ state is written in terms of the ϵp $[3/2]_2$, ϵp $[1/2]_0$, and ϵf $[5/2]_2$ channel wave functions, which we will denote with indices $\mathfrak{D}$, $\mathfrak{E}$, and $\mathfrak{H}$, respectively. With each of these channel functions, a dipole amplitude and a phase shift are associated. This results in a double set of expressions for β_2 and β_4, one for each of the two values of J_c. The asymmetry parameters are written as quotients $\beta_2 = \mathcal{A}_2/\mathcal{A}_0$ and $\beta_4 = \mathcal{A}_4/\mathcal{A}_0$ [3], where for $J_c = 1/2$

$$\mathcal{A}_0 = \mathcal{Q}_{\mathfrak{D}\mathfrak{D}} + \mathcal{Q}_{\mathfrak{E}\mathfrak{E}} + \mathcal{Q}_{\mathfrak{H}\mathfrak{H}}, \tag{9a}$$

$$\mathcal{A}_2 = \frac{1}{4}\mathcal{Q}_{\mathfrak{D}\mathfrak{D}} + \frac{\sqrt{2}}{2}\mathcal{Q}_{\mathfrak{D}\mathfrak{E}} - \frac{\sqrt{6}}{14}\mathcal{Q}_{\mathfrak{D}\mathfrak{H}} - \frac{\sqrt{3}}{2}\mathcal{Q}_{\mathfrak{E}\mathfrak{H}} + \frac{2}{7}\mathcal{Q}_{\mathfrak{H}\mathfrak{H}}, \tag{9b}$$

$$\mathcal{A}_4 = \frac{3}{14}\mathcal{Q}_{\mathfrak{H}\mathfrak{H}} - \frac{3\sqrt{6}}{7}\mathcal{Q}_{\mathfrak{D}\mathfrak{H}}, \tag{9c}$$

and for $J_c = 3/2$

$$\mathcal{A}_0 = \mathcal{Q}_{\mathfrak{A}\mathfrak{A}} + \mathcal{Q}_{\mathfrak{B}\mathfrak{B}} + \mathcal{Q}_{\mathfrak{C}\mathfrak{C}} + \mathcal{Q}_{\mathfrak{F}\mathfrak{F}} + \mathcal{Q}_{\mathfrak{G}\mathfrak{G}}, \tag{10a}$$

$$\begin{aligned}\mathcal{A}_2 = {} & \frac{1}{5}\mathcal{Q}_{\mathfrak{A}\mathfrak{A}} + \frac{3}{10}\mathcal{Q}_{\mathfrak{A}\mathfrak{B}} - \frac{3\sqrt{5}}{10}\mathcal{Q}_{\mathfrak{A}\mathfrak{C}} + \frac{3}{70}\mathcal{Q}_{\mathfrak{A}\mathfrak{F}} + \frac{3\sqrt{6}}{35}\mathcal{Q}_{\mathfrak{A}\mathfrak{G}} - \frac{1}{5}\mathcal{Q}_{\mathfrak{B}\mathfrak{B}} - \frac{\sqrt{5}}{10}\mathcal{Q}_{\mathfrak{B}\mathfrak{C}} \\ & + \frac{3}{10}\mathcal{Q}_{\mathfrak{B}\mathfrak{F}} - \frac{4\sqrt{6}}{35}\mathcal{Q}_{\mathfrak{B}\mathfrak{G}} - \frac{3\sqrt{5}}{10}\mathcal{Q}_{\mathfrak{C}\mathfrak{F}} - \frac{\sqrt{30}}{10}\mathcal{Q}_{\mathfrak{C}\mathfrak{G}} + \frac{1}{5}\mathcal{Q}_{\mathfrak{F}\mathfrak{F}} + \frac{3\sqrt{6}}{35}\mathcal{Q}_{\mathfrak{F}\mathfrak{G}} + \frac{11}{70}\mathcal{Q}_{\mathfrak{G}\mathfrak{G}},\end{aligned} \tag{10b}$$

$$\mathcal{A}_4 = \frac{6}{7}\mathcal{Q}_{\mathfrak{A}\mathfrak{F}} + \frac{3\sqrt{6}}{14}\mathcal{Q}_{\mathfrak{A}\mathfrak{G}} + \frac{3\sqrt{6}}{14}\mathcal{Q}_{\mathfrak{B}\mathfrak{G}} + \frac{3\sqrt{6}}{14}\mathcal{Q}_{\mathfrak{F}\mathfrak{G}} - \frac{3}{28}\mathcal{Q}_{\mathfrak{G}\mathfrak{G}}. \tag{10c}$$

We have introduced $\mathfrak{Q}_{\xi\xi'} = D_\xi D_{\xi'} \cos(\pi\kappa_\xi - \pi\kappa_{\xi'} + \sigma_\ell - \sigma_{\ell'})$, where ξ and ξ' stand for the channel indices ($\mathfrak{A}$, $\mathfrak{B}$, etc.), D_ξ and $D_{\xi'}$ for the real dipole transition amplitudes, $\pi\kappa_\xi$ and $\pi\kappa_{\xi'}$ for the short-range phase shifts, and σ_ℓ and $\sigma_{\ell'}$ for the Coulomb phase shifts. Since the excess energy considered here is low ($\epsilon \lesssim 100$ meV), the dipole matrix elements and phase shifts are obtained by extrapolating density of oscillator strength and quantum defects of the $2p^5n'p$ and $2p^5n'f$ states across the ionisation thresholds. The reader is referred to [3] for details.

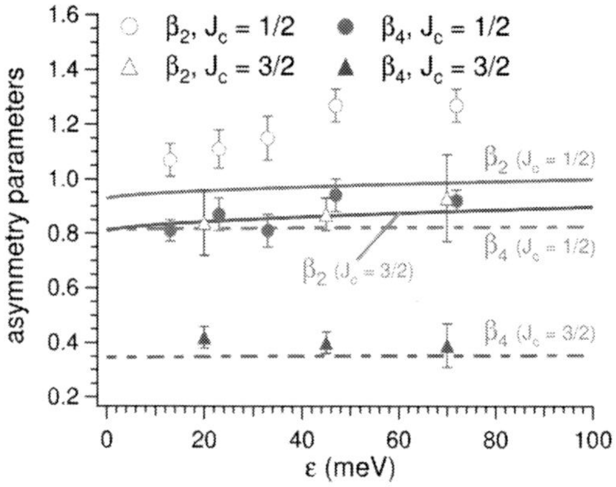

Figure 6. The measured and calculated asymmetry parameters for resonant two-photon ionisation of Ne: the $2p^5$ ($^2P_{1/2}$) $[3/2]_1$ intermediate state (filled and empty circles – experiment, upper solid and upper dashed red line – theory); the $2p^5$ ($^2P_{3/2}$) $[3/2]_1$ intermediate state (filled and empty triangles – experiment, lower solid and lower dashed blue line – theory).

An approximation often used for parametrisation of photoionisation cross sections is to neglect the spin-orbit interaction in the continuum and to assume that the shapes of the orbitals and the phase differences do not depend on the total angular momentum J of continuum channels (e.g., see [16]). If this is assumed in our case, the core-electron ("inner") part and the asymptotic ("outer") part of the wave function can be uncoupled. Simplified expressions are obtained if it is assumed that the phase shifts and the single-electron dipole matrix elements of various channels depend only on the orbital angular momentum ℓ of the continuum electron for a fixed intermediate state [3]. In the present case, however, this approach gives incorrect asymmetry parameters for $J_c = 3/2$ because the value of β_4 is constrained to zero. This means that in order to describe the PADs correctly, the dependence of the transition matrix elements on the channel quantum numbers should be considered.

The energy dependence of the asymmetry parameters given by (9a)–(10c) is shown in figure 6. An important observation is that the asymmetry parameters vary smoothly with ϵ also in the autoionisation region ($J_c = 3/2$ results). As can be seen, the calculated β_4 values for $J_c = 3/2$ are in very good agreement with the experiment. On the other hand, the values of β_2 extracted from the measured PADs deviate systematically from the calculated β_2 for $J_c = 1/2$. Nevertheless, generally speaking, the experimental and the calculated values agree well, especially for the angular momentum $J_c = 3/2$.

3. Conclusions

We have presented analyses of photoelectron angular distributions from polarised He and Ne atoms. These atoms are prepared in polarised excited states with linearly polarised synchrotron light and then ionised with a tunable Ti:Sapphire laser. We have considered photoionisation of He atoms in the singlet $1snp$ states with $4 \le n \le 6$ and Ne atoms in the $2p^5$ ($^2P_{J_c}$) 3d $[3/2]_1$ states with $J_c = 1/2$ or $J_c = 3/2$.

It has been shown that for He, a simpler parametrisation of the asymmetry parameters may be used, where β_2 and β_4 are written in terms of the ratio of the single-electron dipole matrix elements of the accessible $1s\epsilon s$ and $1s\epsilon d$ partial waves and the difference of the corresponding phase shifts. The parameters extracted from the measured data have been found to be in good agreement with our calculations.

For Ne, the measured and calculated asymmetry parameters agree reasonably well. It has been shown that in this case, it is not possible to reproduce the measured asymmetry parameters for the $2p^5$ ($^2P_{3/2}$) 3d $[3/2]_1$ intermediate state if the spin-orbit interaction in the continuum is neglected and it is assumed that the single-electron dipole transition amplitudes and phase differences are independent of the total angular momentum of the channel wave functions.

Acknowledgments

This research was partly supported by the Italy-Slovenia joint research project "Dynamics at nanoscale" and by PRIN 2009W2W4YF and 2009SLKFEX. A. M. and M. Ž. acknowledge the support of the Slovenian Research Agency (research programme P1-0112).

References

[1] Klar H and Kleinpoppen H 1982 *J. Phys.* B **15** 933–950
[2] Sukhorukov V L, Petrov I D, Schäfer M, Merkt F, Ruf M W and Hotop H 2012 *J. Phys.* B **45** 092001
[3] O'Keeffe P, Bolognesi P, Mihelič A, Moise A, Richter R, Cautero G, Stebel L, Sergo R, Pravica L, Ovcharenko E, Decleva P and Avaldi L 2010 *Phys. Rev.* A **82** 052522
[4] O'Keeffe P, Mihelič A, Bolognesi P, Žitnik M, Moise A, Richter R and Avaldi L 2012 Submitted to *New J. Phys.*
[5] Cowan R D 1981 *The theory of atomic structure and spectra* (Berkeley: University of California)
[6] Blyth R R *et al* 1999 *J. Electron Spectrosc. Relat. Phenom.* **101-103** 959–64
[7] Moise A, Alagia M, Banchi L, Ferianis M, Prince K C and Richter R 2008 *Nucl. Inst. Methods Phys. Res.* A **588** 502–8
[8] O'Keeffe P, Bolognesi P, Coreno M, Moise A, Richter R, Cautero G, Stebel L, Sergo R, Pravica L, Ovcharenko Y and Avaldi L 2011 *Rev. Sci. Instrum.* **82** 033109
[9] Garcia G A, Nahon L and Powis I 2004 *Rev. Sci. Instr.* **75** 4989–96
[10] Froese Fischer C 1977 *The Hartree-Fock method for atoms* (New York: John Wiley & Sons)
[11] Haber L H, Doughty B and Leone S R 2009 *Phys. Rev.* A **79** 031401(R)
[12] Lagmago Kamta G, Piraux B and Scrinzi A 2001 *Phys. Rev.* A **63** 040502(R)
[13] Chang T N and Fang T K 1995 *Phys. Rev.* A **52** 2638–44
[14] Chang T N and Fang T K 1995 *Phys. Rev.* A **52** 2052–56
[15] Chan W F, Cooper G, Guo X and Brion C E 1992 *Phys. Rev.* A **45**(3) 1420–33
[16] Schohl S, Cherepkov N A, Petrov I D, Sukhorukov V L, Baier S and Hotop H 1998 *J. Phys.* B **31** 3363–85

Breakdown and discharge regimes in standard and micrometer size dc discharges

N Škoro[1]
Institute of Physics, University of Belgrade, Pregrevica 118, 11080 Belgrade, Serbia

E-mail: nskoro@ipb.ac.rs

Abstract. In this paper, an overview of our recent experimental studies of the breakdown and operation of non-equilibrium discharges in centimetre and micrometer size geometries is presented. In the centimetre size geometries, we focused on elementary processes and phenomenology in gases used for some of the currently most attractive applications of low temperature plasmas - fluorocarbon gases (CF_4, $CHClF_2$) and water vapour. Measurements were performed at electrode separation d = 1.1 cm, in a pressure range from 0.1 to 5 Torr. In the case of micro-discharges, the emphasis was on testing the validity of standard scaling laws and proper determination of scaling parameters pd and j/p^2. This was done at electrode gaps of 200 and 500 μm and pressures of 50 and 20 Torr in argon. The investigation is based on measurements of breakdown potentials (Paschen curves) and Volt-Ampere characteristics, supported by simultaneous ICCD imaging of the discharges.

1. Introduction

Owing to many unique properties, low temperature non-equilibrium discharges have been used in a large number of applications. Fundamental investigations lead to new applications but, recently, in many cases, the needs of applications led to numerous fundamental results. On all accounts, the complexity of these discharges call for full understanding of discharge processes in order to make efficient and reliable devices. Moreover, along with numerous non-equilibrium discharge sources in centimetre size [1,2], many sources with micrometre characteristic dimensions have been introduced recently [3,4]. Hence, an ongoing direction of research about discharges in molecular gases and gas mixtures of interest for applications [5] has been appended with new challenges of discharges taking place between electrodes separated by few tens or hundreds of micrometers – micro-discharges [6]. Additionally, a new type of non-equilibrium discharges taking place in or near liquids has emerged [7], bringing new possibilities for applications together with a whole new set of issues to resolve.

Accordingly, results presented in this paper are divided in two sections. The first one deals with measurements conducted with cm-size chamber. Using previously gathered experience on standard size non-equilibrium discharges in rare gases at low pressures [8-12], our research is continued with experimental measurements in gases of interest for applications: fluorocarbons (CF_4 and $CHClF_2$) which are widely used for plasma etching (e.g. [13]) and water vapour that is present as working environment of numerous commercial devices [14]. Measurements in these gases aim to provide reliable and systematic data sets required for plasma models that may be extended to applications involving those gases.

[1] To whom any correspondence should be addressed.

In the second part, results obtained in case of micro-discharges, were used for comparison with the existing data on breakdown and discharge regimes of cm-size Ar discharges [15]. Our knowledge on basic processes in standard size discharges provided us with a possibility to investigate basic properties of micro-discharges from the point of view of scaling laws [16,17]. This type of non-equilibrium discharges with rising number of applications is inaccessible for most standard diagnostic tools due to its small size. Scaling laws, in general, should provide comparison of discharges with different dimensions or electrical characteristics [18]. Therefore, an approach through the analysis of scaling laws and the use of existing knowledge of processes in standard size discharges becomes valuable way of diagnostics of micro-discharges.

2. Experimental setup

Discharge was ignited in two different chambers, with centimetre and micrometer size electrode gaps. In both cases, two parallel-plate electrodes are placed inside a tightly fitting tube made of dielectric. For micrometer-size discharges, stainless steel cathode and a transparent anode with conductive ITO (Indium Tin Oxide) film were separated by $d = 200$ μm and $d = 500$ μm distances. Radial profiles are recorded from an end-on view, through the transparent anode. In the cm-size discharge chamber, copper cathode and transparent anode, with a conductive film of platinum deposited on a quartz window, were tightly surrounded by a quartz tube. In this way, along with radial discharge profiles, axial emission profiles could be recorded. Diameter of electrodes was $D = 5.4$ cm, while the gap for all measurements presented in this paper was $d = 1.1$ cm. Emission from the discharge, integrated in visual spectral range and in time, were recorded by ICCD camera (Andor IStar DH720-18U-03).

Details of electrical circuit and technique of measurement are presented in more details elsewhere [e.g. 11,15]. Apart from establishing low-current discharge, the electrical circuitry of experiment is able to impose a current pulse of desired length in addition to the running dc discharge. This technique allows us to avoid long breakdown delay times as well as gas and cathode heating and electrode conditioning in high-current discharge regimes [10]. Camera gating is synchronized with the development of voltage and current signals from the discharge in order to record emission from the quasi steady-state part of the pulse.

Before measurements, both vessels were pumped down to low pressure (10^{-6} Torr). In order to create the same cathode surface conditions each time [10], in cm-size chamber, the cathode was treated by moderately high-current discharge in hydrogen (30 μA) until a stable breakdown voltage is achieved. The same treatment but using Ar was done in the case of micrometer chamber since hydrogen discharge may damage ITO film on the anode [19].

3. Properties of standard size discharges in molecular gases important for applications

One of the fundamental properties of the dc discharge is the dependence of breakdown voltage (V_b) on the product of gas pressure and electrode gap (pd), i.e. Paschen curve. In Fig. 1 we show Paschen curves for some of the gases that we studied: water vapour, CF_4 and $CHClF_2$. For comparison, Paschen curve for Ar is also shown. All curves exhibit typical, sharply rising voltages with pd values decreasing to the left from the Paschen minimum and slow increment of breakdown voltages in the right-hand side branch. Curve for Ar has minimum around $pd = 1$ Torrcm with breakdown voltage of around 210 V, well below the ones for molecular gases shown here. The minimal breakdown voltages for $CHClF_2$ and CF_4 are 390 V and 445 V respectively. In water vapour, minimal value required for breakdown is around 530 V. Moreover, positions of minimum of the curves for all molecular gases investigated are shifted towards lower pd values in comparison with Ar. This feature has been noticed in case of other electronegative discharges (SF_6, CO_2, NO, air) [1].

At every point of the Paschen curve a corresponding axial emission profile of the low-current discharge was recorded ($i = 1$ μA). Spatial emission profiles recorded for different pd, ie. reduced electric field (E/N) values, are shown in Fig. 2a for CF_4 and $CHClF_2$ and Fig.2b for water vapour. All profiles are normalized so that peak intensities near the anode are the same. In this way, profile slopes can be compared easily.

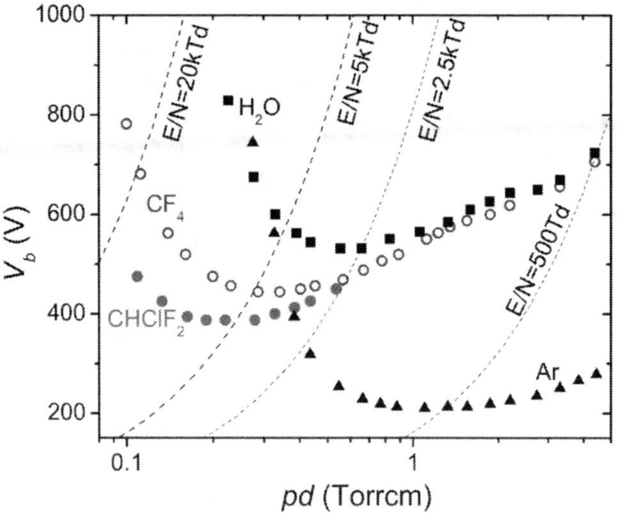

Figure 1. Paschen curves for water vapour – squares, CF_4 – empty circles, $CHClF_2$ – full circles and Ar – triangles. The cathode is made of copper. Dashed lines show characteristic values of reduced electric field (*E/N*) parameter.

Generally, profiles of emission in Townsend regime of discharge exhibit an exponential increase from the cathode towards the anode, when processes of excitation and ionization are predominantly induced by electrons. This kind of behaviour is exhibited in all gasses covered by this study at low *E\N*. Slope of the exponential increase is directly connected to the ionization coefficient [12].

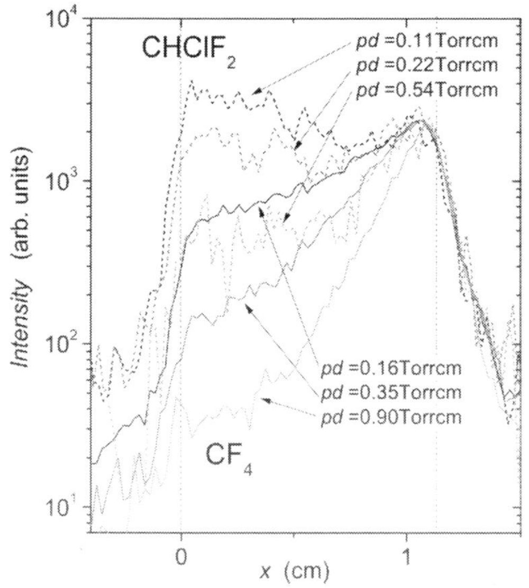

Figure 2a. Low-current discharge profiles in CF_4 (full lines) and $CHClF_2$ (dashed lines) for different values of *pd* parameter. Corresponding *E/N* values: CF_4 0.16 Torrcm / 10 kTd, 0.35 Torrcm / 4 kTd, 0.9 Torrcm / 1.8 kTd; $CHClF_2$ 0.11 Torrcm / 13 kTd, 0.22 Torrcm / 5.5 kTd, 0.54 Torrcm / 2.6 kTd

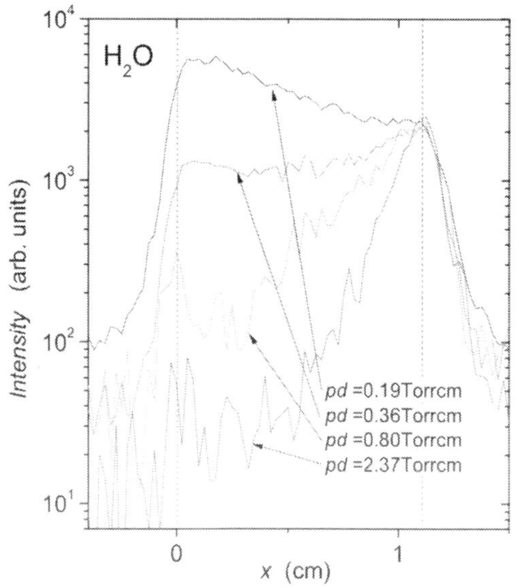

Figure 2b. Low-current discharge profiles in water vapour for different values of *pd* parameter. Corresponding *E/N* values: 0.19 Torrcm / 23 kTd; 0.36 Torrcm / 4 kTd; 0.8 Torrcm / 2 kTd; 2.4 Torr cm / 850 Td

However, for sufficiently high *E/N*, heavy particles, such as positive ions and fast atoms, can also contribute. These processes are revealed through the additional peak of emission near the cathode [11]. As it can be observed in Fig. 2a, heavy particles contribute to the breakdown in CF_4 only at the highest *E/N*-s (~10 kTd), with fast F atoms and positive F^+ ions produced in CF_4 dissociation as probable candidates [20]. On the other hand, $CHClF_2$ and H_2O show notable contribution of heavy particle excitation starting from *E/N* ~ 2 kTd. For highest reduced electric fields, heavy particle excitation becomes even dominant over electron induced excitation – in the case of $CHClF_2$ this is achieved at lower *E/N* values (around 13 kTd) in comparison with H_2O (around 20 kTd) [21]. The apparent candidates for heavy particles that induce significant excitation and ionization in these two gases are hydrogen ions and fast atoms. Further spectrally resolved measurements are taking place to determine exact processes that participate in breakdown of these gases [22].

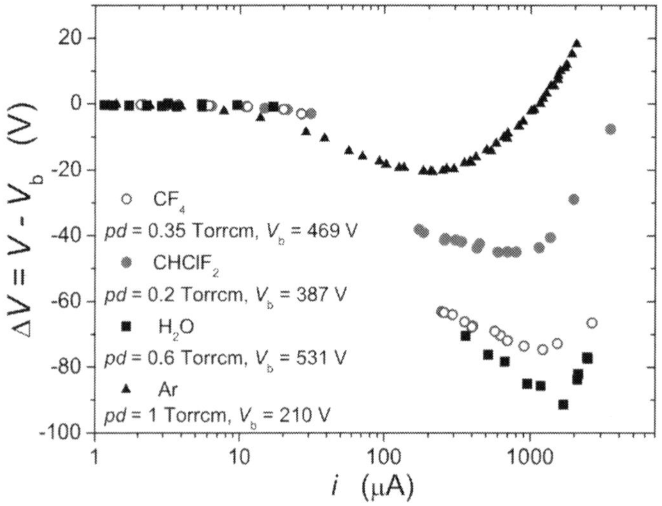

Figure 3. Volt-Ampere characteristics in freons, water vapor and Ar in the chamber with copper cathode. Electrode gap is d = 1.1 cm.

In Fig.3 we show Volt-Ampere characteristics of dc discharges in fluorocarbons ($CHClF_2$ and CF_4), water vapour and Ar, for comparison. All Volt-Ampere characteristics were recorded for *pd* values at the minimum of Paschen curves of particular gases, where the production and losses of the charged particles are well balanced. In all cases the electrode distance was d = 1.1 cm. The voltage is shown as the difference between the discharge and the breakdown voltage for the sake of comparison. Characteristics have a standard shape for low pressure dc discharges, where different regimes of the discharge can be clearly distinguished [23]. In the low-current range, small negative slope of the characteristics can be observed. The slope is larger in Ar discharge in comparison to discharges in molecular gases. At higher currents, fluorocarbon and water vapour discharges enter the current range where oscillations occur (gap in the descending part of Volt-Ampere characteristics). On the other hand, at these conditions, oscillations in Ar discharge exist only in a very small current interval. After the regime of oscillations, decrease in discharge voltage continues as it changes regime of operation to normal glow. Comparison in Fig.3 shows different efficiencies of discharges in different gases. The highest difference between working and breakdown voltage is achieved in water vapour discharge (around 90 V), followed by discharge in fluorocarbons while in Ar difference does not exceed 20 V. However, one should bear in mind that these are relative differences. Positive slope of the characteristics, when a discharge voltage increases with current, is the property of discharge running in a high-current regime, abnormal glow.

Journal of Physics: Conference Series **399** (2012) 012017 doi:10.1088/1742-6596/399/1/012017

4. Properties of micrometer size discharges – scaling laws

Almost all applications of micro-discharges are based on the capability to produce non-equilibrium plasma at atmospheric pressure. So far, non-equilibrium plasma has been comprehensively studied at low pressures and in standard size devices [11]. Since mutual correspondence between cm-size and micro-discharges is indubitably present, it is expected that the employment of scaling laws and data comparison should provide an insight to properties of micro-discharges. To test the applicability of scaling laws and provide description of basic processes governing breakdown and different discharge regimes of micro-discharge, we performed measurements on micro-discharges ignited between parallel electrodes. Although this configuration is more prone to instabilities, compared to hollow cathode-like discharges [3], establishing micro-discharge in plane parallel electrode geometry makes an advantage when it comes to the analysis and interpretation of the results. All measurements presented here were done in pure Ar.

First, we tested the validity of *pd* scaling. Generally, the dependence of breakdown voltage on *pd* parameter, i.e. Paschen curve, should be the same for all electrode gaps unless processes that void scaling take place. In micro-discharges, apart from well known processes that cause violation of scaling (gas and electrode heating, stepwise processes and three-body collisions at high pressures), one may expect processes such as field emission and quantum tunnelling of electrons [24].

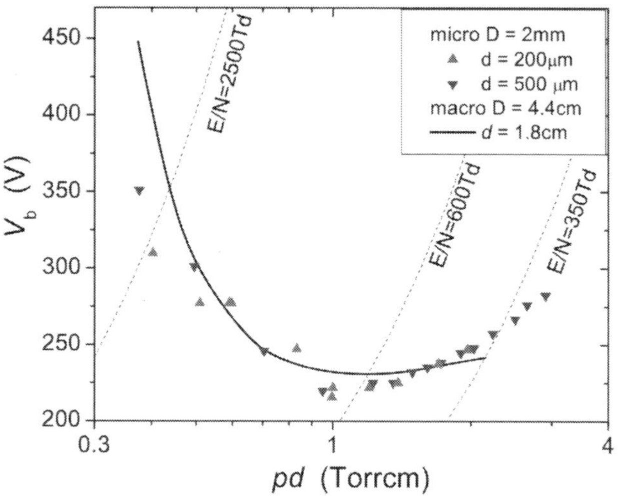

Figure 4. Paschen curves of micro-discharge (triangles) and standard size discharge (line) with stainless steel cathode in pure Ar.

In Fig.4 Paschen curves of micro-discharges with $d = 200$ μm i 500 μm electrode gaps are shown (triangles). For comparison, Paschen curve recorded in a standard size discharge (electrode gap $d = 1.8$ cm, electrode diameter $D = 4.4$ cm) is also shown (line). Breakdown voltages in micro-discharges at different gaps have small differences (up to 20 V) but the curve minimum and shape of the curves agree well in whole *pd* range investigated. In the left hand-side branch of the 200 μm curve the discrepancies are probably due to small differences in cathode surface conditions for the two discharge set-ups. In addition, this branch of Paschen curve is generally more sensitive to variation in gas pressure, due to the steep slope of the curve. The minimum of micro-discharge curves agrees well with the minimum of cm-size curve although the curve taken in cm-size discharge has less steep increase of breakdown voltage in the right-hand branch. There is a very good agreement between Paschen curves of micro-discharges and cm-size discharge. Hence, for the electrode gaps examined no new processes appear in the micro-discharge, i.e. the phenomenology is the same as in standard size discharges. Departures from the standard shape of Paschen curves for this range of electrode gaps, which exists in the literature, in most cases are influenced by discharge ignition at distances larger than electrode gap i.e. by the long path breakdown [15].

The second scaling that has been tested is j/p^2, which is connected to the effect of space charge [17]. It was shown that for standard size discharges both pd (and E/N) and j/p^2 scalings are valid in parameter region around Paschen minimum [11]. However, it is important to note that one needs to know realistic value of the current density i.e. to take into account effective area covered by the discharge, which may be considerably smaller than the area defined by the electrode diameter [25]. Micro-discharges operate at pressures higher than pressures used in standard size discharges, so that the occurrence of constriction in micro-discharge is even more pronounced [24]. Therefore, it is essential to record radial profiles of emission.

Scaled Volt-Ampere characteristics of micro-discharges for two electrode gaps and $pd = 1$ Torrcm are shown in Fig.5. As before, the voltage is given as the difference between the discharge and breakdown voltages. For comparison, the same dependence obtained from the standard size discharge with copper cathode is also shown. Radial profiles of appropriate regimes (indicated by numbers in Fig.5) for both electrode gaps of micro-discharges are shown in Fig.6.

Negative slope of the Volt-Ampere characteristics in low-current regime is significantly smaller in the case of micro-discharge, as the negative differential resistivity scales with d^2. Profile of micro-discharge at $d = 500$ μm is diffuse in this regime (Fig.6, no.1). Still, at higher pressure at the distance of $d = 200$ μm, radial profile is very constricted and discharge occupies only a part of the electrode area. Sometimes, apart from larger area with emission, smaller emission area detached from the main discharge could be observed. The explanation of the second discharge channel's existence due to a high pressure and consequently small diffusion length seems valid [26].

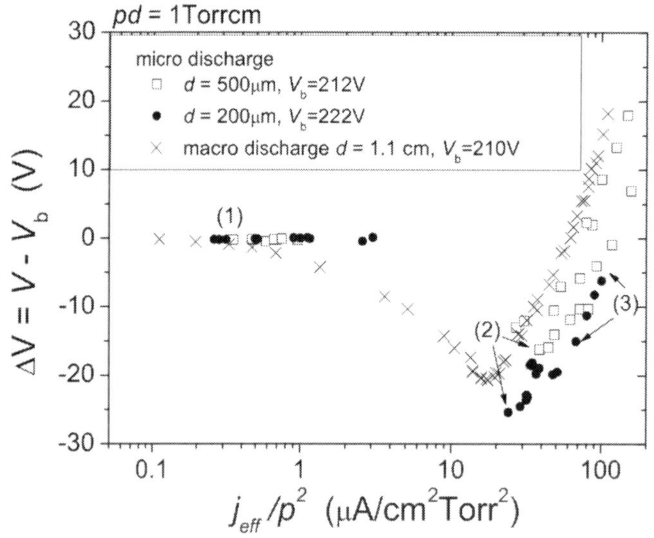

Figure 5. IV characteristics of micro-discharges (squares and circles) and standard size discharge (crosses) in Ar. Numbers correspond to images in Fig.6.

At higher currents, discharge profiles at higher pressure (shorter gap) stays constricted while at lower pressure (larger gap) micro-discharge takes up almost the entire electrode area (Fig.6, no.2). In this current range, the value of j/p^2 parameter is smaller for 200 μm discharge although the current is twice as higher than for 500 μm case because of smaller effective discharge area and higher pressure. Discharges operate in normal glow. After oscillations (interval without points in Fig.5), both characteristics have positive slopes. In both cases, at high-current regime, discharges occupy the whole electrode area (Fig.6, no.3). This current range corresponds to the abnormal glow. Existing discrepancy between micrometer and centimetre size characteristics may be the consequence of different cathode materials but also more pronounced edge effects in micro discharges.

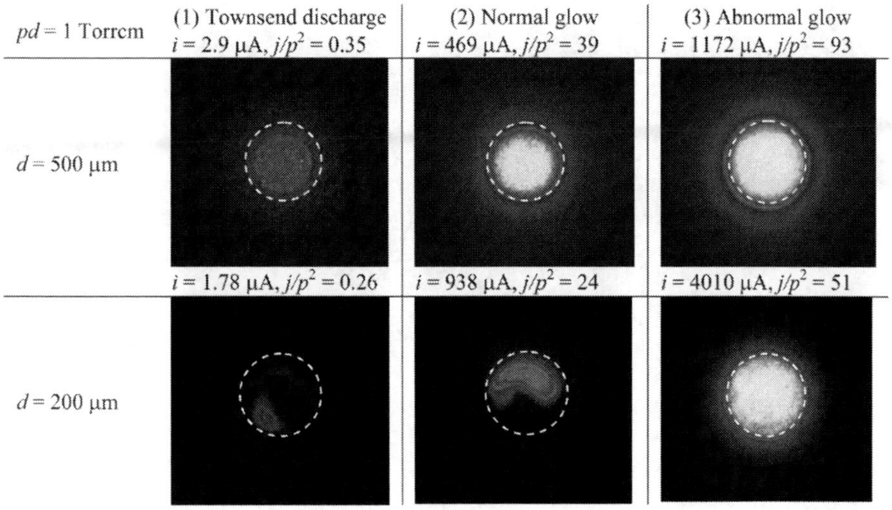

Figure 6. Radial profiles of micro-discharges for two different electrode gaps at different points of IV characteristics (shown in Fig.5).

5. Conclusion

Measurements of breakdown voltages and electrical and emission properties of different discharge regimes were performed in simple plane-parallel standard size discharge with gases used in applications. Simultaneous recordings of Paschen curves and emission profiles in low-current limit allowed the analysis of dominant processes governing breakdown in CF_4, $CHClF_2$ and H_2O gases. It was shown that at higher reduced electrical field values heavy particles have important and even dominant role. Comparison of Volt-Ampere characteristics recorded in aforementioned molecular gases and in Ar revealed differences in electrical properties of operating regimes and efficiencies of discharges in different gases. Further on, we presented results obtained in parallel plate geometry discharge with electrode gap in micrometer range - micro-discharge, that are used in many applications as non-equilibrium discharge sources at high or atmospheric pressures. Since diagnostic of these discharges is difficult due to their small dimensions, we investigated possibility of using pd and j/p^2 scaling laws for comparison with cm-size low-pressure discharges. Comparison of the results show the validity of both scaling laws for micro-discharge electrode gap domain investigated. However, it was shown that a special care should be devoted to the proper determination of scaling parameters.

Present results could be extended with temporal development of the profiles which were also recorded [27]. Besides, in case of micro-discharges metastable densities were recorded [28]. Measurements in oscillatory and steady-state regimes of Volt-Ampere characteristics indicate compatibility of in micro-discharges results [29] with the secondary electron yield model [8]. This provides a support to the expansion of the use of the model to RF plasmas [30]. In addition, it would be interesting to expand the measurements to RF breakdown and thus test the existing theories [31]. Finally, one could also pay closer attention to the Townsend regime where absolute calibration may provide data to model heavy particle effects at high E/N [32] or provide data on ionization coefficients [12] and kinetics of non-local discharges [33] and runaway electrons [34], even data required to study streamers and ionization fronts [35].

6. References

[1] Lieberman M A and Lichtenberg A J 2005 Principles of Plasma Discharge and Materials Processing, (Wiley:Hoboken)
[2] Makabe T and Petrovic Z Lj 2006 Plasma Electronics (Taylor and Francis:New York)
[3] Becker K H, Schoenbach K H and Eden J G 2006 *J. Phys. D: Appl. Phys.* **39** R55

[4] Eden J G and Park S-J 2005 *Plasma Phys. Control. Fusion* **47** B83
[5] Ilic V, Saponjic Z, Vodnik V, Lazovic S, Dimitrijevic S, Jovancic P, Nedeljkovic J M and Radetic M 2010 *Ind Eng Chem Res* **49** 7287
[6] Lazović S, Puač N, Miletić M, Pavlica D, Jovanović M, Bugarski D, Mojsilović S, Maletić D, Malović G, Milenković P and Petrović Z 2010 *New Journal of Physics* **12** 083037
[7] Bruggeman P and Leys C 2009 *J. Phys. D: Appl. Phys.* **42** 053001
[8] Phelps A V, Petrović Z Lj, Jelenković B M 1993 *Phys. Rev. E* **47** 2825
[9] Petrović Z Lj and Phelps A V 1997 *Phys. Rev. E* **56** 5920
[10] Phelps A V and Petrović Z Lj 1999 *Plasma Sources Sci. Technol.* **8** R21
[11] Marić D, Hartmann P, Malović G, Donko Z, Petrović Z Lj 2003 *J. Phys. D: Appl. Phys.* **36** 2639
[12] Malović G, Strinić A, Živanov S, Marić D, Petrović Z Lj 2003 *Plasma Sources Sci. Technol.* **12** S1
[13] Nagatsu M, Ganashev I, Morita S, Sugai H. 1998 *Jpn. J. Appl. Phys.* **37** 2406
[14] Stalder K R, Mcmillen D F, Woloszko J 2005 *J. Phys. D: Appl. Phys.* **38** 1728
[15] Marić D, Škoro N, Maguire P D, Mahony C M O, Malović G, Petrović Z Lj 2012 *Plasma Sources Sci. Technol.* **21** 035016
[16] Francis G 1956 Handbuch der Physik (Berlin: Springer)
[17] Phelps A V 2001 *Plasma Sources Sci. Technol.* **10** 329
[18] Marić D, Malović G, and Petrović Z Lj 2007 *Journal of Physics: Conference Series* **86** 012009
[19] Raniero L, Ferreira I, Pimentel A, Goncalves A, Canhola P, Fortunato E, Martins R 2006 *Thin Solid Films* **511** 295
[20] Rao M V V S and Olthoff J K 1999 *Phys. Rev. E* **60** 7449
[21] Škoro N, Marić D, Malović G, Graham W G, Petrović Z Lj 2011 *Phys. Rev. E* **84** 055401
[22] Sivoš J, Marić D, Petrović Z Lj, unpublished
[23] Petrović Z Lj, Donko Z, Marić D, Malović G, Živanov S 2002 *IEEE Trans. Plasma Sci.* **30** 136
[24] Petrović Z Lj, Škoro N, Marić D, Mahony C M O, Maguire P D, Radmilović-Rađenović M, Malović G 2008 *J. Phys. D: Appl. Phys.* **41** 194002
[25] Škoro N, Marić D, Petrović Z Lj 2008 *IEEE Trans. Plasma Sci.* **36** 994
[26] Sivoš J, Škoro N, Marić D, Malović G, Petrović Z Lj, 2012 *Proc. 26th SPIG (Zrenjanin)* (Novi Sad: University of Novi Sad, Faculty of Sciences, Department of Physics) p 273
[27] Marić D, Malović G, Petrović Z Lj 2009 *Plasma Sources Sci. Technol.* **18** 034009
[28] Kuschel T, Niermann B, Stefanović I, Boke M, Škoro N, Marić D, Petrović Z Lj, Winter J 2011 *Plasma Sources Sci. Technol.* **20** 065001
[29] Stefanović I, Kuschel T, Škoro N, Marić D, Petrović Z Lj, Winter J 2011 *J. Appl. Phys.* **110** 083310
[30] Bojarov A, Radmilović-Radjenović M, Petrović Z Lj 2009 *Hem. ind.* **63** 2333
[31] Savić M, Radmilović-Radjenović M, Šuvakov M, Marjanović S, Marić D, Petrović Z. Lj. 2011 *IEEE Trans. Plasma Sci.* **39** 2556
[32] Stojanović V, Nikitović Ž, Petrović Z Lj 2011 *IEEE Trans. Plasma Sci.* **39** 2592
[33] Petrović Z Lj, Dujko S, Marić D, Malović G, Nikitović Ž, Šašić O, Jovanović J, Stojanović V, Radmilović-Radenović M 2009 *J. Phys. D: Appl. Phys.* **42** 194002
[34] Nikitović Ž, Stojanović V, Petrović Z Lj 2012 *Eur. Phys. Letters* **99** 35003
[35] Puač N, Maletić D, Lazović S, Malović G, Đorđević A, Petrović Z Lj 2012 *Appl. Phys. Lett.* **101** 024103

Acknowledgments

Results presented here are part of a PhD thesis which was completed under the supervision of Dr Dragana Marić and Prof. Zoran Lj. Petrović, to whom author is also thankful for suggestions during writing of this paper. The work was supported by Projects No. ON 171037 and III 41011 of the MES of Serbia.

On the role of electron-driven processes in planetary and cometary atmospheres

Laurence Campbell[1], Oddur Ingólfsson[2] and Michael J. Brunger[1,3]
[1]ARC Centre for Antimatter-Matter Studies, Flinders University, GPO Box 2100, Adelaide, SA 5001, Australia
[2]University of Iceland, Science Institute, Dunhaga 3, IS-107 Reykjavik, Iceland
[3]Institute of Mathematical Sciences, University of Malaya, Kuala Lumpur, 50603, Malaysia
E-mail: Laurence.Campbell@flinders.edu.au

Abstract.
Electron impact excitation plays an important role in the upper atmospheres of planets and their satellites and also in the comae of comets, in both energy transfer and chemistry. Emissions produced by electron impact excitation are vital in remote sensing of these atmospheres. Modeling of the processes and emissions requires knowledge of both the plasma parameters and accurate cross sections or rates for the electron impact processes. This modeling is illustrated by four examples: electron cooling by CO_2 in the atmosphere of Mars, infrared emissions from CO in the atmosphere of Venus, fourth positive emissions from CO in the coma of comet Hale-Bopp and emissions due to excitation of the higher-energy states of molecular oxygen in the atmosphere of Europa. In each case the assembly of the plasma parameters and accurate electron impact cross sections is described, together with the modeling techniques applied and the significant results. These include a possible explanation for low temperatures in part of the upper atmosphere of Mars, evidence that the abundance of CO in comet Hale Bopp has been overestimated and identification of a useful parameter for remote sensing of Europa.

1. Introduction

Sunlight produces photoionization and hence a low-density plasma in upper planetary atmospheres and in the comae around comets. Electrons in the solar wind, often acccelerated in magnetospheric processes, produce localised ionisation. In both cases a flux of secondary electrons is produced, leading to further ionization, plus dissociation and excitation, before the electrons recombine with the ions. The excited atoms and molecules can release energy in radiative decay (as aurora, dayglow or nightglow) or by taking part in chemical reactions. Hence plasma processes drive energy transfer and the composition of minor constituents in cometary and planetary atmospheres.

In order to model these processes, it is necessary to know the composition and temperature of the neutral species, the electron impact cross sections for the ionization, dissociation, and excitation processes, recombination rates and the plasma conditions. The latter are usually characterised either by the electron temperature and density of a Maxwell-Boltzmann distribution, or by a differential electron flux spectrum.

The simplest of such calculations is of "electron cooling", i.e. energy transfer from a Maxwell-Boltzmann distribution of ionospheric electrons to neutral species by electron impact excitation of the neutrals [1]. If there are multiple paths for excitation or deexcitation, such as radiative cascades through different excited states, collisional processes and chemical reactions, then a "statistical equilibrium" calculation [2] is required.

2. Methods

Using the formulation of Pavlov [3] the electron energy transfer rate $Q_{0\nu}$ per molecule is:

$$Q_{0\nu} = E_\nu \{8kT_e(\pi m_e)^{-1}\}^{0.5} \int_0^\infty \sigma_{0\nu}(x)\exp(-x)\mathrm{d}x, \qquad (1)$$

where $x = E(kT_e)^{-1}$, E is the electron energy, E_ν is the energy of vibrational level ν, m_e is the electron mass, T_e is the electron temperature, k is Boltzmann's constant, and $\sigma_{0\nu}$ is the integral cross section (ICS) for excitation from ground(0) to vibrational level ν. If excitation leads to only one possible radiative transition, the emission rate per molecule is $Q_{0\nu}/E_\nu$. The electron cooling rate for a unit volume due to a particular molecule is determined by multiplying $Q_{0\nu}$ by the electron density n_e and the density of the molecule.

The statistical-equilibrium calculation assumes that the gain and loss rates are equal for each excited species, producing equilibrium populations that are determined by the processes involved and thus are generally not in thermal equilibrium [4]. Continuity equations are written for each species and then the whole set is solved iteratively [2]. An example that includes radiative cascade among excited electronic states of a molecule is represented by the statistical-equilibrium equation [5]:

$$k^\alpha_{\nu'0} n^X_0 + \sum_\beta \sum_i A^{\beta\alpha}_{i\nu'} n^\beta_i = \left\{ \sum_\gamma \sum_r A^{\alpha\gamma}_{\nu' r} + Q^\alpha_{\nu'} \right\} n^\alpha_{\nu'}, \qquad (2)$$

where $A^{\alpha\beta}_{ij}$ is the Einstein spontaneous transition probability (s^{-1}) for the radiative transition from the ith vibrational level of electronic state α to the jth vibrational level of electronic state β; n^α_k is the number density of the kth vibrational level of state α; and Q^α_i is the quenching rate (s^{-1}) of the ith level of state α. k^α_{ij} is the electron impact excitation rate (s^{-1}) of the ith level of state α from the jth vibrational level of the ground electronic state. The volume emission rate (VER) from a unit volume due to a particular transition is $A^{\alpha\gamma}_{\nu' r} n^\alpha_{\nu'}$ photons s^{-1}.

Photo- and auroral electrons produce secondary electron distributions which are not Maxwell-Boltzmann and so need to be specified as a function of E. For ionospheric calculations the most common approach is to specify a differential flux distribution $F(E)$, being the number of electrons with a particular energy crossing an area per unit time. The excitation rate $k^\alpha_{\nu'0}$ is then given by:

$$k^\alpha_{\nu'0} = \int_0^\infty \sigma^\alpha_{\nu'0}(E)F(E)dE, \qquad (3)$$

where $\sigma^\alpha_{\nu'0}$ is the rotationally averaged electron impact excitation cross section from the ground state to the ν' level of electronic state α. The higher-energy electrons lose energy and end up as a population of "thermal" electrons in a Maxwell-Boltzmann distribution. They can be included by converting the speed distribution of thermal electrons to a differential flux spectrum, using $F(E) = n_m(E)v = n_m(E)\sqrt{2E/m_e}$, where $n_m(E)$ is the number of electrons with speed v.

3. Applications

Calculations of electron cooling by CO_2 and CO in the atmospheres of Mars and Venus are described below. For CO more detailed calculations using statistical equilibrium were made to include stepwise vibrational transitions, quenching and cascade from excited electronic states. Applications of the statistical-equilibrium method are then described for electron-driven processes in comet Hale-Bopp and in the atmosphere of the Galilean moon Europa.

Journal of Physics: Conference Series **399** (2012) 012018 doi:10.1088/1742-6596/399/1/012018

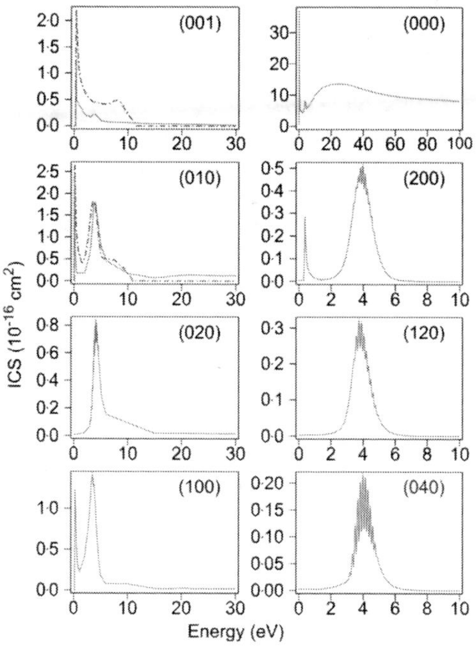

Figure 1. Electron impact integral cross sections for elastic scattering and the vibrational modes (as labelled) of CO_2: (— · —) values used by Morrison and Greene [6] and (——) compilation by Campbell *et al.* [7].

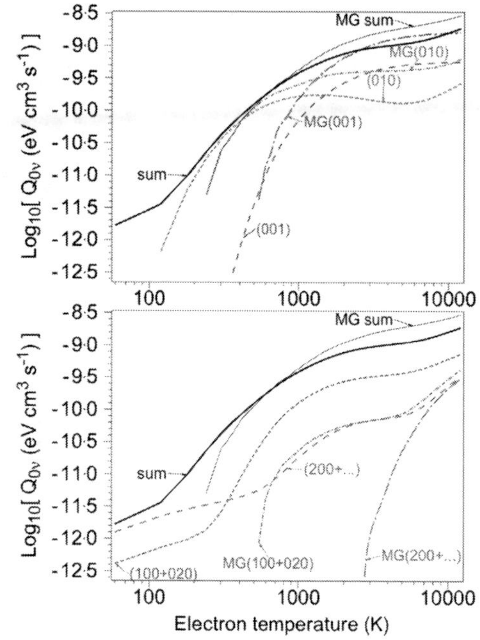

Figure 2. Comparison of the updated electron energy transfer rates (as labeled) with previous calculations by Morrison and Greene [6] (as labeled with prefix "MG"). "(200+...)" represents "(200+120+040)" and the "sum" in both panels is the sum for all 7 vibrational modes.

3.1. Electron cooling by CO_2 in the atmosphere of Mars

As the neutral atmosphere of Mars is dominated by CO_2, Morrison and Greene [6] calculated electron energy transfer rates ($Q_{0\nu}$) due to excitation of CO_2. A more accurate calculation [7] of these rates became possible due to new experimental measurements [8, 9, 10] and theoretical [11] cross sections for electron impact vibrational excitation of CO_2. These new cross sections were combined with others from a compilation [12] of earlier values [13, 14, 15] to produce the set of cross sections [7] shown by solid lines in Fig. 1. The cross sections [16, 17] used by Morrison and Greene are shown in the two cases where they reported transfer rates for individual modes. As expected and as shown in Fig. 2 the larger cross sections used by Morrison and Greene produced larger transfer rates for electron temperatures above ~1000 K, but the larger values of the updated cross sections at very low energies produced larger transfer rates at lower temperatures.

The plasma conditions in the upper atmosphere of Mars are shown by electron densities in Fig. 3 and electron temperatures in Fig. 4. Due to the limited height range of the measured electron temperatures, a calculated profile by Choi *et al.* [18] was used for T_e in equation (1) to calculate energy transfer rates. The electron densities in the atmosphere of Mars were assembled from *in situ* measurements by the Viking I lander [19] and maxima and minima of profiles measured remotely by the Mars Global Surveyor (MGS) satellite [20], to give the "Fit to measurements" [7] shown in Fig. 3. These were combined with CO_2 densities [21] to calculate the cooling rates due to CO_2 as a function of altitude for Mars, as shown in Fig. 5.

It can be seen in Fig. 5 that calculated cooling rates with the more recent cross sections are

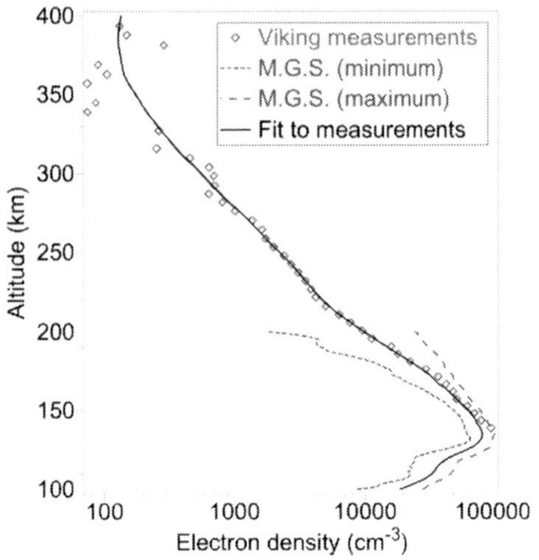

Figure 3. Martian electron densities as a function of altitude: (◇) *in situ* measurements by the Viking I lander [19], (— — —) maximum and (- - - -) minimum of values measured by the MGS [20] and (———) the fit to all these measurements by Campbell *et al.* [7].

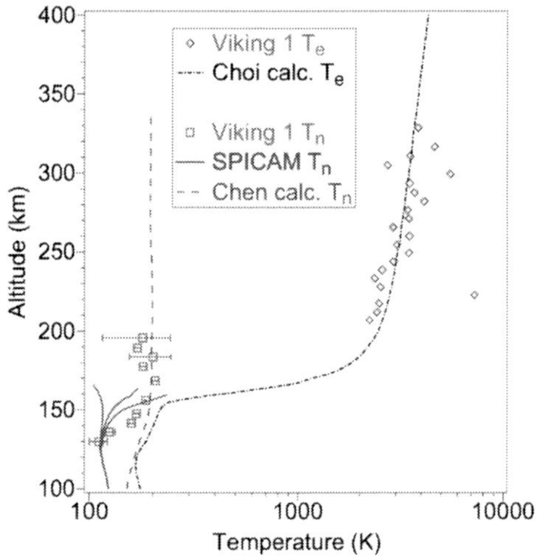

Figure 4. Martian electron temperatures: (◇) measured by Viking I [19] and (— · —) calculated by Choi *et al.* [18]. Neutral temperatures: measured by (□) Viking I [21] and (———) Mars Express [22], and (— — —) calculated by Chen *et al.* [23].

considerably larger than those calculated with the cross sections used by Morrison and Greene [6] at altitudes below 140 km. That difference when using the updated cross sections is larger than the difference due to day-to-day variability in electron density, shown by the maximum and minimum values in Fig. 5. These higher values of electron cooling by CO_2, which imply higher rates of infrared radiation going to space, may explain the very low neutral temperatures observed below 140 km in altitude by both the "SPICAM" (Spectroscopy for the Investigation of the Characteristics of the Atmosphere of Mars) spectrometer on Mars Express [22] and Viking I, shown in Fig. 4, which are much lower than the temperatures calculated by Chen *et al.* [23].

3.2. Infrared emissions from CO in the atmosphere of Venus

A study of electron cooling by CO in the atmospheres of Mars and Venus [27] found that rates by CO were greater than by CO_2 at upper altitudes. This motivated new absolute integral cross section (ICS) measurements [24] for electron impact excitation of the ground state vibrational levels of CO, as plotted in Fig. 6. Also plotted are previous experimental measurements [25] which were put on an absolute scale using the value at 1.91 eV from an earlier study [26]. The main differences between the two sets of measurements are the larger cross sections for the ($0 \rightarrow 1$) excitation around 2 eV and below 0.6 eV in the most recent measurements. The new cross sections [24] were applied in a study [28] of electron-driven infrared emissions from CO in the upper atmospheres of Mars and Venus.

While for Venus a calculation of a photoelectron flux spectrum was available [29], it was necessary to assemble an auroral spectrum for Venus, based on measurements by Spenner *et al* [30]. These measurements, consisting of flux spectra at various altitudes, were subject to considerable time variation, which masked the variation with altitude. However, we found that taking averages of spectra for 2 lower heights and for 4 upper heights gave base spectra at 168 km and 600 km (as plotted in Fig. 7) that were similar in shape to each other. To interpolate

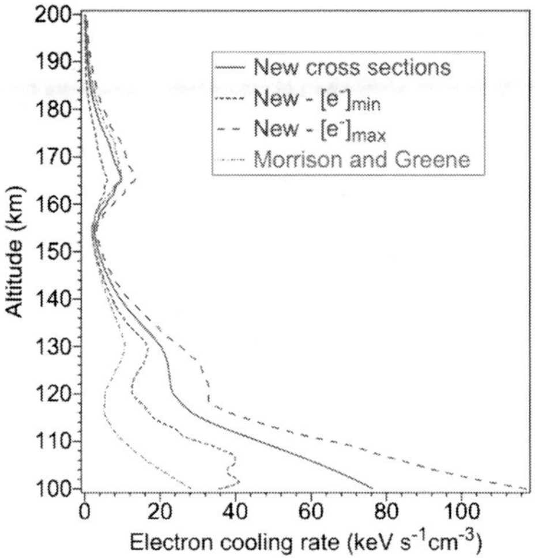

Figure 5. Electron cooling rates as a function of altitude in the atmosphere of Mars, calculated with (— · —) the cross sections used by Morrison and Greene [6] and with the new cross sections for (——) the fit to electron density measurements, and the (- - - -) minimum and (– – –) maximum densities observed by the MGS [20].

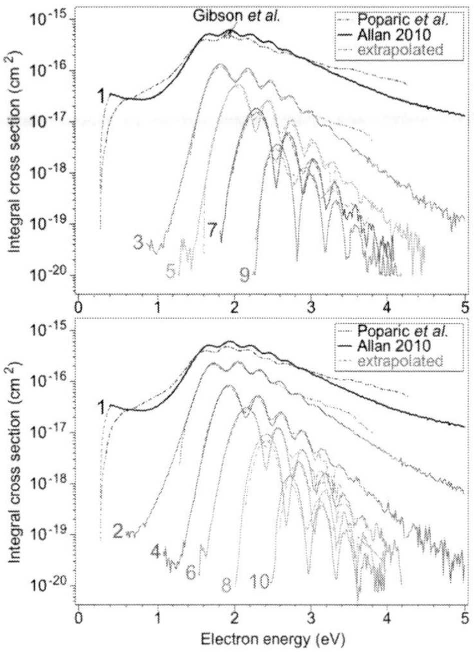

Figure 6. Measurements of ICSs for electron impact of vibrational levels of CO by (——) Allan [24], (— · —) Poparić *et al.* [25] and (□) Gibson *et al.* [26]. (- - - -) Extrapolation at low energies for ν'=1.

or extrapolate from these two "base" spectra to any other height, we assumed that the rate of flux reduction at any height at each energy was proportional to the sum of the densities of O, CO and N_2 at that height. Examples of spectra produced by this procedure are shown for 150 km and 139 km in Fig. 7. The maximum rate of flux loss is predicted at 145 km, which is close to the value of 139 km found by Gérard *et al.* [29], providing some validation for this procedure. As the spectra of Spenner [30] had no values below 8 eV, the shape of the low-energy part of the spectrum was taken from the calculated spectrum by Gérard *et al.* [29] (shown in Fig. 7) and spliced to the 168-km "base" spectrum as shown in Fig. 7, by scaling it to match its slope in the range 8–10 eV to that in the measured spectra. For other heights the added low-energy section is scaled to be proportional to the loss rate in the higher-energy part of the spectrum.

Calculated electron-driven VERs for daytime in the atmosphere of Venus for the (2→1) and (3→2) transitions in the ground electronic state of CO are plotted in Fig. 8. The "Gronoff day" cases apply equation (1) to thermal electrons with parameters as specified by Gronoff *et al.* [32]. Comparison with previous calculations by Crovisier *et al.* [31] for emissions produced by fluorescence and photolysis show that electron-driven (3→2) emissions dominate above 200 km. Results for a "full model", being a statistical-equilibrium calculation incorporating quenching, cascade from excited states, step-wise transitions in the ground state and a flux spectrum including auroral input, are shown by symbols. The agreement with the thermal-equilibrium model for (3→2) emissions at higher altitudes (where auroral input is small) confirms the method of converting the Maxwell-Boltzmann distribution to a flux distribution. The higher values of (2→1) emissions in the full model are due to input to level $\nu' = 2$ from the (3→2) transition.

Other interesting results shown in Fig. 8 are that auroral excitation dominates over that due

Journal of Physics: Conference Series **399** (2012) 012018 doi:10.1088/1742-6596/399/1/012018

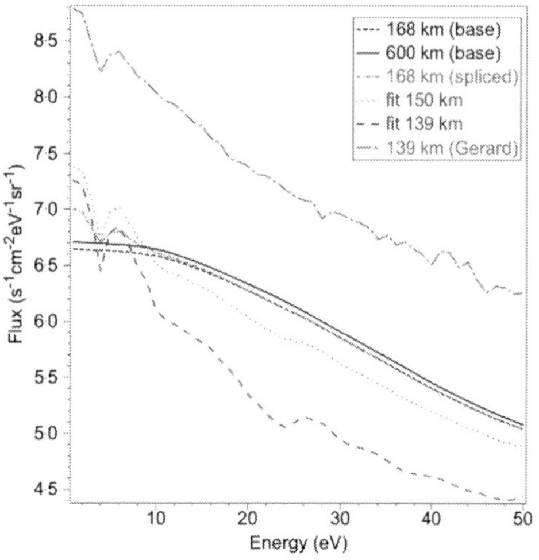

Figure 7. Spectra of auroral electrons for Venus. Base spectra at (- - - -) 168 km and (——) 600 km deduced from the profiles of Spenner *et al.* [30]. The shape at low energy of a (— · —) spectrum at 139 km of Gérard *et al.* [29] is spliced onto the (-·-·-) base spectrum at 168 km. Calculated spectra are shown for (— — —) 139 km and (· · · · · ·) 150 km .

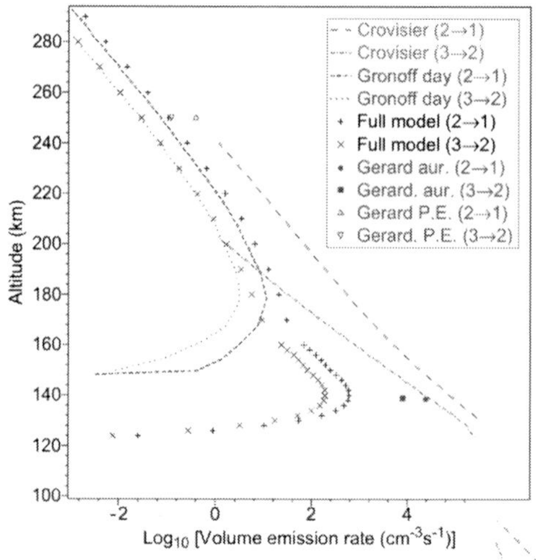

Figure 8. Comparison of electron-driven CO (2→1) and (3→2) VERs to those of Crovisier *et al.* [31] for Venus. Calculations are labelled: "Gronoff day" for using equation (1), "Full" for statistical-equilibrium, "Gerard aur." for the auroral spectrum of Gerard *et al.* [29] and "Gerard P.E." for the photoelectron spectrum of Gerard *et al.*

to thermal electrons below about 180 km, particularly using the auroral spectrum of Gérard *et al.*, and photoelectron-driven emissions dominate over fluorescence plus photolysis at 250 km.

3.3. Fourth positive emissions from comet Hale-Bopp

McPhate *et al.* [33] measured a far ultraviolet spectrum from comet Hale-Bopp, identified particular lines as CO fourth positive emissions ($A\,^1\Pi \rightarrow X\,^1\Sigma^+$) and, assuming these were due to sunlight fluorescence, calculated the abundance of CO in the comet. No contribution to the emissions from electron-impact excitation was included. The determination [34] of electron impact cross sections for excitation of levels 0–7 of the $A\,^1\Pi$ state of CO allowed a calculation [35] of the electron-impact component of the fourth positive emissions.

The plasma environment for the comet was based on a calculated flux spectrum for another comet [36], multiplied by 7.63 to scale the electron density implied by this flux (26.2 electrons cm^{-3}) to that of 200 cm^{-3} calculated for 160000 km from Hale-Bopp [37]. This calculated flux spectrum (shown in Fig. 9) was applied in a statistical-equilibrium calculation of excitation of the $^5\mathrm{S}^\circ$ state of O atoms and vibrational levels 0–7 of the $A\,^1\Pi$ state of CO. The O I emission at 1356 Å is believed to be produced by electron impact [33].

Electron impact excitation cross sections from the ground state of $\mathrm{O}(2p^4\,^3\mathrm{P})$ were given by Laher and Gilmore [38] for $\mathrm{O}(4p\,^5\mathrm{P})$, while those for $\mathrm{O}(3s\,^5\mathrm{S}^\circ)$, $\mathrm{O}(3p\,^3\mathrm{P})$ and $\mathrm{O}(3p\,^5\mathrm{P})$ are averages of values calculated by Zatsarinny and Tayal [39] and Barklem (preprint, P. S. Barklem, 2006, http://arxiv.org/abs/astro-ph/0609684v1). Using transition probabilities from the NIST data base [40], a statistical equilibrium calculation was applied to determine the VERs due to the transitions $3s\,^5\mathrm{S}^\circ \rightarrow 2p^4\,^3\mathrm{P}(0)$ and $3s\,^5\mathrm{S}^\circ \rightarrow 2p^4\,^3\mathrm{P}(1)$, giving the calculated brightness for the 1356 Å and 1358.5 Å lines plotted in Fig. 10. The calculated brightness at 1356 Å compares fairly well with the measurement. While this is somewhat fortituous considering the approximations involved, it shows that the model of the plasma environment in comet Hale-Bopp is physically plausible. The assumed flux was then scaled up by a factor of 1.17 to match the calculated O I emissions with the observations and applied in the calculation of the CO fourth positive emissions. It was found that electron impact produced 40% of the emissions and fluorescence 60%. This is illustrated in Fig. 11 for the subset of CO fourth positive lines selected by McPhate *et al.* [33], comparing their fit to the measurements (solid line) with the cumulative sum of the electron-driven and fluorescent emissions calculated by Campbell *et al.* [35].

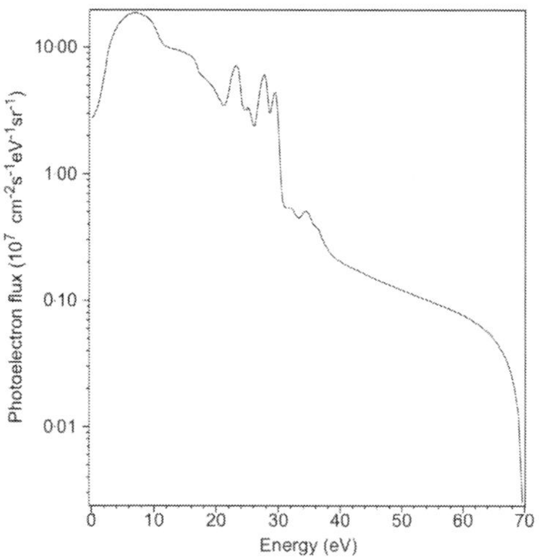

Figure 9. Calculated flux of photoelectrons as a function of energy in the coma of comet Hale-Bopp.

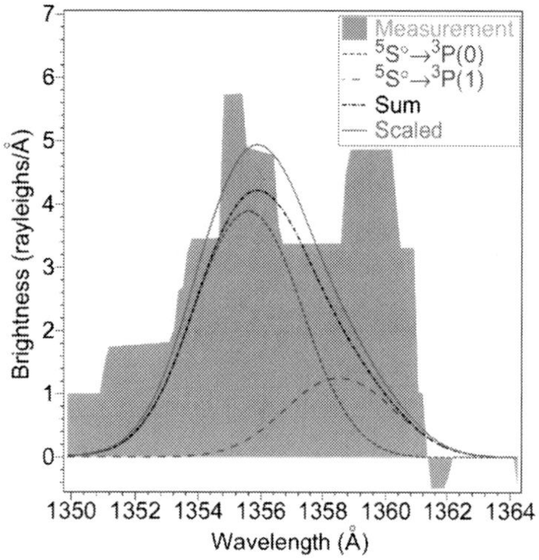

Figure 10. Calculated brightness of O I emissions for (- - - -) $^5S^\circ \rightarrow ^3P(0)$ and (— — —) $^5S^\circ \rightarrow ^3P(1)$ transitions, (— · —) their sum and (——) the sum scaled to the (shading) observations.

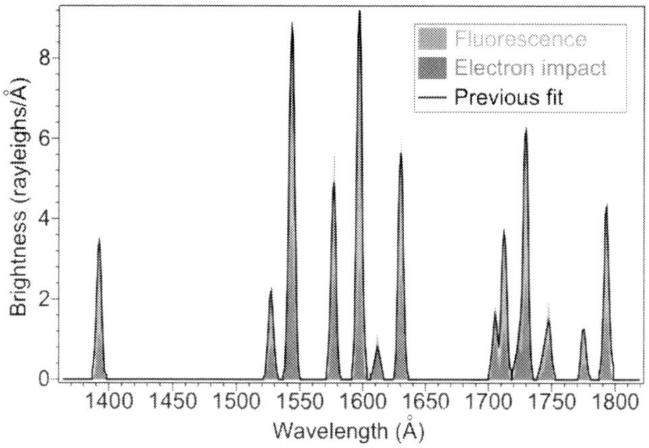

Figure 11. Brightness as a function of wavelength of CO fourth positive emissions for (——) the fit to measurements made by McPhate *et al.* [33] and calculations by Campbell *et al.* [35] of (red shading) electron impact excitation and (green shading) fluorescence.

3.4. Excitation of molecular oxygen in the atmosphere of Europa

Europa, a Galilean moon of Jupiter, has a thin transitory atmosphere believed to be produced predominantly by sublimation and ion impact on its icy surface [44], with O_2 the dominant

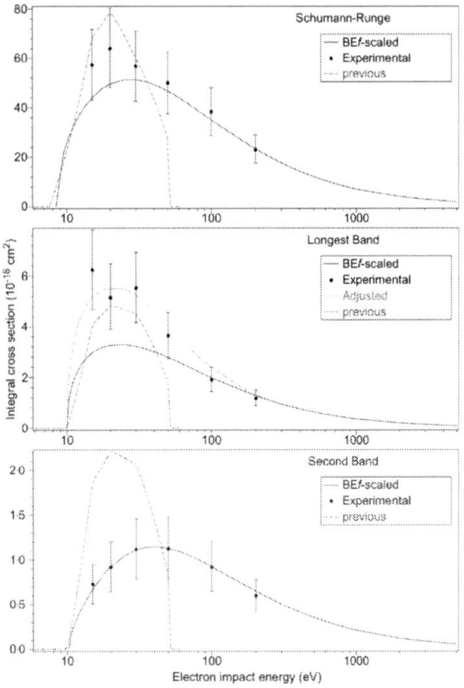

Figure 12. ICSs for excitation of the SR continuum, LB and SB of O_2: (I) measurements [41], (——) BEf-scaling calculations [41], (— · —) adjusted BEf and (- - - -) previous values [42, 43].

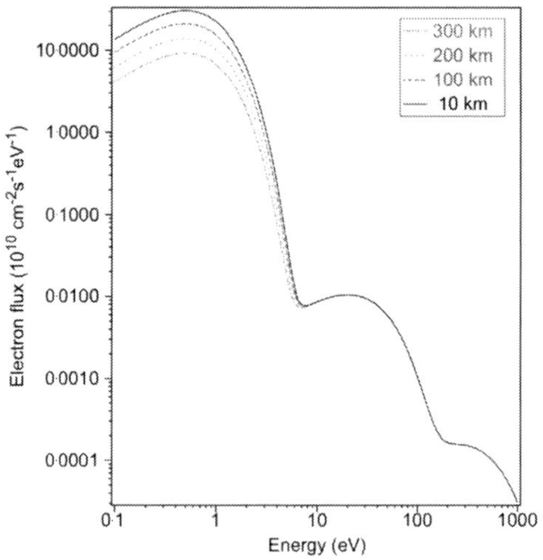

Figure 13. Electron flux spectrum for Europa, at (— · —) 300 km, (· · · · · ·) 200km, (- - - -) 100 km and (———) 10 km.

species below 150 km [45]. As new measurements [41] for electron impact excitation of the higher-energy states of O_2 had become available, a study [46] was made of such excitation in the atmosphere of Europa. The new measurements were for electron impact excitation of the Schumann-Runge (SR) continuum, longest band (LB) and second band (SB) of O_2. These ICSs are shown in Fig. 12 along with BEf-scaled theoretical calculations [41] and earlier measurements [42, 43]. In the calculations the BEf-scaled cross sections were used for the SR continuum and SB, but were adjusted upwards to fit the measurements for the LB.

The electron fluxes used for Europa were sums of 3 Maxwell-Boltzmann distributions, as illustrated in Fig. 13. As described by Saur et al. [47] and Hall et al. [44], electrons in the magnetosphere of Jupiter at the orbit of Europa can be modeled as the sum of two Maxwell-Boltzmann distributions with characteristic temperatures of 20 and 250 eV and densities of 38 and 2 electrons/cm^3 respectively. As only half of this flux impacts on the atmosphere (with the other half shielded by the planet) the quoted densities were halved in the conversion to an isotropic flux spectrum. In addition, ionisation of O_2 produces a distribution of secondary electrons in the atmosphere which is assumed to be isotropic and has a temperature of 0.5 eV [47]. The density of these secondary electrons is given by a fit [48] to experimental measurements of electron density. The fluxes deduced for the sum of these three distributions are plotted for various heights. The reduction in the low-energy distribution with increasing height is a result of the decreasing O_2 density with height, although the change in electron (i.e. O_2^+) density is much less than that of O_2, presumably due to ambipolar diffusion of O_2^+ and electrons upwards.

Most of the excitation of the SR continuum, LB and SB results in predissociation, generally including one excited O(^{1}D) atom [49] which radiatively decays to produce a 630.0-nm photon.

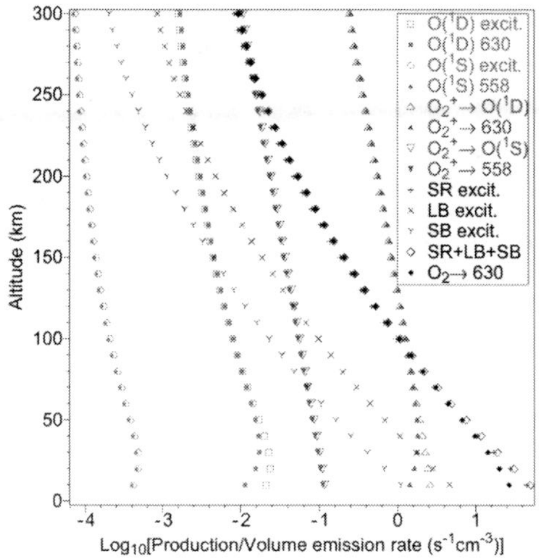

Figure 14. (Open symbols) production rates by excitation or recombination and (solid symbols) VERs for 630.0 and 557.7 nm, as a function of altitude at Europa.

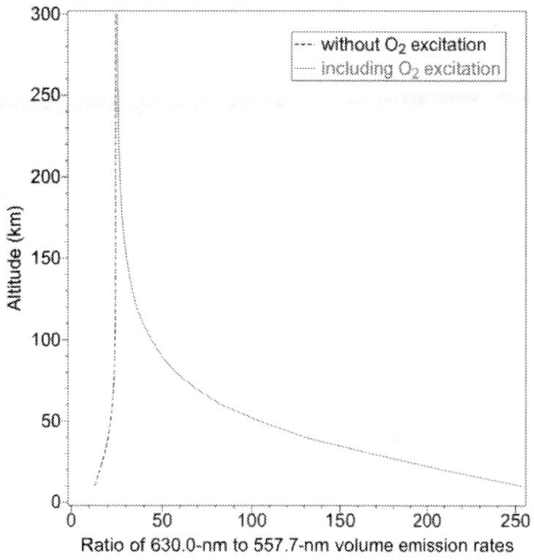

Figure 15. Intensity ratio of 630.0-nm to 557.7-nm VERs versus altitude, (——) including and (- - - -) without electron-impact excitation of O_2.

The other excitation processes (direct electron impact on O atoms and recombination of O_2^+) produce both $O(^1D)$ and $O(^1S)$ atoms and hence both 630.0-nm and, from $O(^1S)\rightarrow O(^1D)$, 557.7-nm radiation . The results of a statistical-equilibrium calculation of these three excitation processes are shown in Fig. 14 as production and VER rates as a function of altitude. We find that 630.0-nm emissions due to excitation of the SR, LB and SB dominate below 90 km.

As the higher-energy states (SR, LB and SB) of O_2 can only be excited by the magnetospheric electrons (above ~8 eV) and as they produce only 630-nm photos, the ratio of 630.0-nm to 557.7-nm radiation will be sensitive to the details of the different electron processes. This ratio is shown in Fig. 15, calculated both with and without the contribution from the O_2 excitation. The ratio is much larger when the contribution from O_2 excitation is included, thus it is a possible indicator for remote sensing of the plasma conditions within the atmosphere of Europa.

4. Conclusions

Examples of calculations of electron impact excitation have been presented, for electron cooling or prediction of emissions in the upper atmospheres of Mars, Venus and Europa and the coma of comet Hale-Bopp. In each case the calculations used recent measurements of accurate cross sections for electron impact excitation and the assembly from multiple sources of a model of the plasma environment. These studies produced significant results, such as a possible explanation for very low neutral temperatures in the atmosphere of Mars, predictions of infrared emissions from CO in the atmosphere of Venus, a reduction of 40% in the estimated abundance of CO in comet Hale-Bopp and identification of a possible indicator for remote sensing of electron parameters in the atmosphere of Europa.

Acknowledgments

This work was supported by the Australian Research Council through its Centres of Excellence Program. One of us (OI) thanks Flinders Uni. for his Visiting Research Fellowship.

5. References

[1] Stubbe P and Varnum W S 1972 *Planet. Space Sci.* **20** 1121–1126
[2] Cartwright D C, Trajmar S and Williams W 1971 *J. Geophys. Res.* **76** 8368–8377
[3] Pavlov A V 1998 *Ann. Geophys.* **16** 176–182
[4] Jefferies J T 1968 *Spectral Line Formation* (Waltham, MA USA: Blaisdell)
[5] Cartwright D C, Brunger M J, Campbell L, Mojarrabi B and Teubner P J O 2000 *J. Geophys. Res.* **105** 20857–20867
[6] Morrison M A and Greene A E 1978 *J. Geophys. Res.* **83** 1172–1174
[7] Campbell L, Brunger M J and Rescigno T N 2008 *J. Geophys. Res.* **113** E08008
[8] Kitajima M, Watanabe S, Tanaka H, Takekawa M, Kimura M and Itikawa Y 2001 *J. Phys. B* **34** 1929–1940
[9] Allan M 2001 *Phys. Rev. Lett.* **87** 033201
[10] Allan M 2002 *J. Phys. B: At. Mol. Opt. Phys.* **35** L387–L395
[11] McCurdy C W, Isaacs W A, Meyer H D and Rescigno T N 2003 *Phys. Rev. A* **67** 042708
[12] Brunger M J, Buckman S J and Elford M T 2003 *Photon and Electron Interactions with Atoms, Molecules and Ions* vol 17/C, chapter 6 ed Itikawa Y (Berlin: Landolt-Börnstein, Springer) pp 118–210
[13] Danner D 1970 *Diplomarbeit* (Germany: University of Freiburg) unpublished
[14] Register D F, Nishimura H and Trajmar S 1980 *J. Phys. B: At. Mol. Phys.* **13** 1651
[15] Antoni T, Jung K, Ehrhardt H and Chang E S 1986 *J. Phys. B: At. Mol. Opt. Phys.* **19** 1377
[16] Lowke J J, Phelps A V and Irwin B W 1973 *J. Appl. Phys.* **44** 4664–4671
[17] Bulos B R and Phelps A V 1976 *Phys. Rev. A* **14** 615–629
[18] Choi Y W, Kim J, Min K W, Nagy A F and Oyama K I 1998 *Geophys. Res. Lett.* **25** 2753–2756
[19] Hanson W B and Mantas G P 1988 *J. Geophys. Res.* **93** 7538–7544
[20] Martinis C R, Wilson J K and Mendillo M J 2003 *J. Geophys. Res.* **108** 1383–1388
[21] Nier A O and McElroy M B 1977 *J. Geophys. Res.* **82** 4341–4349
[22] Forget F *et al European Mars Science and Exploration Conference: Mars Express & ExoMars, ESTEC, The Netherlands, 12–16 November, 2007*
[23] Chen R H, Cravens T E and Nagy A F 1978 *J. Geophys. Res.* **83** 3871–3876
[24] Allan M 2010 *Phys. Rev. A* **81** 042706
[25] Poparić G B, Belić D S and Vićić M D 2006 *Phys Rev A* **73** 062713
[26] Gibson J C, Morgan L A, Gulley R J, Brunger M J, Bundschu C T and Buckman S J 1996 *J Phys B: At Mol Opt Phys* **29** 3197–3214
[27] Campbell L and Brunger M J 2008 *PMC Phys. B* **1** 3
[28] Campbell L, Allan M and Brunger M J 2011 *J. Geophys. Res.* **116** A09321
[29] Gérard J C, Huber B, Shematovich V I, Bisikalo D V and Gladstone G R 2008 *Planet. Space Sci.* **56** 542–552
[30] Spenner K, Knudsen W C and Lotze W 1996 *J. Geophys. Res.* **101** 4557–4563
[31] Crovisier J, Lellouch E, de Bergh C, Maillard J P, Lutz B L and Bézard B 2006 *Planet. Space Sci.* **54** 1398–1414
[32] Gronoff G, Lilensten J, Simon C, Barthélemy M, Leblanc F and Dutuit O 2008 *Astron. Astrophys.* **482** 1015–1029
[33] McPhate J B, Feldman P D, McCandliss S R and Burgh E B 1999 *Astrophys. J.* **521** 920–927
[34] Kato H, Kawahara H, Hoshino M, Tanaka H, Brunger M J and Kim Y K 2007 *J. Chem. Phys.* **126** 064307
[35] Campbell L and Brunger M J 2009 *Geophys. Res. Lett.* **36** L03101
[36] Bhardwaj A, Haider S A and Singhal R P 1990 *Icarus* **85** 216–228
[37] Lovell A J, Kallivayalil N, Schloerb F P, Combi M R, Hansen K C and Gombosi T I 2004 *Astrophys. J.* **613** 615–621
[38] Laher R R and Gilmore F R 1990 *J. Phys. Chem. Ref. Data* **19** 277–304
[39] Zatsarinny O and Tayal S S 2002 *J. Phys. B: At. Mol. Opt. Phys.* **35** 241–253
[40] http://physics.nist.gov/asd3
[41] Suzuki D, Kato H, Ohkawa M, Anzai K, Tanaka H, Limão-Vieira P, Campbell L and Brunger M J 2011 *J. Chem. Phys.* **134** 064311
[42] Shyn T W, Sweeney C J and Grafe A 1994 *Phys. Rev. A* **49** 3680–3684
[43] Shyn T W, Sweeney C J, Grafe A and Sharp W E 1994 *Phys. Rev. A* **50** 4794–4801
[44] Hall D T, Strobel D F, Feldman P D, McGrath M A and Weaver H A 1995 *Nature* **373** 677
[45] Smyth W H and Marconi M L 2006 *Icarus* **181** 510–526
[46] Campbell L, Tanaka H, Kato H, Jayaraman S and Brunger M J 2012 *Eur. Phys. J. D* **66** 26
[47] Saur J, Strobel D F and Neubauer F M 1998 *J. Geophys. Res.* **103** 19947–19962
[48] Kliore A J, Hinson D P, Flasar F M, Nagy A F and Cravens T E 1997 *Science* **277** 355–358
[49] Gibson S T and Lewis B R 1996 *J. Elect. Spectrosc. Rel. Phen.* **80** 9–12

Journal of Physics: Conference Series **399** (2012) 012019 doi:10.1088/1742-6596/399/1/012019

Plasma spectroscopy in the conditions of the ITER tokamak

J Rosato[1], Y Marandet[1], V Kotov[2], D Reiter[2], H Capes[1], L Godbert-Mouret[1], M Koubiti[1], R. Hammami[1] and R Stamm[1]

[1]Laboratoire PIIM, UMR 7345 Université d'Aix-Marseille / CNRS, Centre de Saint-Jérôme, Case 232, F-13397 Marseille Cedex 20, France

[2]IEK-4, Forschungszentrum Jülich GmbH, Trilateral Euregio Cluster, Euratom Association, D-52425 Jülich, Germany

E-mail: joel.rosato@univ-amu.fr

Abstract. The state of art of the line shape modeling techniques involved in tokamak edge plasma spectroscopy is reported, in the context of the preparation for ITER. Hydrogen spectra are calculated assuming a line-of-sight crossing a 2D-plasma background obtained from numerical simulations. The Doppler, Zeeman and Stark effects are retained. The possibility for a line shape-based diagnostic of the ITER divertor plasma is examined through fittings of simulated spectra and comparison with the input plasma fields. The role of turbulent fluctuations on line shapes is also examined.

1. Introduction

ITER is an international project aimed at demonstrating the principle of producing more energy from the fusion process than is used to initiate it. The tokamak, presently under construction in Cadarache (France), will be equipped with an extensive set of plasma diagnostics [1]. Passive spectroscopy of the hydrogen Balmer lines with a low upper principal quantum number n (in particular, the Hα transition) is presently considered to measure fluxes of the hydrogen isotopes and to determine the isotopic proportion of the fuelling gas (N_T/N_D,N_H/N_D) in the divertor region. Furthermore, information on the electron density N_e will be available through the analysis of the Stark broadening of high-n lines. Such lines are already observed in present divertor plasmas in the so-called "detached regime", e.g. in JET [2] and Alcator C-Mod [3]. A well-known difficulty that occurs in the interpretation of an observed spectrum is provided by the non-uniformity of the plasma along the line-of-sight. Recently, this issue has been investigated numerically for plasma conditions relevant to ITER, by using a line shape code and the transport code B2-EIRENE [4]. The possibility for a line shape-based diagnostic has been demonstrated, at least in one typical plasma background [5,6]. An additional difficulty in the interpretation of spectra is provided by the presence of turbulent fluctuations. Efforts have been undertaken in recent years in order to retain low-frequency fluctuations on line shapes [7–9]. In this work, we report on the modeling techniques involved in tokamak edge plasma spectroscopy with a special emphasis on hydrogen lines. Section 2 gives an overview on the physical mechanisms leading to line broadening. Next, we simulate spectra resulting from a line-of-sight crossing the ITER divertor (Sec. 3) and we examine the possibility for a diagnostic by performing fittings of the simulated spectra to recover plasma parameters from the line shape (Sec. 4). An investigation of the role of turbulent fluctuations on spectral lines is presented in Sec. 5.

2. Line broadening mechanisms

The line broadening mechanisms are due to the physical processes that affect the emitting atomic species at the microscopic scale. These processes include the perturbation of the energy levels induced by the plasma's microscopic electric field (Stark effect) and the magnetic field (Zeeman effect), and the thermal motion of the emitters (Doppler effect). The common description of spectral line shapes relies on a quantum statistical formalism [10]. In the atom's frame of reference, a line shape can be written as the Fourier transform of the dipole autocorrelation function

$$I_0(\omega) = \frac{1}{\pi}\mathrm{Re}\int_0^{\infty} dt \mathrm{Tr}\{\rho \vec{d}_\perp \cdot U^+(t)\vec{d}_\perp U(t)\} e^{i\omega t}. \quad (1)$$

Here, ρ is the atomic density operator, $\vec{d}_\perp$ is the dipole projected onto the plane perpendicular to the radiation's propagation direction, the trace is performed over the atomic states of the upper level of the transition, the brackets {...} denote a classical average over the perturber trajectories (semi-classical approximation), and $U(t)$ is the evolution operator. It obeys the time-dependent Schrödinger equation

$$i\hbar\frac{dU}{dt}(t) = (H_0 + V(t))U(t), \quad (2)$$

where $H_0 = H_{\mathrm{at}} - \vec{\mu}\cdot\vec{B}$ is the Hamiltonian including both the atomic energy level structure and the Zeeman effect, and $V(t) = -\vec{d}\cdot\vec{E}(t)$ is the time-dependent Stark effect term resulting from the action of the microscopic electric field $\vec{E}(t)$. The line shape is the laboratory frame (i.e., accounting for the Doppler effect) is obtained by convolution with the atoms' velocity distribution function $f(\vec{v})$:

$$I(\omega) = \int d^3v f(\vec{v}) I_0(\omega - \omega_0 \vec{v}\cdot\vec{n}/c). \quad (3)$$

Here, ω_0 stands for the central frequency and $\vec{n}$ is a unitary vector along the propagation direction.

In tokamak edge plasmas, the typical densities and temperatures are such that the contribution of the electrons to Stark broadening can generally be described with an impact collision operator. Namely, for most of the observed Balmer lines, one has $\rho_{We} << r_0$ and $\tau_{ce} << \tau_i$ where ρ_{We}, r_0, τ_{ce} and τ_i are respectively the electron Weisskopf radius, the mean interparticle distance, the electron collision time and the time of interest at half-maximum (of the Doppler-free profile I_0). This amounts to performing the substitution $V(t) \to -iK_e - \vec{d}\cdot\vec{E}_i(t)$ where $\vec{E}_i(t)$ is the microfield due to the ions only and K_e is a constant operator denoting the electronic contribution, described as a series of short-time binary interactions. The impact approximation ceases to be valid only for very high-n lines, which will not be considered hereafter.

The ion contribution to Stark broadening is strongly dependent on the upper principal quantum number of the line under consideration. The Stark broadening is generally affected by the ion dynamics, which leads to a much broader and less structured line shape than expected from the quasi-static theory (which assumes $\vec{E}_i(t) \approx \vec{E}_i(0)$). The latter is only valid for high-n lines (typically $n >$ 10). In the following, we shall retain the ion dynamics on the low-n lines by using the simulation method, which consists of solving the time-dependent Schrödinger equation numerically with an electric field obtained from particle simulation [11–13]. The code used here has been designed essentially for weakly-coupled magnetized hydrogen plasmas and can be applied to tokamak divertor conditions [14]. The ions are described with a quasi-particle model, and the electron contribution is retained in the Schrödinger equation with the Griem-Kolb-Shen collision operator model [15].

In ITER, a correct estimation of the Stark effect will be critical, because the latter at high density can be of the same order as the Doppler broadening (which is traditionally assumed to be dominant in the case of low-n Balmer lines). This point is illustrated in Fig. 1. The Dα line profile obtained using various degrees of accuracy in the line broadening description is plotted at $N_e = 3.5\times10^{15}$ cm^{-3}, $T_e = T_i = T_{\mathrm{at}} = 1$ eV, B = 8 T and in perpendicular observation. The line has a Lorentz triplet structure due to the magnetic field which, as shown in the figure, for each component has a width at half maximum about two times larger when the Stark effect is retained. For higher principal quantum number, the Stark broadening becomes more important and tends to dominate the Zeeman splitting (Fig. 2).

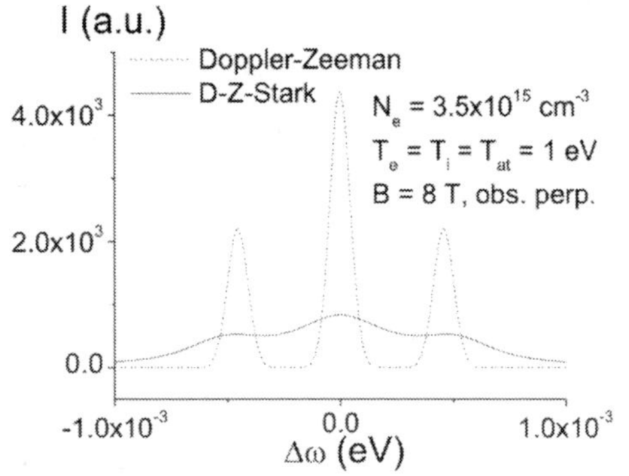

Figure 1. The contribution of the Stark effect to the line broadening is important on Dα and generally dominant on the other lines of the Balmer series. Here, plot of Dα using different degrees of accuracy on the line shape description.

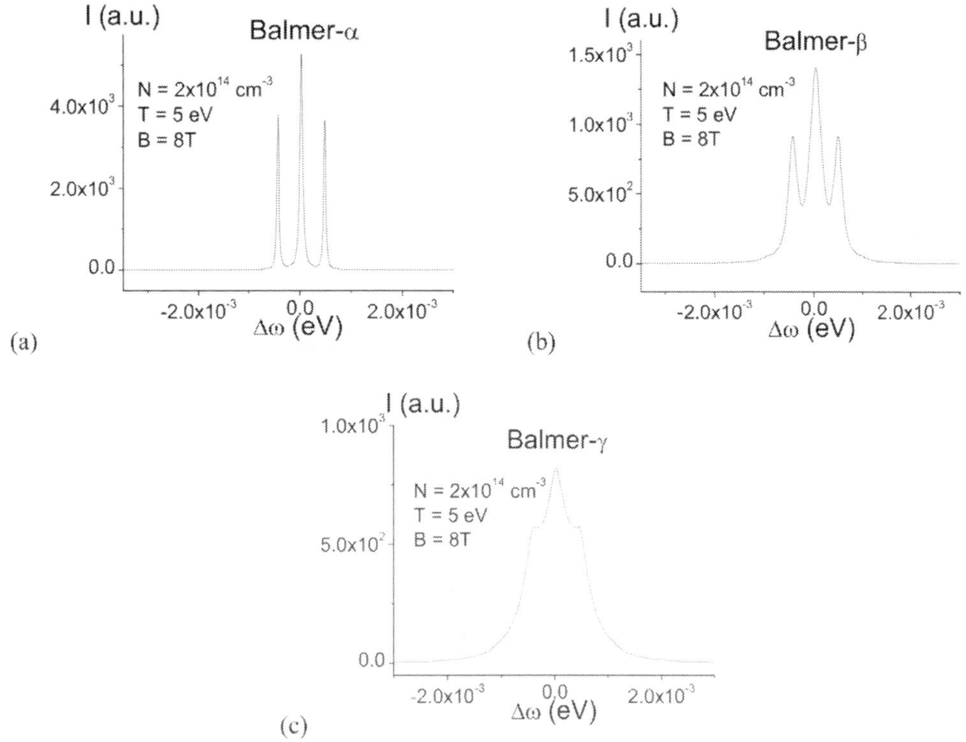

Figure 2. Plot of line shapes with increasing upper principal quantum number: (a) Balmer-α (n = 3), (b) Balmer-β (n = 4), (c) Balmer-γ (n = 5), in the atom's frame of reference, i.e. without Doppler effect. The Stark broadening increases with the principal quantum number and tends to dominate the Zeeman effect.

3. Simulation of an observed spectrum

In order to investigate the role of plasma non-uniformity along the line-of-sight, we have to consider a plasma background relevant to the divertor region and we have to perform a spatial integration. We have done so by using the transport code B2-EIRENE, which is extensively used for design relevant modeling of the ITER edge and divertor plasma [4]. Basically, it is a package formed by a Monte-Carlo solver for kinetic transport (EIRENE) and a two-dimensional plasma fluid code (B2). The EIRENE part solves a linear kinetic Boltzmann equation for the neutrals (atoms, molecules and also photons of the Lyman series) assuming a given plasma background and constructs, by Monte-Carlo method, quantities of interest like spatial distributions of reaction rates and energy and momentum sinks or sources. The presence of various energy levels is retained through the use of an equilibrium collisional-radiative model, accounting for both atoms and molecules. The obtained quantities of interest (e.g. particle sources for charged species) are then used in B2 as source and loss terms so that the fluid equations for charged particles can be solved. Obviously, the system formed by the neutral gas and the plasma is nonlinear, so that the solution is obtained by iterations. In general, the result of B2-EIRENE simulations leads to divertor plasma in recombining regime close to the target plates (see Fig. 3), with a high electron density (up to 3×10^{15} cm^{-3}) and a low temperature (down to 1 eV and smaller). The corresponding atomic density (hence, the emissivity) is concentrated in this zone. The dotted line in Fig. 3 denotes the line-of-sight considered in the present work.

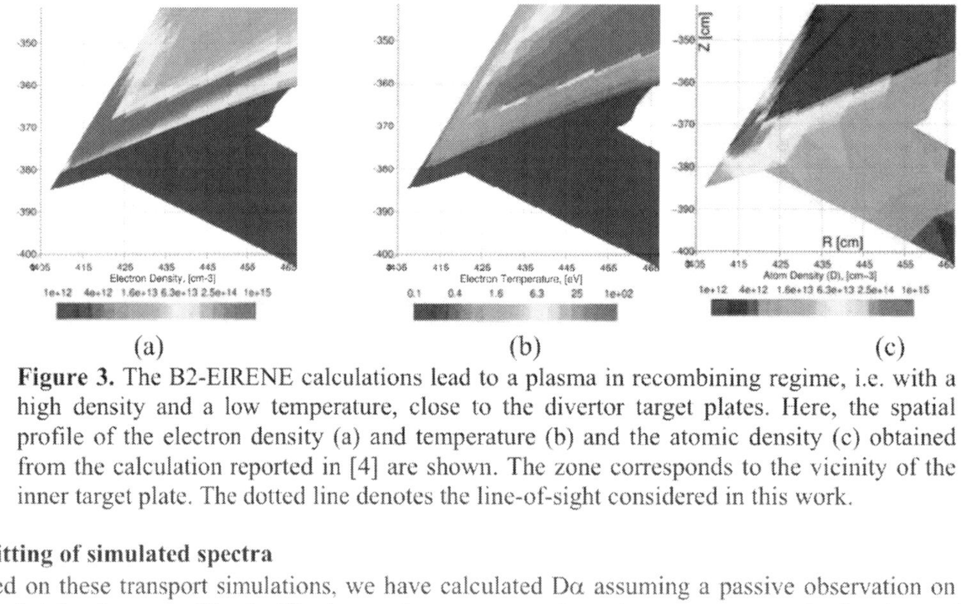

(a) (b) (c)

Figure 3. The B2-EIRENE calculations lead to a plasma in recombining regime, i.e. with a high density and a low temperature, close to the divertor target plates. Here, the spatial profile of the electron density (a) and temperature (b) and the atomic density (c) obtained from the calculation reported in [4] are shown. The zone corresponds to the vicinity of the inner target plate. The dotted line denotes the line-of-sight considered in this work.

4. Fitting of simulated spectra

Based on these transport simulations, we have calculated Dα assuming a passive observation on the line-of-sight shown in Fig. 3. The interesting property of passive spectroscopy is that it provides information on the plasma without perturbing it, provided that a suitable line shape model is employed. Here we present an analysis of Dα using a three-Voigt function model, accounting for Zeeman, Doppler and Stark effects. This model contains four independent parameters, namely the separation between the three components, the Gaussian width, the Lorentzian width and the relative amplitude of the lateral components. We have used a genetic algorithm routine [16]. The obtained result (Fig. 4) indicates that the fitted spectrum is in a quite good agreement with the "observed" spectrum from our modeling. The inferred atomic temperature extracted from the Gaussian width is about 1 eV, which correlates well with the temperature at the densest location on the line-of-sight, i.e. within about 10%. The magnetic field and angle of observation $\theta = (\vec{B},\vec{n})$ extracted from the fitting analysis are 8.4 T and 77°, respectively, which also roughly corresponds to the correct local value at the emission zone, within 10%. As concerning the Lorentzian width, it should be noticed that its interpretation using a Stark model is not obvious, since, as mentioned above, an analytical model cannot be used in general. The use of the impact approximation leads to underestimating the density by one order of magnitude, whereas, on the other hand, the quasi-static approximation provides the correct order of magnitude, but with a discrepancy still of about 50%. The situation is quite different for the lines with a high upper principal quantum number observed in detached regime, whose Stark analysis provides the density with a good accuracy. For example, an analysis of a simulated profile of D_{10} (deuterium Balmer 10 line, $n = 10 \rightarrow 2$ transition) done with the semi-empirical formula $\Delta\omega_{1/2} = C_1N_e + C_2N_e^{2/3}$ (here: $\Delta\omega_{1/2}$ is the half-width at half-maximum, the C_1 and C_2 constants which correspond to the electrons' and the ions' contribution to the Stark broadening, respectively) has provided a density of 3.4×10^{15} cm^{-3}, which correlates well with the densest location on the line-of-sight (Fig. 5).

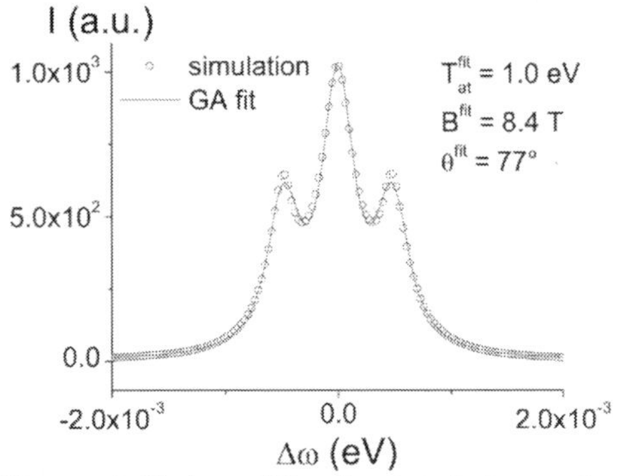

Figure 4. Fitting of the simulated observation (circles) by a three-Voigt function model (line). The fitted model-spectrum is in a good agreement with the simulated one, and the obtained plasma parameters T_{at}^{fit}, B^{fit}, θ^{fit} correspond closely to their value at the densest location on the line-of-sight.

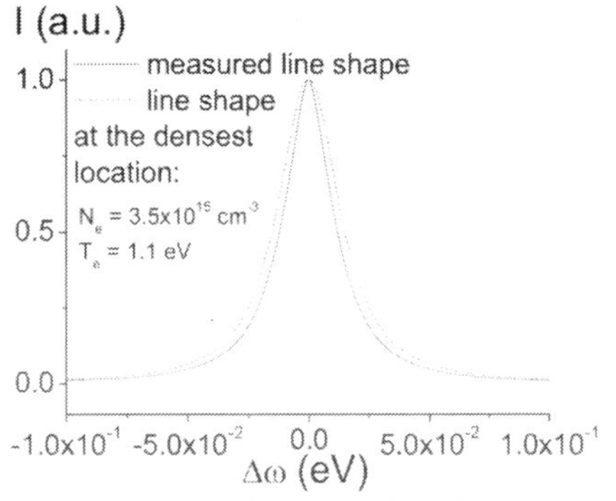

Figure 5. Plot of the D_{10} line shape. Here the measured line shape, i.e. resulting from integration along the line of sight, is close to that at the densest location.

5. Line shapes and turbulence

A major issue in magnetic fusion concerns the control of the instabilities induced by plasma edge turbulence. Recent works have been devoted to the description of passive spectroscopy signals in the presence of low-frequency fluctuations [7–9], aiming at providing a non-intrusive diagnostic capability or for tests of the spectroscopy models neglecting turbulence. The formalism developed in these investigations relies on a statistical modeling of the fluctuations along the line-of-sight. A measured line shape is described with the following integral:

$$I_{\text{mes}}(\omega) = \int dXW(X)B(X)I(\omega, X). \tag{4}$$

Here $X = (N_e, T_e\ldots)$ stands for the fluctuating parameters, $W(X)$ denotes their joint probability density function (PDF), $B(X)$ is the line brightness defined in such a way that $I(\omega,X)$ is area-normalized. This quantity is usually obtained from a collisional-radiative model. As an example, we consider here the D_{10} line, in the presence of density fluctuations. The shape of this line is governed by the Stark effect and therefore is strongly dependent on the density. The brightness is roughly proportional to N^{α} with α close to unity. Figure 6 shows a plot of D_{10}, obtained assuming a Gamma PDF for the density fluctuations with $\langle N_e\rangle = 10^{13}$ cm^{-3} and $\Delta N_e/\langle N_e\rangle = 50\%$ (with ΔN_e being the root mean square). Such a large fluctuation rate has already been detected in edge plasmas [17]. The temperatures T_e, T_i have been set equal to 1 eV. As can be seen, the density fluctuations lead to a weak broadening of the line shape, by less than 20%. This result would suggest that passive spectroscopy diagnostics based on Stark broadening are robust, at least for moderate fluctuation rate. A much stronger deviation is expected in the case of electron temperature fluctuations, because the brightness has a strong, non-monotonic dependence on this parameter [8]. To illustrate it we consider a spectral line broadened due to the Doppler effect only, e.g. Dα at low-density conditions, and assume correlated fluctuations of T_e and T_{at}. We focus on one component of the Zeeman Lorentz triplet and we note $T_e = T_{at} \equiv$ T. Figure 7 shows a profile of Dα with 50% of temperature fluctuations compared to the same profile without fluctuations, (a) for $\langle T\rangle = 20$ eV and (b) for $\langle T\rangle = 1$ eV, obtained with a Gamma PDF for the temperature. The brightness has been obtained from the collisional-radiative code SOPHIA [18]. The line width is strongly sensitive on the average temperature, in particular on its relative value with respect to the brightness' maximum value. The line shape is narrowed if $\langle T\rangle$ is significantly larger than the brightness maximum [case (a)]. This strong dependence on the average temperature is of interest for diagnostics. Note, the system considered here was ideal, with only one fluctuating parameter; a complementary work on the case where several parameters fluctuate at the same time should also be carried out.

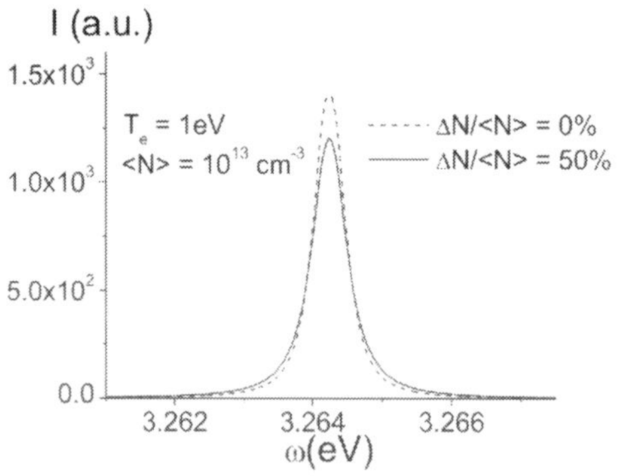

Figure 6. Profile of D_{10} with 50% of density fluctuations compared to the same profile without fluctuations, with $\langle N_e\rangle = 10^{13}$ cm^{-3} and $T_e = T_i = 1$ eV. In this particular case, the spectrum is scarcely affected by the fluctuations.

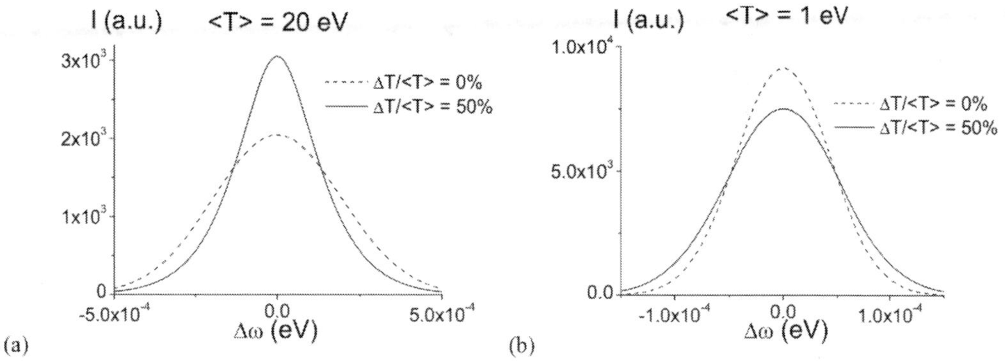

Figure 7. Profile of Dα with 50% of temperature fluctuations compared to the same profile without fluctuations, (a) for <T> = 20 eV and (b) for <T> = 1 eV. The width is strongly sensitive on the average temperature. In the case (a), the fluctuations lead to a narrowing of the profile.

6. Conclusion

The operation of ITER will require passive spectroscopy diagnostics in the edge plasma regions. In this work, we have assessed Balmer lines by using a line shape code and a plasma edge code. As a result, it is shown that low-*n* lines like Hα can be affected by Stark broadening significantly. In general, the Stark effect cannot be described by a simple analytical model. In spite of this, the use of a fitting routine with a three-Voigt function model could permit an extraction of the atomic temperature in a localized region with a quite good accuracy. In a similar way, it has been shown that local information on the electron density can also be obtained from analysis of the width of high-*n* lines. The role of turbulent fluctuations on line broadening has been examined through ideal cases. In particular, it has been shown that temperature fluctuations may lead to a significant modification of Doppler line shapes. The work presented in this paper provides additional information on what can be extracted from passive spectroscopy analysis in ITER. Further work could consist in similar investigations in pure helium discharges. The line radiation due to the impurities (viz. Be, C, W) should be also addressed, both regarding the shapes and the intensities, in equilibrium or in the presence of fluctuations (as done for lithium in [19,20]). Another issue of interest concerns the modeling of resonance lines, affected by opacity effects, in the presence of turbulence. Preliminary investigations have indicated a strong correlation between opaque lines and turbulence properties that could be used for diagnostics [21,22].

Acknowledgements

This work was carried out within the framework of the European Fusion Development Agreement and the French Research Federation for Fusion Studies. It is supported by the European Communities under the contract of Association between Euratom and CEA. The views and opinions expressed herein do not necessarily reflect those of the European Commission. Financial support was also received from the Agence Nationale de la Recherche, project ANR-11-BS09-023 (SEDIBA).

References

[1] Donné A J H *et al* 2007 *Nucl. Fusion* **47** S337–84
[2] Koubiti M, Loch S, Capes H, Godbert-Mouret L, Marandet Y, Meigs A, Stamm R and Summers H 2003 *J. Quant. Spectrosc. Radiat. Transfer* **81** 265–73
[3] Welch B L, Griem H R, Terry J, Kurz C, LaBombard B, Lipschultz B, Marmar E and McCracken G 1995 *Phys. Plasmas* **2** 4246–51
[4] Kotov V, Reiter D and Kukushkin A S 2007 *Forschungszentrum Jülich Report* Jül-4257
[5] Rosato J, Kotov V and Reiter D 2010 *J. Phys. B: At. Mol. Opt. Phys.* **43** 144024

[6] Rosato J, Kotov V, Reiter D, Capes H, Godbert-Mouret L, Koubiti M, Marandet Y and Stamm 2010 *AIP Conf. Proc.* **1290** 63–7
[7] Marandet Y, Capes H, Godbert-Mouret L, Koubiti M and Stamm R 2005 *Europhys. Lett.* **69** 531–7
[8] Rosato J, Capes H, Marandet Y, Rosmej F B, Stamm R, Godbert-Mouret L and Koubiti M 2008 *Europhys. Lett.* **84** 43002
[9] Marandet Y, Rosato J, Capes H, Godbert-Mouret L, Koubiti M, Mekkaoui A, Rosmej F B and Stamm R 2009 *High Energ. Dens. Phys.* **5** 312–9
[10] Griem H R 1974 *Spectral Line Broadening by Plasmas* (New York: Academic)
[11] Stamm R, Smith E W and Talin B 1984 *Phys. Rev. A* **30** 2039–46
[12] Gigosos M A, Fraile J and Torres F 1985 *Phys. Rev. A* **31** 3509–11
[13] Stambulchik E, Alexiou S, Griem H R and Kepple P C 2007 *Phys. Rev. E* **75** 016401
[14] Rosato J, Marandet Y, Capes H, Ferri S, Mossé C, Godbert-Mouret L, Koubiti M and Stamm R 2009 *Phys. Rev. E* **79** 046408
[15] Griem H R, Kolb A C and Shen K Y 1959 *Phys. Rev.* **116** 4–16
[16] Charbonneau P 1995 *Astrophys. J. Suppl. Ser.* **101** 309–34
[17] Boedo J A 2009 *J. Nucl. Mater.* **390–391** 29–37
[18] Rosmej F B, Stamm R, Fritzsche S, Capes H, Koubiti M, Marandet Y, Lisitsa V S, Ohno N, Takamura S and Nishijima D 2005 *J. Nucl. Mater.* **337–339** 1101–5
[19] Rosato J, Capes H, Catoire F, Kadomtsev M B, Levashova M G, Lisitsa V S, Marandet Y, Rosmej F B and Stamm R 2011 *J. Nucl. Mater.* **415** S617–9
[20] Rosato J, Catoire F, Marandet Y, Mekkaoui A, Capes H, Koubiti M, Stamm R, Kadomtsev M B, Levashova M G, Lisitsa V S and Rosmej F B 2011 *Phys. Lett. A* **375** 4187–9
[21] Rosato J, Reiter D, Marandet Y and Mekkaoui A 2009 *ECA* **33E** P5.155
[22] Rosato J, Mekkaoui A, Marandet Y, Reiter D, Kotov V, Capes H, Godbert-Mouret L, Koubiti M and Stamm R 2012 *Contrib. Plasma Phys.* **52** 429–34

Journal of Physics: Conference Series **399** (2012) 012020 doi:10.1088/1742-6596/399/1/012020

Gyrokinetic simulation studies for non-axisymmetric plasma confinement: turbulent transport and entropy transfer

T H Watanabe[1,2], H Sugama[1,2], M Nunami[1,2], and M Nakata[3]

[1]National Institute for Fusion Science, Toki 509-5292, Japan
[2]The Graduate University for Advanced Studies, Toki 509-5292, Japan
[3]Japan Atomic Energy Agency, Rokkasho 039-3212, Japan

E-mail: `watanabe.tomohiko@nifs.ac.jp`

Abstract. Turbulent ion heat transport and entropy transfer in non-axisymmetric toroidal plasma confinement are investigated by means of gyrokinetic simulations. Elongation of fluctuation spectrum into the radial wavenumber space is observed in the ion temperature gradient (ITG) turbulence simulation for a helical plasma configuration reconstructed from experimental data, and is examined by the entropy transfer function. For studying turbulence and zonal flows in a poloidally-rotating helical plasma, the conventional flux tube simulation model is extended to a bundle of flux tubes, where a multi-scale model of zonal flows and turbulence is formulated and tested by simulations of the linear ITG instability and the zonal flow response.

1. Introduction

In fusion, space, and astrophysical plasmas with a high temperature, the conventional fluid approximation is considered to be invalid, because a mean-free-path of particles is often comparable to or longer than the size of interest. Kinetic approaches dealing with the particle distribution function in the phase space are indispensable to theoretical and numerical studies of the collisionless or weakly collisional plasmas. For studying a magnetized plasma turbulence, the nonlinear gyrokinetic equations [1, 2] are derived from the Vlasov equation, where the phase space dimension is reduced by the gyro-phase average while preserving the spatial and time resolutions for drift waves.

Gyrokinetic simulations of drift wave turbulence in fusion plasmas have made a large progress in the last two decades [3], where interaction of ion temperature gradient (ITG) turbulence and zonal flows [4] is one of the main issues. However, application to non-axisymmetric toroidal systems, such as the Large Helical Device (LHD) [5], still remains as one of the challenges, since the complicated equilibrium configuration demands careful numerical treatments and expensive computational costs. The ITG turbulent transport simulations with model LHD configurations have been carried out [6, 7] by means of a gyrokinetic simulation code, GKV [8], where the zonal flow enhanced in an optimized field configuration with less effective helical ripple components leads to reduction of the turbulent ion heat transport. The gyrokinetic simulation for LHD plasmas has been extended so as to incorporate the equilibrium configuration reconstructed from the experimental data [9, 10, 11]. The GKV-X code successfully reproduces the ion heat transport observed in a high ion temperature discharge on LHD [11].

In the δf approach employed in the GKV and GKV-X codes, time evolution of the perturbed distribution function δf from the background is calculated in the three-dimensional real and the two-dimensional velocity spaces, where fluctuations of δf are transferred into different spacial and velocity scales though linear and nonlinear advection terms. In the present study, we examine the nonlinear interactions of turbulence and zonal flows in the LHD configuration reconstructed from experimental data, by means of a transfer function of a quadratic integral of δf, that is, the entropy transfer function [12, 13].

It is known that the neoclassical particle transport in helical systems spontaneously generates an equilibrium scale radial electric field which drives the poloidal $\boldsymbol{E} \times \boldsymbol{B}$ rotation and improves collisionless orbits of helical-ripple-trapped particles. To incorporate the effect of poloidal plasma rotation, however, the field-line label dependence of the confinement field strength is necessary to be introduced, which demands extension of the conventional flux tube model [14]. In the later part of the paper, a novel idea for multi-scale simulations of zonal flows and turbulence will be discussed as generalization of the flux tube model.

This paper is organized as follows. The basic formulation of the δf gyrokinetic equations is explained in the next section. In section 3, results of the flux tube simulations of the ITG turbulent transport in the LHD configuration are analyzed by means of the entropy transfer function. A new multi-scale simulation model using a bundle of flux tubes for non-axisymmetric systems with the equilibrium radial electric field will be discussed in section 4, where simulation results of the linear ITG instability growth and the zonal flow response are discussed. A summary is given in the last section.

2. Basic equations

We consider the electrostatic gyrokinetic equation [1], which is represented as an inhomogeneous advection-diffusion equation for perturbed gyrocenter distribution function, δf, on the five dimensional phase space,

$$\left[\frac{\partial}{\partial t} + v_\parallel \boldsymbol{b}\cdot\nabla + \boldsymbol{v}_d\cdot\nabla - \frac{\mu}{m}(\boldsymbol{b}\cdot\nabla B)\frac{\partial}{\partial v_\parallel}\right]\delta f + \frac{c}{B_0}\{\Phi, \delta f\} = \left(\boldsymbol{v}_* - \boldsymbol{v}_d - v_\parallel \boldsymbol{b}\right)\cdot\frac{e\nabla\Phi}{T_i}F_M + C \quad (1)$$

where the parallel velocity, $v_\parallel$, and the magnetic moment, $\mu = mv_\perp^2/2B$, are chosen as the velocity coordinates. The real space (gyroceter) coordinates are defined by the minor radius (flux surface label), r, the field line lable, α, and the poloidal angle, θ. Here, $\alpha = \zeta - q\theta$ for the toroidal angle, ζ, and the safety factor, q. The equilibrium quantities and their local gradients are assumed to be constant in the vicinity of a flux surface at the minor radius $r = r_0$, where the local turbulence simulation is carried out. The abbreviations of $\boldsymbol{v}_d$, Φ, and F_M denote the magnetic drift velocity, the gyro-averaged potential, and the Maxwellian distribution function, respectively. In the numerical simulation, it is convenient to use the coordinates (x, y, z), such that $x = r - r_0$, $y = (r_0/q_0)[q(r)\theta - \zeta]$, and $z = \theta$ [14] with assumption of the constant magnetic shear, $\hat{s} = (r_0/q_0)dq/dr$, in r. The diamagnetic drift velocity $\boldsymbol{v}_*$ associated with density and temperature gradients is given as

$$\boldsymbol{v}_*\cdot\nabla = -\frac{cT_i}{eL_nB_0}\left[1 + \eta_i\left(\frac{m_iv^2}{2T_i} - \frac{3}{2}\right)\right]\frac{\partial}{\partial y}\ . \quad (2)$$

The linear model collision term is abbreviated as C. Scale lengths of the density and the ion temperature gradients are denoted by L_n and L_T, respectively, providing $\eta_i = L_n/L_T$. Other notations are standard (see Ref.[6] for more details).

3. Flux tube simulations of gyrokinetic plasma turbulence

3.1. Local flux tube model

In general toroidal confinement systems, the field-line-label dependence of the confinement field strength appears in several terms of Eq. (1), such as the magnetic drift and mirror force terms *etc.* Assumption of the toroidal axisymmetry for the confinement field is valid in a core region of tokamaks, while the non-axisymmetric effect should be taken into account in helical plasmas, such as LHD. In either case, the scale length of the confinement field strength $B(\alpha, \theta)$ is assumed to be much longer than those of turbulence. The former is characterized by R_0/M or $q_0 R_0$ in the α and θ directions, respectively, where M means the toroidal periodicity of the non-axisymmetric component of confinement field. The later is supposed to be of the order of thermal gyro-radius ρ_t. Indeed, the gyrokinetic ordering used in derivation of the gyrokinetic equations assumes the smallness parameter of $\rho_* = \rho_t/L \ll 1$ where L means an equilibrium scale length, such as the minor radius of a torus. Therefore, it is natural to assume scale separation of the turbulence and the equilibrium. Fixing the field-line-label α in $B(\alpha, \theta)$ to a constant leads to the flux tube model [14] which is a standard gyrokinetic model for studying the local turbulent transport in tokamak and helical plasmas.

The local turbulence assumption in r enables ones to assume the periodic boundary condition in the radial direction. In the local flux tube geometry, the modified periodic boundary condition for the z (θ) coordinate is applied to the turbulent fluctuations in case with the finite magnetic shear [14]. The normal periodic boundary condition is employed for the zonal flow components of which wavenumbers in the y (α) direction vanish. The locality assumption simplifies the numerical scheme for the flux tube model, where the spectral method is used for fluctuations in the x and y coordinates. Thus, the Fourier transformation is applied to Eq.(1) with a constant value of α in B, where the perpendicular wavenumber vector $\boldsymbol{k}_\perp$ is defined for the (x, y) plane.

By means of the quasi-neutrality condition, the electrostatic potential in the particle coordinates is calculated in the perpendicular wavenumber $\boldsymbol{k}_\perp$ space,

$$\int J_0 f_{\boldsymbol{k}_\perp} d^3v - \frac{e\phi_{\boldsymbol{k}_\perp}}{T_i} n_0 (1 - \Gamma_0) = n_{e,\boldsymbol{k}_\perp} , \tag{3}$$

where $f_{\boldsymbol{k}_\perp}$ denotes the Fourier component of δf. The electron density is given by the adiabatic response model $n_{e,\boldsymbol{k}_\perp}/n_0 = e\phi_{\boldsymbol{k}_\perp}/T_e$ for turbulence components. For zonal flows, no electron density perturbation arises, where no parallel electric field exists. Thus, $n_{e,\boldsymbol{k}_\perp}/n_0 = e(\phi_{\boldsymbol{k}_\perp} - \langle\phi_{\boldsymbol{k}_\perp}\rangle)\phi/T_e$. The flux surface average is denoted by

$$\langle A \rangle = \int_{-N_\theta \pi}^{+N_\theta \pi} \sqrt{g_F} A_{k_y=0} dz \Bigg/ \int_{-N_\theta \pi}^{+N_\theta \pi} \sqrt{g_F} dz \ . \tag{4}$$

The poloidal extent of the simulation domain is defined as $-N_\theta \pi \le z \le N_\theta \pi$ with a positive integer N_θ. Usually, one assumes $N_\theta = 1$. The Jacobian of the flux tube coordinates, (x, y, z), is denoted by $\sqrt{g_F}$.

The simplest flux tube coordinates are given by the circular co-centric flux surfaces with the assumption of a large aspect ratio torus, where the flux surface average is simply given with the Jacobian, $\sqrt{g_F} = 1/B(\alpha_0, z)$ (See Ref. [6, 8] for more detailed implementation of the conventional `GKV` code). In a recent version of our flux tube gyokinetic code, `GKV-X` [9, 10], the Jacobian is obtained from the `VMEC` code [15] which is an equilibrium configuration solver for non-axisymmetric systems. Then, the LHD experimental configurations reconstructed by the `VMEC` code are introduced in the `GKV-X` simulation. In the `GKV-X` code, the derivative operators in Eq.(1) are also defined with metric coefficients given by the `VMEC` equilibrium calculation, while simpler expressions are used in the conventional `GKV` code, such as $\boldsymbol{b} \cdot \nabla \delta f = (1/q_0 R_0)(\partial \delta f/\partial z)$.

Hereafter, unless it is explicitly stated, we use the normalization units of $x = x'/\rho_i$, $t = t' v_{ti}/L_n$, $v = v'/v_{ti}$, $B = B'/B_0$, and $\phi = e\phi' L_n/T_i \rho_i$, where T_i, v_{ti}, and B_0 denote the ion temperature, the ion thermal speed, and the typical magnitude of the confinement field, respectively. Prime means dimensional quantities.

3.2. Entropy balance relations in a flux tube model

Magnitude of kinetic plasma fluctuations in the phase space can be characterized by a quadratic form of the perturbed distribution function, $|\delta f|^2$. The integral of $|\delta f|^2/2F_M$ on the phase space corresponds to the second-order difference of the micro- and macroscopic entropy, $-\int f \ln f dxdv$ and $-\int F_M \ln F_M dxdv$, respectively. The entropy variable δS is defined by $\delta S = \sum_{\boldsymbol{k}_\perp} \delta S_{\boldsymbol{k}_\perp} = \sum_{\boldsymbol{k}_\perp} \langle \int (|f_{\boldsymbol{k}_\perp}|^2/2F_M) d^3v \rangle$. A balance equation for $\delta S_{\boldsymbol{k}_\perp}$ is derived from Eqs. (1) and (3) with a fixed field-line-label, $\alpha = \alpha_0$,

$$\frac{\partial}{\partial t}\left(\delta S_{\boldsymbol{k}_\perp} + W_{\boldsymbol{k}_\perp}\right) = \eta_i Q_{\boldsymbol{k}_\perp} + \mathcal{T}_{\boldsymbol{k}_\perp} + D_{\boldsymbol{k}_\perp} \ , \tag{5}$$

where $W_{\boldsymbol{k}_\perp}$, $Q_{\boldsymbol{k}_\perp}$, and $D_{\boldsymbol{k}_\perp}$ mean the potential energy, the ion heat transport flux, and the collisional dissipation, respectively. The detailed definitions are found in Refs. [13] and [16]. The entropy transfer among different turbulent and zonal flow components is represented by, $\mathcal{T}_{\boldsymbol{k}_\perp}$, which is derived from the $\boldsymbol{E} \times \boldsymbol{B}$ nonlinear term in Eq.(1) [12, 13], such that $\mathcal{T}_{\boldsymbol{k}_\perp} = \sum_{\boldsymbol{q}_\perp} \sum_{\boldsymbol{p}_\perp} \delta_{\boldsymbol{k}_\perp+\boldsymbol{p}_\perp+\boldsymbol{q}_\perp,0} \mathcal{J}\left[\boldsymbol{k}_\perp \,|\, \boldsymbol{p}_\perp, \boldsymbol{q}_\perp\right]$ with

$$\mathcal{J}\left[\boldsymbol{k}_\perp \,|\, \boldsymbol{p}_\perp, \boldsymbol{q}_\perp\right] = \left\langle \boldsymbol{b}\cdot(\boldsymbol{p}_\perp \times \boldsymbol{q}_\perp) \int d^3v \frac{1}{2F_M} \mathrm{Re}\left[\Phi_{\boldsymbol{p}_\perp} h_{\boldsymbol{q}_\perp} h_{\boldsymbol{k}_\perp} - \Phi_{\boldsymbol{q}_\perp} h_{\boldsymbol{p}_\perp} h_{\boldsymbol{k}_\perp}\right] \right\rangle \ . \tag{6}$$

Here, the non-adiabatic part of the perturbed gyrocenter distribution is given by $h_{\boldsymbol{k}_\perp} = f_{\boldsymbol{k}_\perp} + \Phi_{\boldsymbol{k}_\perp} F_M$. The entropy transfer function $\mathcal{T}_{\boldsymbol{k}_\perp}$ is regarded as a kinetic extension of the Reynolds stress [12], and is useful for analyzing interactions among turbulence and zonal flows [13]. The total entropy balance relation in a single flux tube is given by the sum of Eq.(5) over $\boldsymbol{k}_\perp$,

$$\frac{d}{dt}\left(\delta S + W\right) = \eta_i Q + D \ , \tag{7}$$

where each term stems from their counter part in Eq.(5) while $\sum_{\boldsymbol{k}_\perp} \mathcal{T}_{\boldsymbol{k}_\perp}$ vanishes by construction.

The entropy balance relations give us insights how fluctuations of δf are transferred in the phase space. One of the simple examples is the collisionless damping of zonal flows and the geodesic acoustic mode (GAM) [17]. Suppose the initial condition given by an Maxwellian distribution with the poloidal and toroidal mode numbers of $m = n = 0$ but with a finite radial wavenumber so that the zonal flow perturbation is constant in z with $k_x \neq 0$ and $k_y = 0$. As the right hand side of Eq. (7) vanishes ($Q = D = 0$), $\delta S + W$ should be conserved. The zonal flow with $m = n = 0$ couples with the $m = 1$ mode through the toroidicity, and starts to oscillate with the GAM frequency. The GAM oscillation with a $m = 1$ component is damped through the collisionless Landau damping by passing particles which generate fine-scale velocity-space structures [8]. As the GAM is damped, W decreases while δS increases. It means that the entropy is transferred from large to small velocity scales. In a long time limit, the zonal flow amplitude with $m = 0$ remains finite after the initial GAM damping [18], where a coherent structure of $f_{\boldsymbol{k}_\perp}$ associated with the neoclassical polarization is formed in the trapped particle region of the velocity space [8].

In non-axisymmetric systems, the zonal flow damping process is influenced by radial drift motion of helical ripple trapped particles which lead to additional shielding of the zonal flow potential through mixing of the perturbed distribution function [19, 20, 21]. The gyrokinetic simulations of zonal flow damping in the LHD model configuration clearly captures the modulation of $f_{\boldsymbol{k}_\perp}$ by the radial drift which also increases the entropy variable.

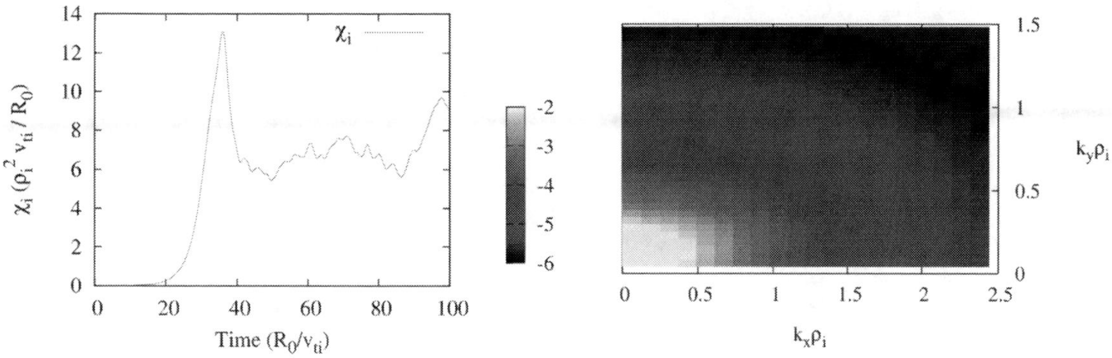

Figure 1. Results of ITG turbulence simulation for the LHD experiments #88343. (Left) Time history of the ion heat transport coefficient χ_i normalized in the gyro-Bohm unit of $\chi_{GB} = \rho_i^2 v_{ti}/R_0$. (Right) Wavenumber spectrum of the electrostatic potential fluctuations observed at $z = 0$ in the ITG turbulence, where the time average is taken for $40R_0/v_{ti} < t < 90R_0/v_{ti}$.

3.3. Turbulent entropy transfer in LHD plasma

A quantitative analysis of the nonlinear entropy transfer among zonal flows and turbulence was first carried out for the tokamak ITG and ETG (electron temperature gradient) turbulence simulations [13], where the successive entropy transfer from low to high radial wavenumber region results in the elongated spectrum of ITG turbulence fluctuations. The same analysis has also been adopted to the model LHD configurations with a simple helical confinement field given by an analytic expression [16], where the conventional `GKV` code is employed. We have been developing a new simulation code package, `GKV+`, with a modern programing structure for higher flexibility in maintenance and extension, where the full geometrical computations used in the `GKV-X` are also incorporated. Then, the entropy transfer analysis can be applied to the ITG turbulence simulations for the LHD experimental configurations.

As the initial test, the same configuration as that used in the `GKV-X` simulation is employed, that is, the `VMEC` equilibrium reconstructed from the LHD experimental data #88343 [11]. Figure 1 (left) shows the time evoluton of the ion heat transport coefficient normalized by the gyro-Bohm unit, $\chi_{GB} = \rho_i^2 v_{ti}/R_0$, where the normalized minor radius $\rho = 0.83$. For saving the computational costs, we have reduced the resolution for z, $v_\parallel$, and μ space by half. Nevertheless, the resultant transport coefficient averaged for $60L_n/v_{ti} < t < 100L_n/v_{ti}$ is $7.19\chi_{GB}$, and agrees with the results of Ref [11] within an error of 10% that is smaller than the fluctuation level of χ_i found in Fig.1 (left). Wavenumber spectrum of the electrostatic potential fluctuations is shown in Fig. 1 (right), where $|\phi_{\boldsymbol{k}_\perp}(z = 0)|^2$ averaged from $t = 40$ to $90L_n/v_{ti}$ is plotted by color contours. It is found that the fluctuation spectrum spreads into the high radial wavenumber (k_x) region in comparison to the k_y direction. This is attributed to the strong interaction of zonal flows and turbulence, where the entropy fluctuations of turbulence are transferred to small radial scales.

The entropy transfer function $\delta_{\boldsymbol{k}_\perp+\boldsymbol{p}_\perp+\boldsymbol{q}_\perp,0}\overline{\mathcal{J}}\,[\boldsymbol{p}_\perp\,|\,\boldsymbol{q}_\perp,\boldsymbol{k}_\perp]$ is plotted in Fig.2 for $\boldsymbol{p}_\perp\rho_i = (0.244, 0.127)$, where the time average $\overline{\cdots}$ is taken for $40L_n/v_{ti} < t < 90L_n/v_{ti}$. The entropy transfer to the mode with $\boldsymbol{p}_\perp\rho_i = (0.244, 0.127)$ is represented by the red color which is typically found at $\boldsymbol{q}_\perp\rho_i = (0, -0.127)$ and $(-0.244, 0)$. The wavenumbers of three modes satisfy the triad interaction condition, $\boldsymbol{k}_\perp + \boldsymbol{p}_\perp + \boldsymbol{q}_\perp = 0$. In contrast, the entropy of $\boldsymbol{p}_\perp\rho_i = (0.244, 0.127)$ mode is transferred to the higher radial wavenumber region, such as $\boldsymbol{q}_\perp\rho_i = (-0.610, -0.127)$ or higher $|q_x|$, where the transfer function has negative values colored by blue. The result confirms that the successive entropy transfer found in the previous works also acts in the LHD experimental

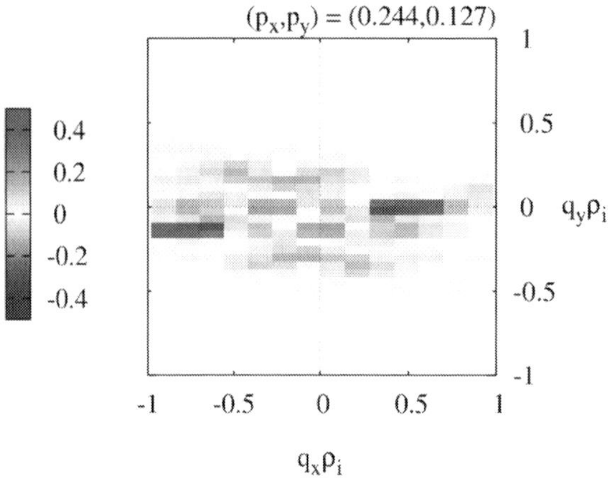

Figure 2. Entropy transfer analysis for the simulation result shown in Fig.1. Color contours of entropy transfer function $\delta_{\boldsymbol{k}_\perp+\boldsymbol{p}_\perp+\boldsymbol{q}_\perp,0}\mathcal{J}\left[\boldsymbol{p}_\perp\,|\,\boldsymbol{q}_\perp,\boldsymbol{k}_\perp\right]$ for $\boldsymbol{p}_\perp\rho_i = (0.244, 0.127)$ are plotted, where the time-averaged is taken for $40R_0/v_{ti} < t < 90R_0/v_{ti}$.

configuration analyzed here. An additional simulation carried out for $\rho = 0.46$ presents a weaker entropy transfer to high k_x region, where the wavenumber spectrum of turbulence fluctuations looks more isotropic (not shown here) [11]. The obtained results suggest that the fluctuation spectrum or the radial correlation length, as well as the zonal flow and turbulence interactions, has a flux surface (or ρ) dependence in the LHD configuration.

4. Flux-tube bundle model for non-axisymmetric systems

In the previous section, we have discussed the entropy transfer process in the local flux tube model for the LHD configuration with no equilibrium-scale radial electric field E_{r0}. In the axisymmetric system, the rigid poloidal $\boldsymbol{E}\times\boldsymbol{B}$ rotation with a small poloidal Mach number causes only the Doppler shift of mode frequencies, but no influence on the turbulent transport. In the non-axisymmetric systems, however, the poloidal rotation by E_{r0} plays a crucial role in determining the long-time response of zonal flows, which was pointed out by the gyrokinetic theory [22, 23, 24, 25]. Gyrokinetic simulations for non-axisymmetric systems have also confirmed enhancement of the residual zonal flow level due to the poloidal flow driven by the macro-scale (uniform and constant) E_{r0} [25, 26, 27, 28]. In the helical systems, thus, the turbulent transport may be influenced by the radial electric field E_{r0} spontaneously formed by the ambipolarity condition of the neoclassical particle transport, leading to coupling of the neoclassical and anomalous transport processes through zonal flows.

For the gyrokinetic simulations of helical plasma turbulence in case with E_{r0}, the conventional local flux tube model should be extended so as to introduce the field-line-label dependence of $B(\alpha,\theta)$ which appears in the magnetic drift, the mirror force, and other terms in Eq.(1). A preliminary simulation for the ITG turbulent transport in case with E_{r0} has been carried out by means of an extended `GKV` code [27], where two different scale lengths (that is, an equilibrium scale associated with helical ripples, R_0/M, and the turbulence scale of ρ_i) are simultaneously involved in a single equation. It is contrast to the conventional flux tube model, and is not preferable from a viewpoint of the gyrokientic ordering assuming separation of equilibrium and turbulence scales. To overcome limitation of the conventional methods, we have developed a novel gyrokinetic simulation model with multiple flux tubes, that is, a "flux-tube bundle" model of which concept was first discussed in Ref. [29].

4.1. Formulation of flux-tube bundle model

Let us start with the gyrokinetic equation in Eq.(1) where the explicit α dependence of B remains. We postulate the field-line-label dependence on slow (α) and fast (y) coordinates, such that $\delta f = \widehat{f}(x, \alpha, z, v_\parallel, \mu) + \widetilde{f}(x, y, z, v_\parallel, \mu; \alpha)$ and $\Phi = \widehat{\Phi}(x, \alpha, z, \mu) + \widetilde{\Phi}(x, y, z, \mu; \alpha) + \Phi_0$, where $y = (r_0/q_0)(q\theta - \zeta) = -(r_0/q_0)\alpha$ in the conventional flux tube model. The zonal flow and turbulence components are represented by $\widehat{\cdots}$ and $\widetilde{\cdots}$, respectively. The equilibrium potential is denoted by Φ_0 where $E_{r0} = -d\Phi_0/dr$. Assuming $[\![\delta f]\!]_\alpha = \widehat{f}$ and $[\![\widetilde{f}]\!]_\alpha = 0$, where $[\![\cdots]\!]_\alpha$ means the average over the y coordinate, one finds the gyrokinetic equation for the zonal flows,

$$\begin{aligned}\left[\frac{\partial}{\partial t} + v_\parallel \boldsymbol{b}\cdot\nabla + v_{dx}\frac{\partial}{\partial x} - \frac{\mu}{m}(\boldsymbol{b}\cdot\nabla B)\frac{\partial}{\partial v_\parallel} + \omega_\theta q_0 \frac{\partial}{\partial \alpha}\right]\widehat{f} \\ = \left(-v_{dx}\frac{\partial\widehat{\Phi}}{\partial x} - v_\parallel \boldsymbol{b}\cdot\nabla\widehat{\Phi}\right)\frac{e}{T_i}F_M + C(\widehat{f}) + S^{ZF}\end{aligned} \quad (8)$$

where S^{ZF} represents the source term driven by the turbulence,

$$S^{ZF} = -\frac{c}{B_0}[\![\{\Phi, \delta f\}]\!]_\alpha \; . \quad (9)$$

The magnetic and diamagnetic drift terms in the α direction is neglected by the ordering of $|\partial/\partial x| \gg |\partial/\partial\alpha|$. This is because the α dependence appears in the equilibrium scale of $\sim R_0/M$, while the radial scale length is characterized by ρ_i. The poloidal $\boldsymbol{E}\times\boldsymbol{B}$ rotation with $\omega_\theta = cE_{r0}/r_0 B_0$ should be included as the fifth term in the square brackets on the left hand side of Eq.(8), since it is usually larger than the magnetic and diamagnetic drift terms in the α directions and modifies the collisionless orbit of ripple-trapped particles.

By subtracting Eq. (8) from Eq. (1) with $\delta f = \widehat{f} + \widetilde{f}$ and $\Phi = \widehat{\Phi} + \widetilde{\Phi} + \Phi_0$, one finds the gyrokinetic equation for $\widetilde{f}(x, y, z, v_\parallel, \mu; \alpha)$,

$$\begin{aligned}\left[\frac{\partial}{\partial t} + v_\parallel \boldsymbol{b}\cdot\nabla + \boldsymbol{v}_d\cdot\nabla - \frac{\mu}{m}(\boldsymbol{b}\cdot\nabla B)\frac{\partial}{\partial v_\parallel} - \omega_\theta r_0\frac{\partial}{\partial y}\right]\widetilde{f} + \frac{c}{B_0}\{\Phi, \delta f\} \\ = (\boldsymbol{v}_* - \boldsymbol{v}_d - v_\parallel \boldsymbol{b})\cdot\frac{e\nabla\widetilde{\Phi}}{T_i}F_M + C(\widetilde{f}) - S^{ZF}\end{aligned} \quad (10)$$

where the slow coordinate, α, is treated as a parameter. It means that the local flux tube model with a fixed values of α can be applied to numerically solving Eq.(10). Interactions between zonal flows and turbulence are formulated though the nonlinear terms of $\{\Phi, \delta f\}$ and S^{ZF}.

The gyrokinetic equaiton for the zonal flows, Eq.(8), can be numerically solved on discretized grids, that is, $\alpha_j = j\Delta\alpha$, where $\Delta\alpha$ is the grid spacing, and j is an integer. We locate a flux tube simulation box at each grid point $\alpha = \alpha_j$ for calculating the turbulence components from Eq.(10). The multi-scale simulation model, thus, consists of the zonal flow solver and multiple flux tubes for turbulence [where a total number of flux tubes is equal to the number of grid points (N_α) for discretizing Eq.(8)], and is called a "flux-tube bundle" model. Here, the same boundary conditions and numerical methods as those for the local flux tube model are applied to each flux tube at $\alpha = \alpha_j$. Finally, we postulate a closure relation, which relates the $k_y = 0$ component of $\widetilde{f}_{j,\boldsymbol{k}_\perp}$ in the jth flux tube to the zonal flow, so that

$$\widetilde{f}_{j,k_y=0} = \widehat{f}(\alpha_j) \; . \quad (11)$$

It greatly simplifies the numerical treatment, since Eq.(8) at $\alpha = \alpha_j$ is the same as that for the $k_y = 0$ components of Eq.(10) except for the poloidal rotation term.

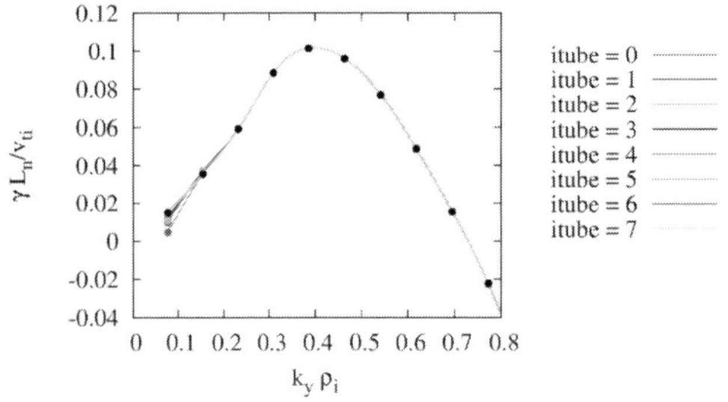

Figure 3. Linear growth rates of ITG mode calculated on different flux tubes placed at $\alpha = \alpha_j$. Plotted eight lines are almost overlapping each other except in a low k_y region.

4.2. Linear ITG instability and zonal flow response

As Eq.(10) for $k_y \neq 0$ modes has the same form as that of the conventional flux tube model, except for the poloidal rotation term causing the Doppler shift of the mode frequency, the linear growth rate of the ITG instability remains unchanged in the flux-tube bundle model. Linear growth rates γ of the ITG mode at different α_j are plotted in Fig. 3, where physical parameters are the same as those used in the inward-shifted case discussed in Ref. [21]. The numerical simulation is carried out using the grid points of $(256, 64, 24)$ in the $(z, v_\parallel, \mu)$ directions, respectively, where $-\pi \leq z \leq \pi$, $-5v_{ti} \leq v_\parallel \leq 5v_{ti}$, and $0 \leq \mu \leq 12.5m_i v_{ti}^2/B_0$. Eight flux tubes are set in a range of $-\pi/M \leq \alpha \leq +\pi/M$ for $M = 10$ of the LHD configuration, and hence, $\Delta\alpha = \pi/40$. Weak α dependence of γ is found only in the low k_y region, which is consistent with the previous linear stability analysis [30].

The zonal flow response to a given source term is also calculated by linear gyrokinetic simulations using the flux-tube bundle model. Figure 4 shows the time-history of the real part of the zonal flow potential for cases with and without E_{r0}. Physical and numerical parameters are the same as those for the simulation of the ITG instability shown above. One can clearly find that the residual zonal flow level after the initial GAM damping is enhanced in the case of $M_p = 0.3$ where the poloidal Mach number is defined as $M_p = |\omega_\theta| R_0 q_0/v_{ti} = (R_0 q_0/r_0)|cE_{r0}/B_0 v_{ti}|$. The residual level after the GAM damping is consistent with the results in Ref. [27]. As the Lenard-Bernstein collision operator is introduced with the collision frequency of $\nu = 2 \times 10^{-3} v_{ti}/L_n$, the residual zonal flow slowly decays in time.

5. Summary

The gyrokinetic simulations towards quantitative evaluation and prediction of the turbulent transport in magnetic fusion plasmas have been applied to a non-axisymmetric toroidal fusion system, such as LHD. The ITG turbulence simulation using the `GKV-X` code has demonstrated the capability for reproducing the ion heat flux in the high ion temperature discharge of LHD experiment [11]. In this paper, we have shown the application of a new code package, `GKV+`, to the entropy transfer function analysis [13, 16] for the ITG turbulence simulation on the LHD experimental configuration. Analysis of the nonlinear entropy transfer function $\delta_{\boldsymbol{k}_\perp+\boldsymbol{p}_\perp+\boldsymbol{q}_\perp,0}\mathcal{J}[\boldsymbol{p}_\perp|\boldsymbol{q}_\perp,\boldsymbol{k}_\perp]$ confirms that the successive entropy transfer process also works in the LHD equilibrium configuration for the high ion temperature discharge.

In the latter part of this paper, a novel idea for the multi-scale simulation of turbulence and zonal flows is formulated for the non-axisymmetric system with poloidal rotation, and is

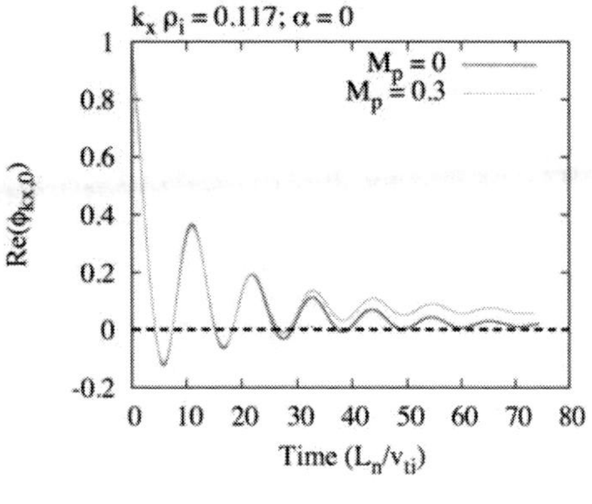

Figure 4. Zonal flow response at $\alpha = 0$ for $k_x\rho_i = 0.117$ in cases with the poloidal Mach number $M_p = 0$ (red) and 0.3 (green), where the zonal flow potential is normalized by the initial amplitude.

implemented with the flux-tube bundle model. The linear ITG instability and the zonal flow response are tested by means of the new simulation model. Application of the multi-scale simulation to the turbulent transport and the zonal flow generation is currently in progress, and the results will be published elsewhere.

Acknowledgments

This work is supported in part by grants-in-aid of the Ministry of Education, Culture, Sports, Science and Technology (No. 21560861, 22760660, and 24561030), and in part by the National Institute for Fusion Science (NIFS) Collaborative Research Program (KNTT014, KNST043, UNTT002, and KNTT015). Numerical simulations are carried out by use of the Plasma Simulator system at National Institute for Fusion Science, and by use of Helios system at International Fusion Energy Research Center (Project code: GTNAXIS). A part of this work is carried out during one of the authros (T.-H. W.) stays at the gyrokinetic theory working group meeting held at EURATOM-CIEMAT.

References

[1] Frieman E A, Chen L 1982 *Phys. Fluids* **25** 502
[2] Brizard A J, Hahm T S 2007 *Rev. Mod. Phys.* **79** 421
[3] Garbet X, Idomura Y, Villard L and Watanabe T H 2010 *Nucl. Fusion* **50** 043002
[4] Diamond P H, Itoh S I, Itoh K and Hahm T S 2005 *Plasma Phys. Control. Fusion* **47** R35
[5] Yamada H et al. 2011 *Nucl. Fusion* **51** 094021
[6] Watanabe T H, Sugama H. and Ferrando-Margalet S. 2007 *Nucl. Fusion* **47** 1383
[7] Watanabe T H, Sugama H and Ferrando-Margalet S 2008 *Phys. Rev. Lett.* **100** 195002
[8] Watanabe T H and Sugama H 2006 *Nucl. Fusion* **46** 24
[9] Nunami M, Watanabe T H, and Sugama H 2010 *Plasma Fusion Res.* **5** 016
[10] Nunami M, Watanabe T H, Sugama H, and Tanaka K 2011 *Plasma Fusion Res.* **6** 1403001
[11] Nunami M, Watanabe T H, Sugama H and Tanaka K 2012 *Phys. Plasmas* **19** 042504.
[12] Sugama H, Watanabe T H and Nunami M 2009 *Phys. Plasmas* **16** 112503
[13] Nakata M, Watanabe T H and Sugama H 2012 *Phys. Plasmas* **19** 022303
[14] Beer M A, Cowley S C and Hammett G W 1995 *Phys Plasmas* **2** 2687
[15] Hirshman S P, Betancourt O 1991 *J Comput Phys* **96** 99
[16] Watanabe T H, Sugama H, Nunami M, Tanaka K and Nakata M 2012 *Plasma Phys. Control. Fusion* (in press)
[17] Winsor N, Johnson J L and Dawson J J 1968 *Phys. Fluids* **11** 2448

[18] Rosenbluth R M and Hinton F L 1998 *Phys. Rev. Lett.* **80** 724
[19] Sugama H and Watanabe T H 2005 *Phys. Rev. Lett.* **94** 115001
[20] Sugama H and Watanabe T H 2006 *Phys. Plasmas* **13** 012501
[21] Ferrando-Margalet S, Sugama H and Watanabe T H 2007 *Phys. Plasmas* **14** 122505
[22] Mynick H E and Boozer A H 2007 *Phys. Plasmas* **14** 072507
[23] Sugama H, Watanabe T H and Ferrando-Margalet S 2008 Plasma Fus. Res. **3** 041
[24] Sugama H and Watanabe T H 2009 *Phys. Plasmas* **16** 056101
[25] Sugama H, Watanabe T H and Nunami M 2010 *Contrib. Plasma Phys.* **50**, 571-575
[26] Kleiber R, Hatzky R and Mishchenko A 2010 *Contrib. Plasma Phys.* **50**, 766-769
[27] Watanabe T H, Sugama H and Nunami M. 2011 *Nucl. Fusion* **51** 123003
[28] Sánchez E et al. in *The Joint 18th International Stellarator/Heliotron Workshop and 10th Asia Pacific Plasma Theory Conference* (Sydney, Australia, 2012) S2.3Tu
[29] Sugama H et al. 2012 *Plasma Fusion Res.* **7** 2403094
[30] Kuroda T, Sugama H 1998 *J Phys Soc Jpn* **67** 3787

The structure of the Si IV region in Be stars: a study of Si IV spectral lines in Be 68 stars

A Antoniou[1, 3], E Danezis[1], E Lyratzi[1, 2] and D Stathopoulos[1, 2]
[1]University of Athens, Faculty of Physics Department of Astrophysics, Astronomy and Mechanics, Panepistimioupoli, Zographou 157 84, Athens, Greece
[2] Address: Eugenides Foundation, 387 Sygrou Av., 17564, Athens, Greece
[3]Address: University of Peloponnese, Faculty of Science and Technology, Department of Telecommunications Science and Technology, Karaiskakis Str. 22100 Tripolis, Greece

Abstract. Using the GR model, we analyze the UV Si IV resonance lines in the spectra of 68 Be stars of different spectral subtypes, in order to detect the structure of Si IV region. We calculated the values of the rotational, radial and random velocities as well as the absorbed energy of each one of the λλ 1393.755, 1402.778 Å Si IV resonance lines and we present the relation between some of these parameters and their variation as a function of the spectral subtype.

1. Introduction

The Si IV resonance lines in the Be stellar spectra have a peculiar and complex profile. Additionally many researchers have observed the existence of absorption Si IV components shifted to the violet or red side of the main spectral line (e.g. Refs [1-5]). We named these components Discrete or Satellite Absorption Components (DACs: [6], SACs: [5]). DACs or SACs originate in separate density clouds, in the Si IV region and have different rotational and radial velocities. In any case, the whole observed feature of the Si IV resonance lines is not the result of a uniform atmospherical region, but results from different Si IV clouds with different physical parameters which create a number of components.

Using the Gauss-Rotation (GR) model ([3]), we analyze the UV Si IV λλ 1393.755, 1402.778 Å resonance lines in the spectra of 68 Be stars of different spectral subtypes, taken with IUE, in order to investigate the presence of Satellite Absorption Components (SACs) and Discrete Absorptions Components (DACs). From this analysis we can calculate the values of a group of physical parameters, such as the apparent rotational and radial velocities, the random velocities of the thermal motions of the ions, the Full Width at Half Maximum (FWHM), the optical depth, as well as the absorbed energy and the column density of the independent Si IV clouds of matter which produce the main and the satellite components of the studied spectral lines. In this paper, we present the variation of the rotational, radial and random velocities as well as the absorbed energy of each one of the λλ 1393.755, 1402.778 Å Si IV resonance lines, as a function of the spectral subtype. In some cases we give the relation among some of the calculated parameters. In these cases we calculate the Linear Regression and the Linear Correlation Coefficient R^2.

2. Data

The spectrograms of the 68 Be stars have been taken by IUE satellite, with the Long Wavelength range Prime and Redundant cameras (LWP, LWR) at high resolution (0.1 to 0.3 Å). The list of the studied stars contains six B0, four B1, twenty five B2, seventeen B3, seven B5, two B6, one B7, three B8 and four B9 stars.

Published under licence by IOP Publishing Ltd

3. Variation of the physical parameters of the Si IV regions, as a function of the spectral subtype

In Fig. 1, we present the Si IV doublet of the B2 Ve star HD 58050, and its best fit. The best fit has been obtained with two SACs. The graph below the profile indicates the difference between the fit and the real spectral line.

In Figures 2-4 we present the variation of the values of the rotational and the radial velocities of the independent Si IV clouds as well as the random velocities of the ions of the Si IV resonance lines, as a function of the spectral subtype. We fitted the observed profiles with one, two or three components. In the case of the rotational velocities we detect three levels of values. The first level corresponds to the 1st component and has a mean value about 455 km/s, the second one corresponds to the 2nd component with mean value 221 km/s and the third one corresponds to the 3rd component with mean value 85 km/s. In the case of the radial velocities we found two levels of values. The first level corresponds to the 1st component and has values between -236 and 90 km/s (mean value -83 km/s) and the second one corresponds to the 2nd and 3rd component and has values between 360 and 10 km/s (mean value 137 km/s). Finally, in the case of the random velocities we detected an almost constant trend of the random velocities for each component. We found mean values of about 90 km/s for the first component, 51 km/s for the second one and 31 km/s for the third one.

In Figures 5 and 6 we present the variation of the absorbed energy (Ea) in eV, of each one of the λλ 1393.755, 1402.778 Å Si IV resonance lines as a function of the spectral subtype. We see a slightly increasing linear trend of the rotational velocities.

In Figure 7 we see the relation among the rotational velocities and the absorbed energy of the λ 1393.755 Å Si IV spectral line and finally, in Figure 8 one can see the relation among the rotational and radial velocities of the Si IV independent density clouds. In both cases we see a slightly increasing linear trend and a "good" linear correlation (R^2=0.6752 and R^2=0.553 respectively). We note that, if R^2=1 the linear correlation is ideal, if 0.5 <R^2<1 the linear correlation is considered "good" and if 0.3< R^2<0.5 the linear correlation is considered "weak".

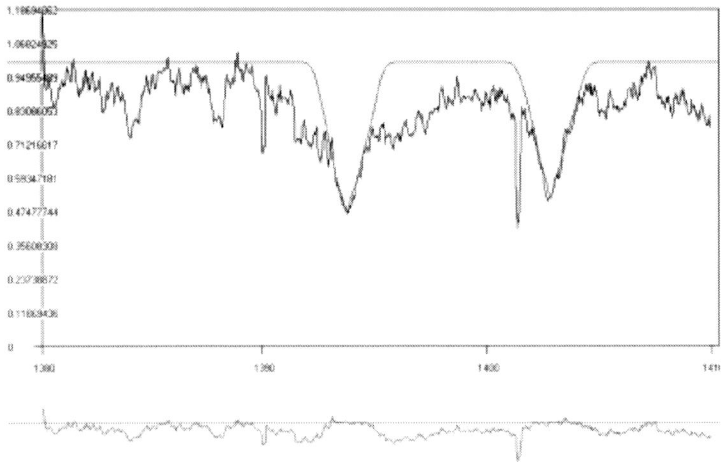

Figure 1. The Si IV λλ 1393.755, 1402.778 Å resonance lines in the spectrum of HD 58050. Each of Si IV spectral lines consists of two components. The graph below the profile indicates the difference between the fit and the real spectral line.

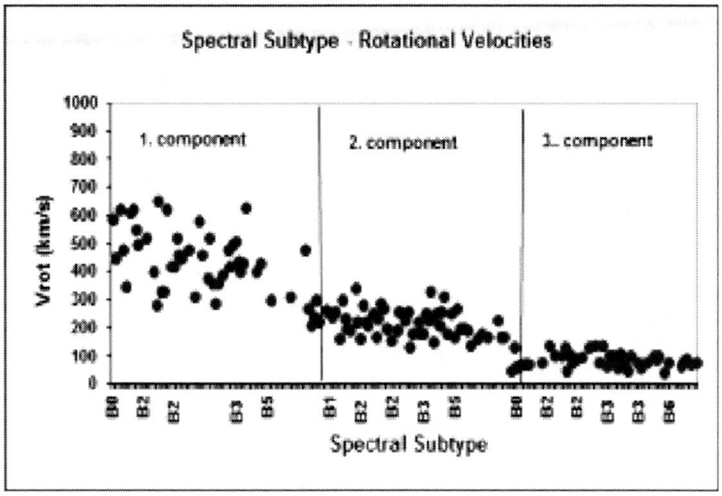

Figure 2. Variation of the values of the rotational velocities of the density clouds, which create the absorption components of the Si IV doublet (λλ 1393.755, 1402.778 Å) as a function of the spectral subtype. We detect three levels of values. The first level corresponds to the 1st component and has a mean value about 455 km/s, the second one corresponds to the 2nd component with mean value 221 km/s and the third one corresponds to the 3rd component with mean value 85 km/s.

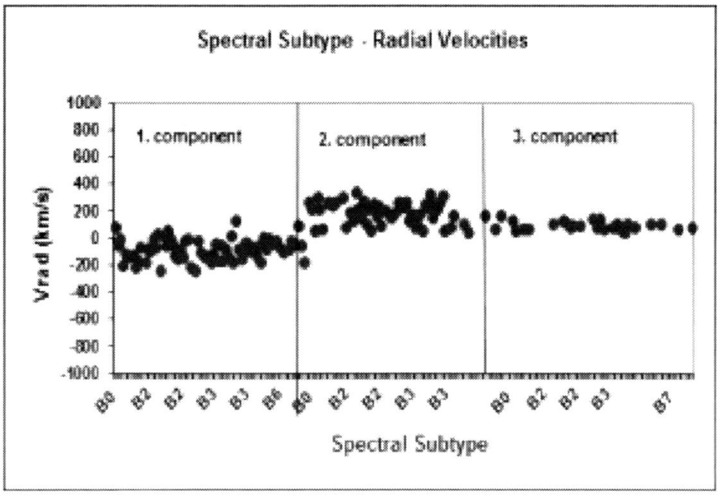

Figure 3. Variation of the radial velocities of the density clouds, which create the absorption components of the Si IV doublet (λλ 1393.755, 1402.778 Å) as a function of the spectral subtype. Here we found two levels of values. The first level corresponds to the 1st component and has values between -236 and 90 km/s (mean value -83 km/s) and the second one corresponds to the 2nd and 3rd component and has values between 360 and 10 km/s (mean value 137 km/s).

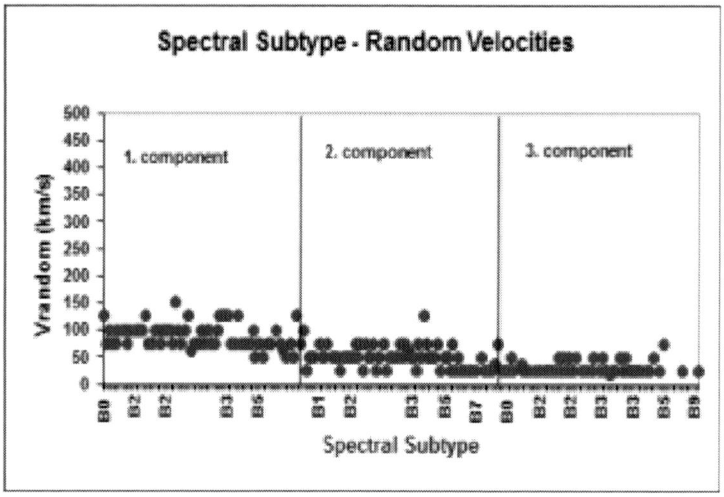

Figure 4. Variation of the random velocities of the Si IV ions of the density clouds, which create the absorption components of the Si IV doublet (λλ 1393.755, 1402.778 Å) as a function of the spectral subtype. We detected an almost constant trend of the random velocities for each component. We found mean values of about 90 km/s for the first component, 51 km/s for the second one and 31 km/s for the third one.

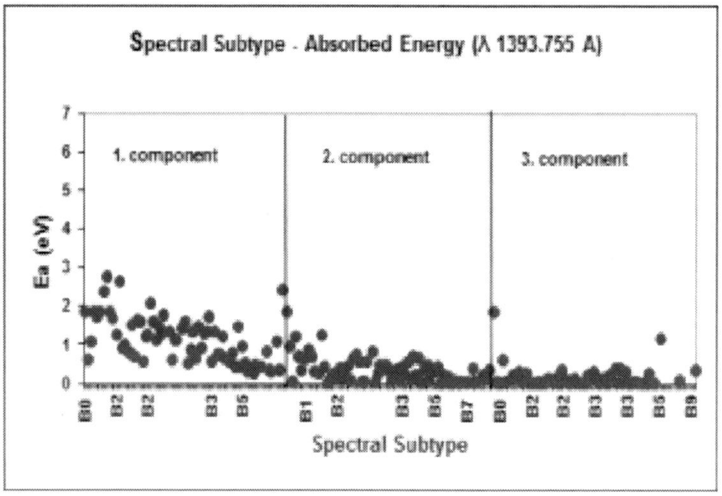

Figure 5. Variation of the absorbed energy (Ea in eV) of the Si IV line λ 1393.755 Å, for the independent density clouds, which create the absorption components as a function of the spectral subtype.

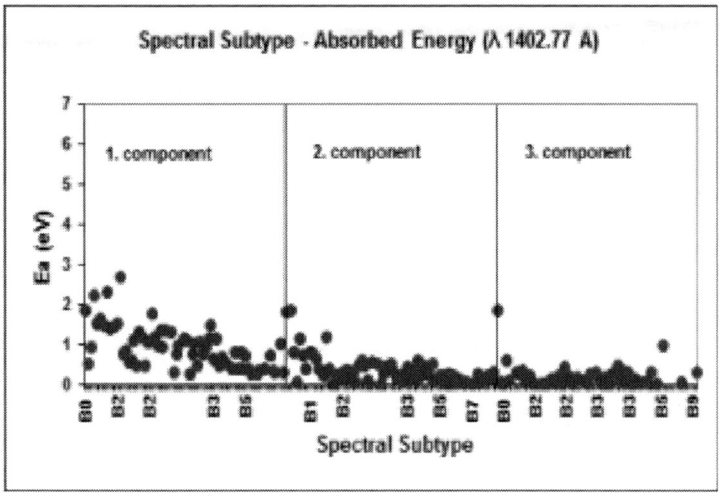

Figure 6. Variation of the absorbed energy (Ea in eV) of the Si IV line λ 1402.77 Å, for the independent density clouds, which create the absorption components as a function of the spectral subtype.

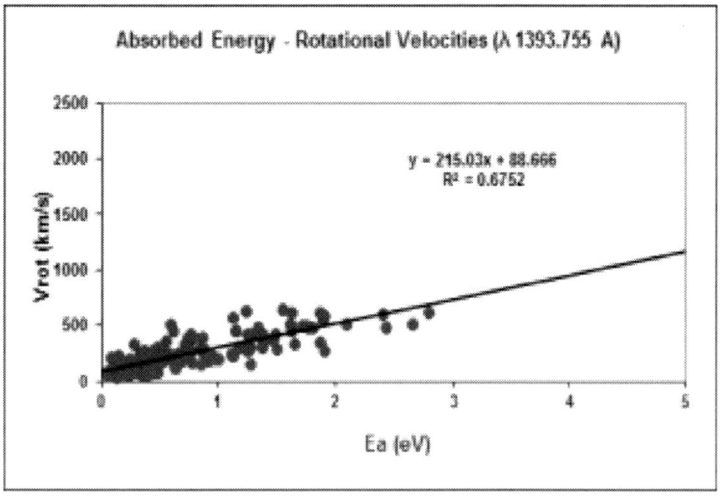

Figure 7. Variation of the rotational velocities of the independent density clouds which create the absorption components of the λ 1393.755 Å Si IV spectral line as a function of the absorbed energy. We see a slightly increasing linear trend of the rotational velocities.

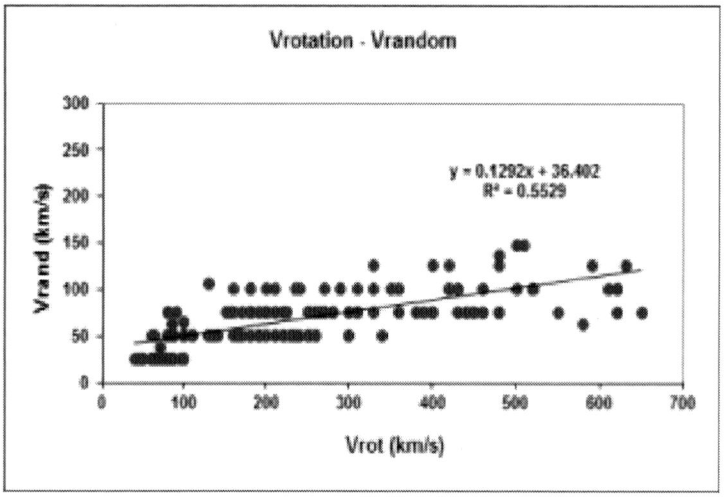

Figure 8. Variation of the rotational velocities of the independent density clouds which create the absorption components of the Si IV resonance lines (λλ 1393.755, 1402.778 Å) as a function of the random velocities. Here also we see a slightly increasing linear trend of the random velocities and a "good" linear correlation (R^2=0.553).

4. Discussion and conclusions

Our general comments are the following: Firstly, the values of all the calculated parameters are in agreement with the physical theory. Secondly, we accomplished the best fit using one absorption component in eleven of the 68 studied stars, with two components in twenty eight of them and with three components in twenty nine of them.

Rotational velocities: The kinematical independent clouds rotate with different mean velocities. We detected three levels of values. The first level corresponds to the 1st component and has values between 620 and 310 km/s (mean value 455 km/s), the second one corresponds to the 2nd component and has values between 300 and 210 km/s (mean value 221 km/s) and the third one corresponds to the 3rd component and has values between 140 and 40 km/s (mean value 85 km/s). We also note that the rotational velocities of the three independent regions present a uniform fluctuation with the spectral subtype.

Radial velocities: Here we detected three components but two levels of values. The first level corresponds to the 1st component and has values between -236 and 90 km/s (mean value -83 km/s) and the second one corresponds to the 2nd and 3rd component and has values between 360 and 10 km/s (mean value 137 km/s).

Random velocities: We detected an almost constant trend of the random velocities for each component. We found values from about 110 to 70 km/s for the first component, from about 70 to 50 km/s for the second one and from about 50 to 30 km/s for the third one. This is expected because the random motions of the ions depend on the temperature of the region in which the specific spectral lines are created.

Absorbed Energy: We note that this parameter has the same behavior as the rotational velocities. We detected a descending trend from the first to the third component.

Absorbed Energy- Rotational Velocities Vrot=f(Ea): We detected a slightly increasing linear trend of the rotational velocities in the Si IV spectral line and a "good" linear correlation (R^2=0.675). This means that high values of the absorbed energy indicate high values of rotational velocities.

Rotational Velocities – Random Velocities Vrand=f(Vrot): We also detected a slightly increasing linear trend of the random velocities as a function of the rotational velocities and a "good" linear correlation (R^2=0.558).

In the cases of *Vrot=f(Ea)* and *Vrand=f(Vrot)* the fact that the calculated linear correlation is "good" indicates that in Be stars, if we know the value of one of them, we could estimate the other one. This idea must be confirmed by a much greater sample of Be stars and it is part of our future work.

Acknowledgments
This research project is progressing at the University of Athens, Department of Astrophysics, Astronomy and Mechanics, under the financial support of the Special Account for Research Grants, which we thank very much.

5. References

[1] Danezis E, Theodossiou E and Laskarides P G 1991 *Ap&SS* 179 111
[2] Danezis E, Nikolaidis D, Lyratzi E, Popović L Č, Dimitrijević M S, Antoniou A and Theodosiou E 2007 *PASJ* 59 827
[3] Lyratzi E, Danezis E, Popović L Č, Dimitrijević M S, Nikolaidis D and Antoniou A 2007 *PASJ* 59, 357
[4] Doazan V, Sedmak G, Barylak M and Rusconi L 1991 *A Be Star Atlas of Far UV and Optical High-Resolution Spectra* (ESA SP-1147, Paris: ESA Sci. Publ.)
[5] Danezis E, Lyratzi E, Stathopoulou M, Theodossiou E, Nikolaidis D, Drakopoulos C and Soulikias A 2003 *Spectral Lines Shapes: Proc. IV SCSLS (Arandjelovac, Serbia)*
[6] Bates B and Halliwell D R 1986 *MNRAS* 223 673

Stellar Population in the Sample of Type 2 Active Galactic Nuclei

Nataša Gavrilović Bon[1,2], **Edi Bon** [1] **and Luka Č. Popović** [1]

[1]Astronomical Observatory, Volgina 7, 11060, Belgrade, Serbia
[2] Observatoire de Lyon, 9 avenue Charles André, Saint-Genis Laval cedex, F-69561, France

E-mail: nbon@aob.bg.ac.rs

Abstract. We analyse simultaneously stellar populations (SPs) in active galactic nuclei (AGNs), AGN featureless continuum and emission lines, using NBurst code, adopted to fit the total spectrum of an active galaxy. In this paper we present the investigations of the accuracy, limitations and applications of the code in a research of stellar populations in narrow emission line galaxies (Type 2). To validate the method, we simulated several thousands of line-of-sight integrated spectra of Type 2 galaxies. We fitted simulated spectra with NBurst and found that code can extract SP and AGN parameters with a high precision. Additionally, we used the method to investigate SP and AGN spectral parameters of a sample of Type 2 AGNs. Especially we investigate the properties of stellar population in the first kiloparsec.

1. Introduction

The nuclear spectra of Seyfert galaxies can be composed of two components: a stellar component characterized by absorption lines from stellar atmospheres and an active galactic nucleus component composed of strong emission lines and a featureless continuum (FC). In the standard model for active galactic nuclei the UV/optical continuum is produced by non-thermal processes, following a power-law of the form $f_\lambda \sim \lambda^\alpha$ with a typical value of α in the range α = [-1.5,2] (see e.g. [1, 2, 3]). However, hot, young stars also produce a featureless continuum that attenuates the absorption lines of the older stellar population. Therefore, any reliable measurement of the emission-line spectrum of galactic nuclei has to properly account for the starlight contamination.

A typical approach to investigate gas properties in the center of an AGN is to dissociate the spectrum constituents in order to remove the starlight from an integrated spectrum. In order to minimize the degeneracies between the spectrum components, it is desirable not to subtract the SP from the integrated spectrum before the analysis of the AGN emission, but to fit simultaneously all constituents that contribute to the total spectrum. One of techniques for the analysis of an unresolved component from stellar populations of galaxies in an integrated spectrum is the spectral synthesis, which consists in the decomposition of an observed spectrum in terms of single stellar populations (SSPs) of various ages and metalicities, producing as output the star formation and chemical histories of a galaxy, together with its extinction and velocity dispersion (see e.g. [4, 5, 6, 7]). This is achieved by a full spectrum fitting including the continuum shape and absorption features. Sarzi et al. developed the ameliorate technique to fit in the same time the stellar and the gas kinematics [8].

In this work we presented and tested the advanced method to analyse simultaneously gas and stellar kinematics and continuum radiation coming from AGNs in the spectra of Type 2 galaxies, using NBurst full spectrum fitting routine [9].

2. Method for the simultaneous analysis of the starlight and AGN emission

In order to analyse AGN spectra, we used NBurst [9] full spectrum fitting package that we modified and adopted to analyse simultaneously all components of the integrated light from an active galaxy. Single stellar population (SSP)[1] models that we used are spline interpolated over an age-metallicity grid of models, generated with PEGASE.HR code[2] [10]. Therefore, by fitting the spectra of interest with NBurst, we reconstracted the SSP-equivalent age and metallicity.

Following [11] we defined a model M(x) of the integrated AGN spectrum, consisted of a stellar template spectrum T(x) convolved with a line-of-sight velocity broadening function G(x), a model for the AGN continuum C(x), here assumed to be a single power law, and a sum of Gaussian/Gauss-Hermit series S(x), that represent AGN emission lines:

$$M(x) = P(x)([T(x) \otimes G(x)] + C(x) + \sum S(x)), \quad (1)$$

were P(x) is a multiplicative polynomial.

We assumed a Gaussian velocity broadening function G(x) for simplicity, but technically it is possible to solve for higher order moments of the velocity profile (see e.g. [12, 13]).

We incorporated a multiplicative polynomial P(x), into the fit to remove large-scale shape differences between the observed stellar and galactic spectra.

In the original version of NBurst, the optional additive continuum C(x) is represented by Legendre additive polynomial. We added a single power low component in the fit to represent the featureless continuum. The AGN continuum is well described with the power-law function ($F \sim \lambda^{\alpha}$) over all electromagnetic spectrum [14]. The spectral index α depends on the continuum slope, thus in different spectral domains it has different values. For the optical domain, that we analyzed for the moment, expected value for spectral index is between -1.5 and 2.

When we made a model of emission lines, we defined each emission line component separately with Gaussian or Gauss-Hermit series.

To test the accuracy of the method, we simulated line-of-sight integrated spectra of low luminous AGNs. We assumed that it is composed of the featureless AGN continuum, narrow emission lines from the narrow line region and underlying stellar population (SP) from the host galaxy. We made a grid of 7200 spectra with Solar-type stellar population, combined with different fraction and slopes of featureless continuum, different intensities and widths of emission lines, various spectral ranges, signal-to-noise ratio (SNR) and degrees of Legendre multiplicative polynomial (more details will be given in [15]).

To evaluate the ability of the method to restore characteristics of the gas and stars in the nucleus and in the host galaxy, we fitted simulated spectra with described model of integrated AGN spectra.

Figure 1 represents the restored SSP ages in a response of FC contribution to the total spectrum (10%-90%) in the cases of SNR=20 (top panel) and SNR=40 (bottom panel). In both cases the method succeeds to restore SP age, but if the SP contributes with less than $\sim$ 10% to the total spectrum, the reliability of the result is very low, according to the wide error bars. Figure 2 represents the response of the FC continuum contribution and SSP age on the SNR variability. One can notice that method restors the FC contribution highly precise independently on the noise level. Only the errors for restored FC contribution computed by the program from the

[1] SSP is population with single age and metallicity.

[2] http://www2.iap.fr/pegase/pegasehr/

covariance matrix are low, and probable not realistic. We can expect to derive higher discrepancies from Monte Carlo simulations. On the other hand, in the case when SNR is lower that 20, the restored SP age is not reliable.

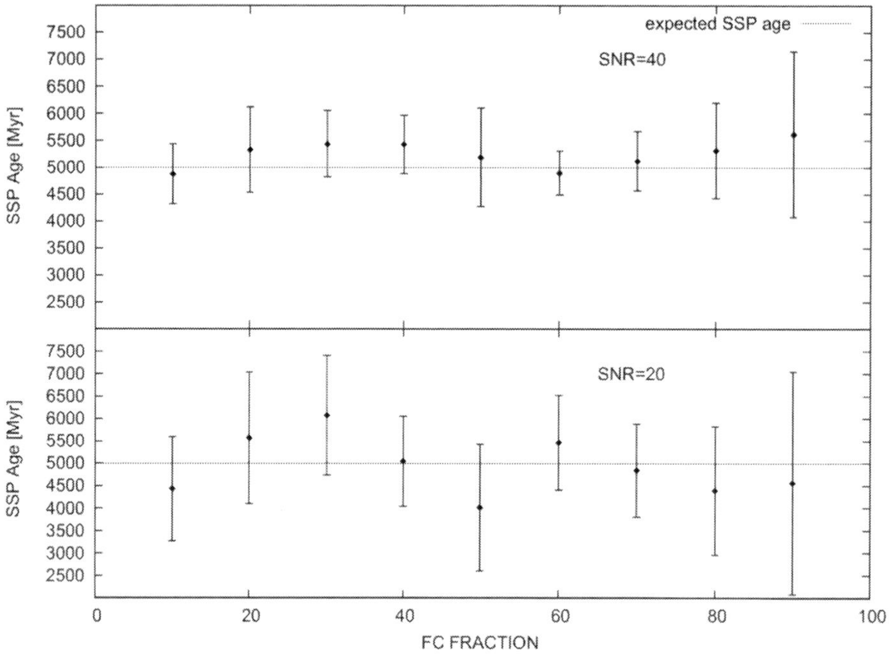

Figure 1. The restored SSP ages obtained from the single best fit for different FC contribution to the total spectrum (10%-90%) in the cases of SNR=20 (top panel) and SNR=40 (bottom panel).

We can conclude that the result depends mainly on the AGN continuum contribution to the total spectrum and on the signal-to-noise ratio. Parameters of the gas and stars in the galactic spectrum can be well restored if the spectra have SNR > 15 and if the stellar population fraction is higher than 20 %. Method is not able to recover SP characteristics when the SP contribution to the total observed spectrum is less than 10%.

3. Application of the method on a sample of Type 2 objects

We used described method for the statistical analysis of stellar population in the first kpc of Type 2 galaxies. We selected ~ 4000 SDSS Type 2 galaxies. Considering that we were interested only in the stellar population inside the inner kpc of the galaxies, and since galaxies were observed through a fixed angular fiber, we constrained redshift of the galaxies between 0.01 and 0.05. To firmly confine the field-of-view (FOV) of the fiber only on the inner part of the galaxy, we constrained the size of the SDSS objects, using the isophotes in the r-band, that represents the size of the major axis of the galaxy. Therefore, we bounded major axis to be larger than FOV of the galaxy. In this way, we prevent the influence of the stars from host galaxy to the observed spectrum.

Since narrow emission lines can be used to classify the dominant energy source in emission line galaxies, we used well known Hα/N[II]6548,6583Å vs. Hβ/O[III]4959,5007Å BPT diagnostic

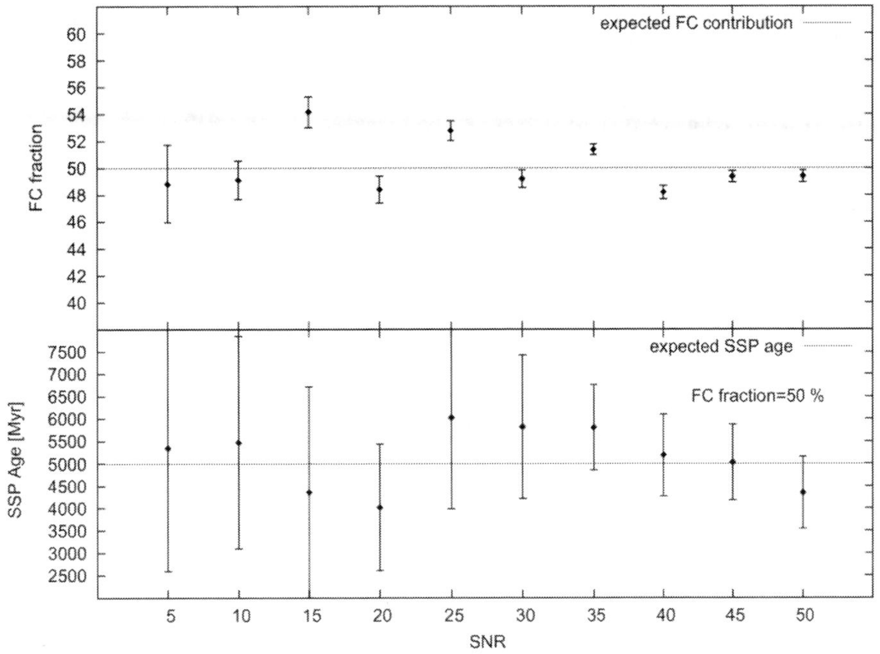

Figure 2. The restored FC continuum contribution (top panel) and SSP age (bottom panel) from the single best fit for different signal-to-noise ratio (SNR=5-50) in the case of 50% of FC contribution.

diagram [16] to separate four classes of objects - Seyfert 2, LINERs, HII regions and so called "composite objects".

To analyse the stellar population parameters we presented the results of the fit in the form of histograms. Here we will briefly disscuss the results of the stellar population ages.

3.1. Discussion of the stellar population ages in the first kpc of Type 2 galaxies

Figure 3. represents distributions of analysed Type 2 galaxies with redshift between 0.01 and 0.05 over the mean stellar population ages. As expected, the youngest stellar populations have been found in starburst galaxies. Moreover, there was just a few HII objects with SP older that 6 Gyr. In the case of only three Sy2 galaxies have been found stellar population younger than 120 Myr. The possible explanation for such result could be that the young stellar population is quenched in the inner kpc of Sy2 galaxis by the central engine. Very interesting result appeared in comparing restored SP ages in the samples of Sy2, LINERs and composite objects. Namely, even the number of galaxies in three samples is very different, one can notice similar distribution of SP ages. We made cross-correlation between Sy2 - LINER, and Sy2 - composite objects histograms and obtained very high correlation, with cross-correlation coefficient of $r = 0.96$ and P-value that describes the probability of no correlation $P(null) = 3.87 \times 10^{-26}$ in the first case, and $r = 0.89$ and $P(null) = 3.15 \times 10^{-14}$ in the second case. In addition, the SP age distribution in all selected Type 2 galaxies has maximum at ~ 2 Gyr, that suggests the dominance of an intermediate-age stellar population. To verify obtained result, we analysed the SP age distribution separately in each redshift interval, and we obtain very similar distribution in the redshift intervals in the case of Sy2, LINERs and composite objects. So, it seems that the result does not depend on the cosmological scale (we emphasize that we investigate only local redshift objects $z < 0.05$). We can conclude that apart from the case of starburst galaxies,

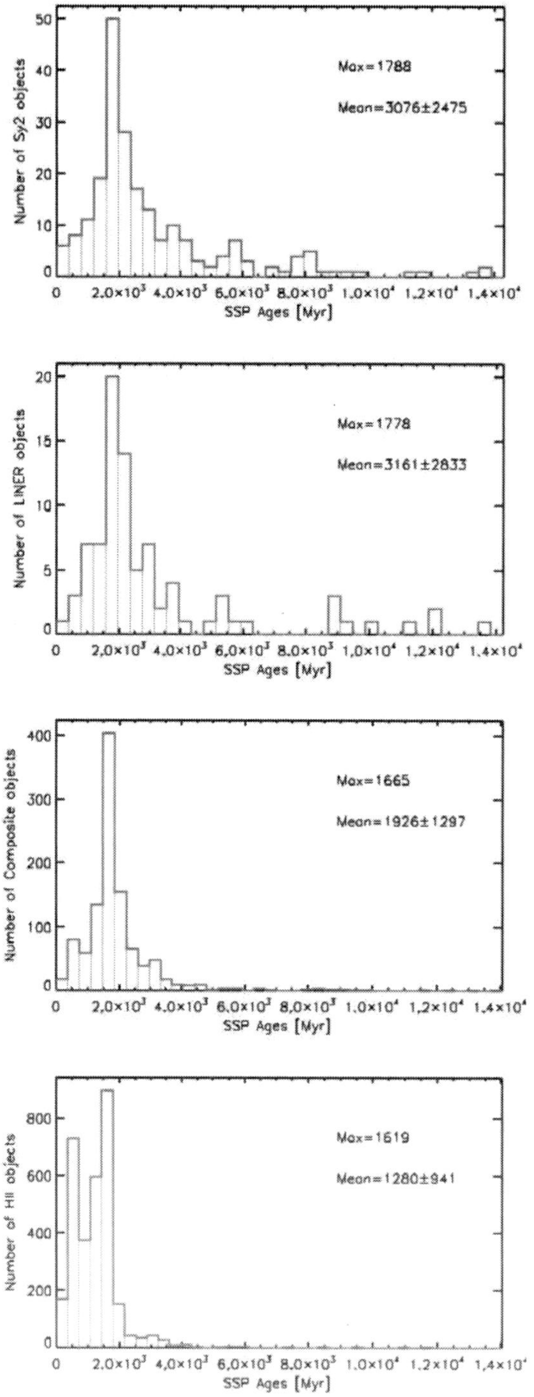

Figure 3. Stellar population age in the integrated spectra of Sy2, LINER, composite objects and HII, respectively. On the graph are presented the maximum, mean and standard deviation of the distribution.

stellar population ages in the first kpc of Type 2 objects do not depend on activity type, and have similar distribution.

4. Conclusions

In order to analyse the properties of the starlight and its contribution to the spectra of AGN, we used NBurst, full spectrum fitting package. We upgraded the NBurst code with the intention of simultaneously analysing all components that contribute to the integrated AGN spectra along a line of sight: stellar population, AGN featureless continuum, and emission lines. In the original version, NBurst uses stellar populations synthesis models for analysing galaxy spectra. We added new components in the model: a featureless continuum, represented with a power law function, and emission lines, represented with Gaussian or Gauss Hermit series.

In order to assess the possibilities of restoring the properties of gaseous and stellar component in the spectrum of interest, we made numerical experiments. These experiments show that the NBurst code efficiently restores the information about the shape and light fraction of the AGN continuum, as well as the kinematics, age, and metallicity of the underlying stellar population in AGN spectra. Analysis revealed that the method is not able to restore the SP characteristics in the case when the SP contribution to the total observed spectrum is less than 10%.

We applied our proposed method to analyse the SP in the first kpc of Type 2 galaxies. We concluded that Sy2, LINERs and composite objects have the same distribution of stellar population ages in the first kpc.

Overall, our results validated spectral synthesis as a powerfull tool to study the history of galaxies and show some interesting characteristics of stellar population in the first kiloparsec of Type 2 galaxies from our sample.

Acknowledgments

This research is part of the project 176001 "Astrophysical spectroscopy of extragalactic objects" supported by the Ministry of Education, Science and Technological Developement of the Republic of Serbia.
We are grateful to Philippe Prugniel for his useful suggestions.

References

[1] Boroson, T. A., & Green R. F. 1992, *ApJS*, **80**, 109
[2] Winkler, H., Glass, I. S., van Wyk, F., Marang, F., Jones, J. H. S., Buckley, D. A. H., Sekiguchi, K. 1992, *MNRAS*, **257**, 659
[3] Garcia Rissmann, A. 2004 *IAUS*, **222**, 93
[4] Cid Fernandes, R., Mateus, A., Sodré, L., Stasíska, G., Gomes, J. M., 2005, *MNRAS*, **358**, 363
[5] Cid Fernandes, R., Delgado, G., Rosa M. 2010, *MNRAS*, **403**, 780
[6] Koleva, M., Prugniel, P., Bouchard, A., Wu, Y. 2009 *A&A*, **501**, 1269
[7] MacArthur, L. A., González, J. J., & Courteau, S. 2009, *MNRAS*, **395**, 28
[8] Sarzi, M., Falcón-Barroso, J., Davies, R. L., Bacon, R., Bureau, M., Cappellari, M. et al. 2006, *MNRAS*, **366**, 1151
[9] Chilingarian I. V., Prugniel P., Sil'Chenko O. K., Afanasiev V. L., 2007, *MNRAS*, **376**, 1033
[10] Le Borgne D., Rocca-Volmerange B., Prugniel P., Lancon A., Fioc M., Soubiran C., 2004, *A&A*, **425**, 881
[11] Barth, A. J., Ho, L. C., Sargent, W. L. 2002, *ApJ*, **124**, 2607
[12] Rix, H.-W., & White, S. D. M. 1992, *MNRAS*, **254**, 389
[13] van der Marel, R. P. 1994, *MNRAS*, **270**, 271
[14] Osterbrock, D. E. 1989, *Astrophysics of Gaseous Nebulae and Active Galactic Nuclei* (Mill Valle: University Science Press)
[15] Bon, N, Bon, E, Prugniel, P., Popović, L. Č. 2012, in preparation
[16] Baldwin, J. A., Phillips, M. M. & Terlevich, R. 1981, *PASP*, **93**, 5

The properties of emission lines and their correlations in spectra of Active Galactic Nuclei

Jelena Kovačević and Luka Č. Popović

Astronomical Observatory, Volgina 7, 11060 Belgrade, Serbia

E-mail: jkovacevic@aob.bg.ac.rs

Abstract. Active Galactic Nuclei (AGNs) are the most luminous objects in the Universe. It is assumed that an AGN consists of a supermassive black hole in the center, surrounded by accreating gas. In that process, a large amount of energy is produced, resulting in a very complex spectrum that shows a number of strong emission lines, which arise in the plasma located close to the supermassive black hole.

Here we analyse the line profile shapes, which are signature of the geometry, physical and kinematical properties of the AGN emission gas. Especially we analysed the optical Fe II lines, in order to answer to some unexplained questions about their origin. We found correlations between line properties which physical background is still not explained. Possible influence of starbursts to AGN spectral properties is discussed.

1. Introduction

It is estimated that about 5-10% of all galaxies have Active Galactic Nucleus (AGN) in their center. AGNs are the most luminous objects in the Universe and their luminosity may be a few thousands times higher than luminosity of nuclei of normal galaxies. It is assumed that an AGN consists of a supermassive black hole in the center, surrounded by accreating gas. In the process of accreation, a large amount of energy is produced which photoionize the gas in surrounding of the black hole. It results in rich and complex spectra, which consist of the continuum emission and strong broad and narrow emission lines.

The emission region closer to the black hole (<0.1 pc) is called the Broad Line Region (BLR), because it produces the broad emission lines. It has a higher density (10^8-10^{11} cm^{-3}) than plasma which is further (<100 pc) from the black hole and accretion disk, which is called the Narrow Line Region (NLR). In this region, narrow lines arise and plasma has a very low density, in some cases even lower than the vacuum which is usually produced in laboratories. The density of the Narrow Line Region is assumed to be 10-1000 cm^{-3}, and temperatures of both regions are estimated to be: 10000-25000 K.

The densities of the Narrow and Broad Line Regions are estimated according the presence or absence of some forbidden or semi-forbidden lines. Namely, some lines which arise in the Narrow Line Region, and which are very strong in AGN spectra, could be strongly forbidden, and almost impossible to be produced in laboratory conditions. These lines (as [O III]$\lambda\lambda$4959, 5007 Å, [S II]$\lambda\lambda$6716, 6731 Å, [N II]$\lambda\lambda$6548, 6583 Å, etc.) are collisionally excited and their ratios could be used for determination of the temperature and density in the NLR. These forbidden lines could be very strong in AGN spectra probably because of very low density of the NLR –

which allow the radiative deexcitation from metastable level, and presence of very large amount of the low density gas.

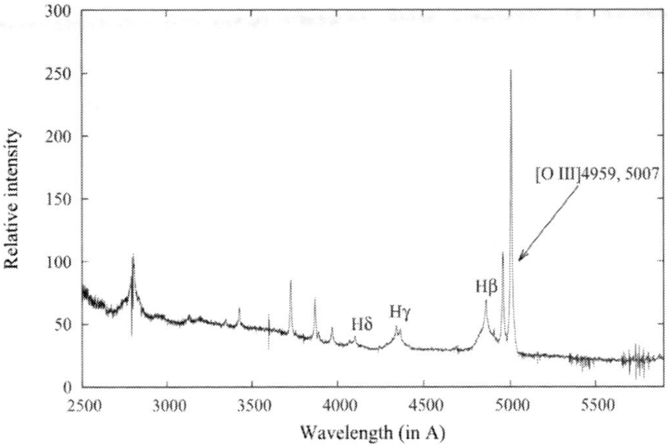

Figure 1. Example of an AGN spectrum (object SDSS 041210.17-051109.1), where the forbidden [O III]λλ4959, 5007 Å lines are very prominent.

The most prominent forbidden emission lines in AGN spectra are the [O III]λλ4959, 5007 Å lines which arise in magnetic dipole and partly in electric quadrupole transitions (see [1]). In some spectra, they may have higher intensity than allowed lines in the optical part. The example of strong [O III]λλ4959, 5007 Å lines is shown in Fig 1.

While the average width of narrow lines is FWHM $\approx$ 500 km/s, the broad lines have more complex shapes and larger widths. Namely, the BLR probably consists of different layers of the plasma, which have different kinematical properties. It results with complex and very broadened line profiles, which have an average FWHM $\approx$ 5000 km/s, but it may go up to 15000 km/s. The broad line profile composed from kinematicaly different components is shown in Fig 2. Also, the line emission from the accretion disk may contribute to the final line profile, as for example of an extreme case Hα line in spectrum of object SDSS J0942+0900 [3], for which FWHM Hα $\approx$ 40000 km/s (see Fig 3).

Shapes of the emission lines, their flux ratios, as well as their shifts and widths, are signature of the geometry, physical and kinematical properties of the emission gas. Therefore, we investigated relationships between different spectral properties in large AGN sample, in order to get better insight of AGN nature.

2. The sample and analysis

For this investigation we use the sample of 302 Type 1 AGNs (with the broad and narrow emission lines), selected with criteria described in paper [4]. The spectra are taken from Sloan Digital Sky Survey (SDSS) database.

Basic assumptions we made are: 1) that emission lines arise by superposition of the radiation from different emission regions, which have different physical and kinematical properties. These regions are: the Narrow Line Region, Broad Line Region and Intermediate Line Region (ILR) – which is between these two. 2) The complex profiles of emission lines may be described by a summ of Gaussians, which represents emission from different emission regions. Their widths, shifts and intensities reflect the physical and kinematical properties of emission regions.

The detailed procedure of fitting of spectra is described in [4]. All Balmer lines are fitted with three Gaussians, which represent the emission from the NLR, BLR and ILR. We assume

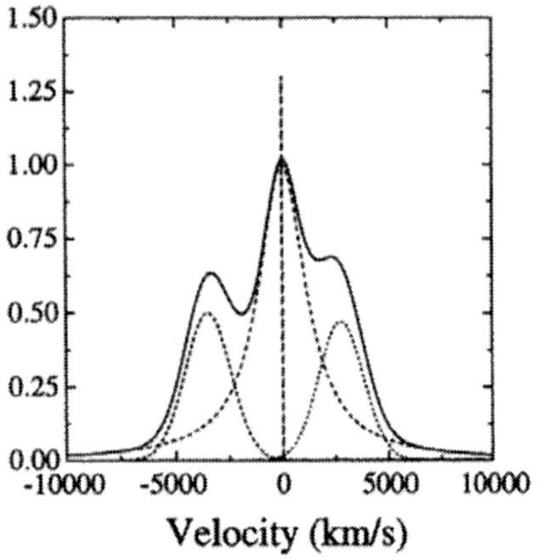

Figure 2. Example of complex shape of Hβ line in AGN spectra. It is caused by kinematical properties of emitting gas [2].

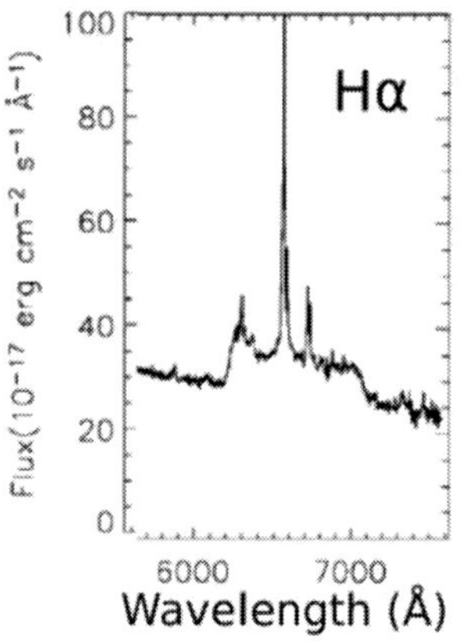

Figure 3. Extreme example of very broad emission line: Hα line from spectrum of object SDSS J0942+0900 [3]. H$\alpha \approx 40000$ km/s.

that all narrow lines originate from the same emission region, and consequently they have the same widths and shifts. The iron lines within 4400 – 5500 Å range are fitted with calculated template given in [4]. Example of a fit is shown in Fig 4.

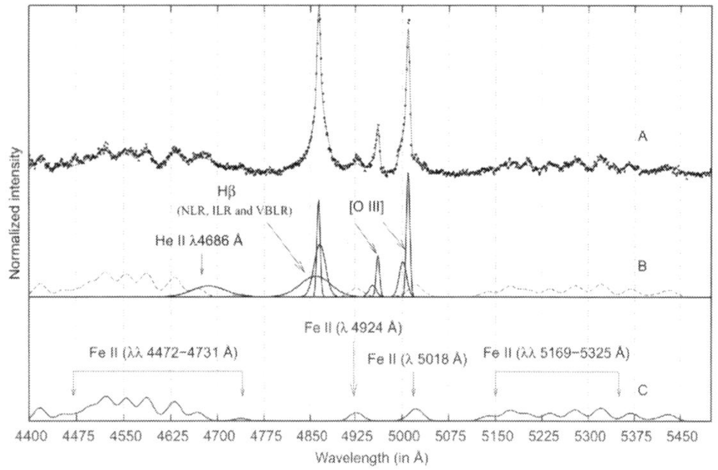

Figure 4. Example of the fit of an AGN spectrum in range 4400 – 5500 Å.

There are some indications that evolution of AGNs is probably related with starburst regions (see [5], [6], [7]). Namely, it is possible that AGNs in an earlier phase of their evolution are composed of starburst (star-forming) regions and the central engine - accretion disk and supermassive black hole ([6], [8], [9]). In order to investigate the influence of starbursts to AGN spectra, we separated the total sample to two subsamples: pure AGNs - objects in which radiation from AGN dominates, and starburst dominant subsample, which are probably objects with composite structure (AGN associated with starbursts). For separation of the sample we accepted a criteria of R=log([O III]/Hβ NLR) = 0.5 (see [10]), where pure AGNs are with $R > 0.5$, and starburst dominant objects with $R < 0.5$.

3. Results

After we found all line parameters from the best fit of spectra, we performed correlations between measured spectral properties for the total sample and for two subsamples. We focus our investigation to: find the place in the AGN structure where the Fe II lines arise, to analyse the influence of starbursts to AGN spectra and to investigate Baldwin effect correlations, which physical cause is still unknown.

3.1. Location of the Fe II emitting region in AGNs

The iron lines in AGN spectra are very interesting for investigation, because there are still many open questions connected to them and they are assumed to be related with the AGN evolution. The complex iron ion in the AGN gas produces many lines, which overlap and make irregular features in AGN spectra. Therefore, the line identification, as well as calculation of the relative intensities of the Fe II lines are very difficult. Namely, it seems that mechanism of excitation of all Fe II lines is still unknown because the observed lines can not be well explained with simple photoionization models, and therefore some additional mechanisms are requested. Also, many correlations are observed between the Fe II lines and other spectra properties, which do not have physical explanation yet.

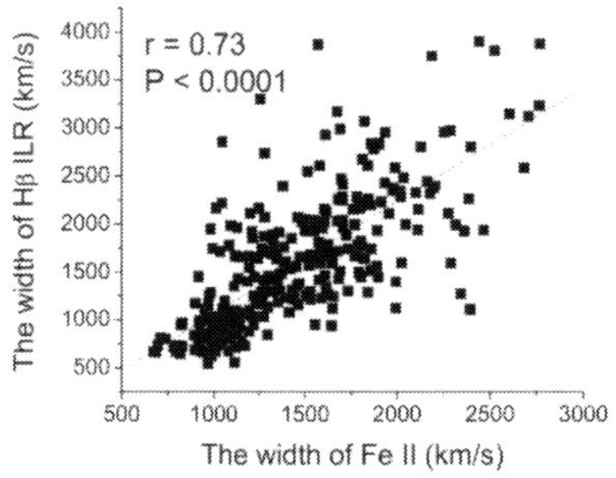

Figure 5. The correlation between the width of Fe II and Hβ ILR component.

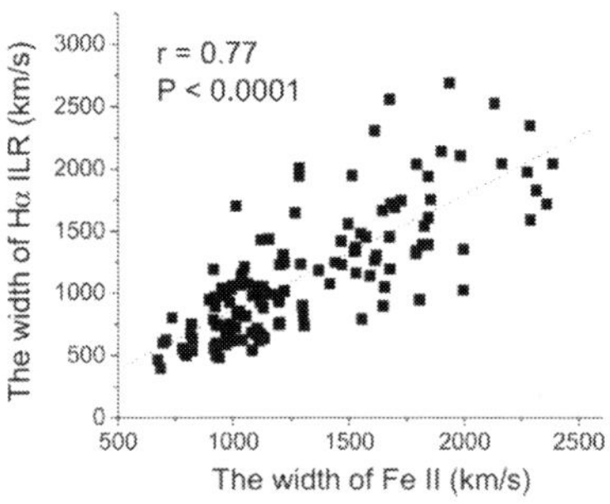

Figure 6. The same, but for the Hα ILR component.

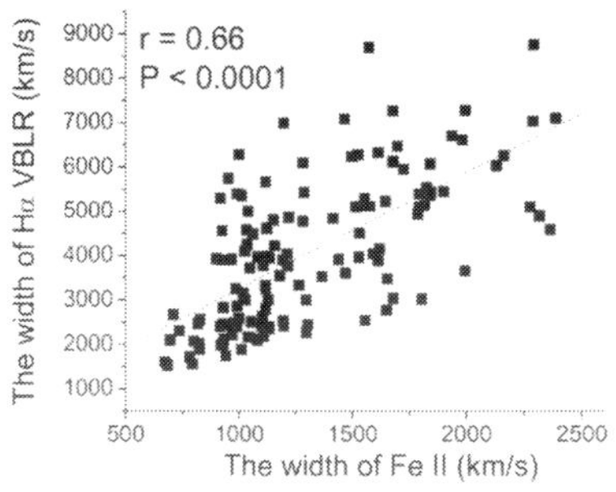

Figure 7. The same, but for the Hα BLR component.

In this investigation we try to find the place in the AGN structure where the Fe II lines are produced, which is still open problem. There are many propositions by different authors: accretion disk, BLR, ILR, or partly in the BLR and partly in the NLR (see references in [4]).

We compared the widths of the Fe II optical lines and the widths of different components of Balmer lines, which are coming from different emission regions (NLR, ILR, BLR). Since the widths of the line components reflect the kinematical properties of the emission regions, the correlation between them imply the kinematical connection.

We found significant correlations between the widths of the Fe II and ILR components of Balmer lines, and trend with BLR components of Balmer lines. Correlations are shown in Figs 5, 6 and 7. No correlation is found between the widths of NLR components and Fe II. This imply that optical iron lines probably arise in the ILR region, and partly in the BLR [4].

3.2. Contribution of starbursts to AGN spectra (in type 1 AGNs)

We performed correlations between spectral properties (continuum luminosity, broad and narrow line widths, equivalent widths of lines) for two subsamples (pure AGN and starburst dominant) and found significant differences in some correlations. We found that the width of the broad Hβ is in a significant correlation with the FWHM of narrow [O III] lines (see Fig 8), as well as with continuum luminosity (L_{5100}) (see Fig 9), but only for the starburst dominant subsample. There are no any correlations between these parameters for the AGN dominant subsample (see [10]). One of the possible explanations may be influence of the starburst to the emission, not only narrow emission lines, but also partly broad ones.

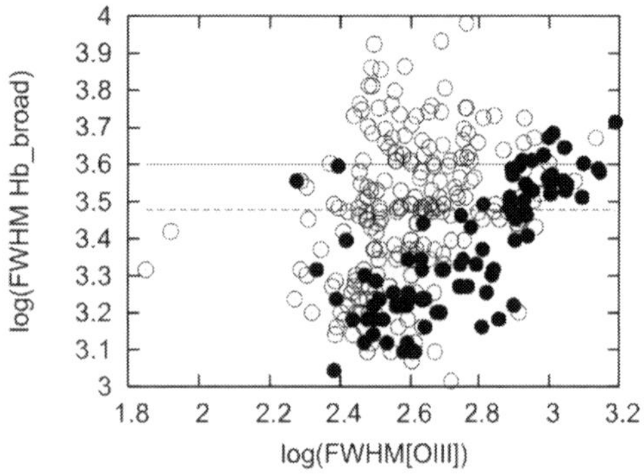

Figure 8. FWHMs of the broad Hβ vs. the narrow [O III] lines. Filled circles denote the starburst dominant subsample, and open circles, pure AGNs.

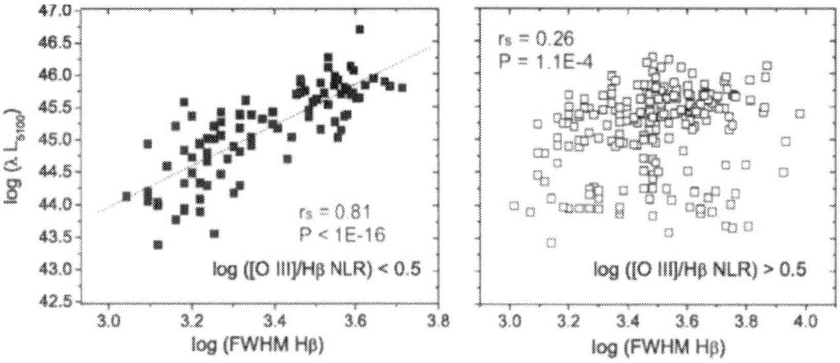

Figure 9. Correlations between the continuum luminosity and FWHM Hβ for starburst dominant (left) and AGN dominant subasmple (right).

3.3. Baldwin effect correlations

Anticorrelations between EWs (equivalent widths) of the emission lines and continuum luminosity are commonly called the Baldwin effect (see [11]), and their physical cause is still unclear, because they can not be explained by a simple photoionization model. Here, we investigated the Baldwin effect correlations for different line components, and we perform these correlations also separately to the pure AGN and starburst dominant subsample.

We found that different components of the same line (for example Hβ), have different relationships with the continuum luminosity (as the continuum luminosity increases, EW Hβ NLR decreases, but with broad Hβ no correlation!).

From the other hand, when we analyse separately different subsamples, we find correlation between the EW of Hβ broad and L_{5100} for the starburst dominant subsample (r=0.56, P=9×10^{-9}), while there is no correlation for the AGN dominant subsample (r=-0.03, P=0.67). It is even more interesting that for the starburst dominant subsample, EW of the broad Hβ increases, as continuum luminosity increases, which is Inverse Baldwin effect.

The previous results imply that Baldwin effect correlations depend on dominant source of ionization in a sample (accretion disc in pure AGN or starburst).

Future work is needed to explain physical background of observed correlations.

4. References

[1] Dimitrijević M S, Popović L Č, Kovacević J, Dačić M, Ilić D 2007 *Monthly Notices of the Royal Astronomical Society* **374** 118
[2] Gaskell C M, Snedden S A 1997 *Emission Lines in Active Galaxies: New Methods and Techniques; Astronomical Society of the Pacific Conference Series; San Francisco: edited by Bradley M. Peterson, Fu-zhen Cheng, and Andrew S. Wilson* **113** 193
[3] Wang T-G, Dong X-D, Zhang X-G, Zhou H-Y, Wang J-X & Lu, Y-J 2005 *Astrophysical Journal Letters* **625** 35.
[4] Kovačević J, Popović L Č, Dimitrijević M S 2010 *Astrophysical Journal Supplement* **189** 15
[5] Lípari S L & Terlevich R J 2006 *Monthly Notices of the Royal Astronomical Society* **368** 1001
[6] Mao Y-F, Wang J, Wei J-Y 2009 *Astronomy & Astrophysics* **9** 529
[7] Sani E, Lutz D, Risaliti G, Netzer H, Gallo L C, Trakhtenbrot B, Sturm E, Boller T 2010 *Monthly Notices of the Royal Astronomical Society* **403** 1246
[8] Wang J, Wei J 2006 *Astrophysical Journal* **648** 158
[9] Wang J, Wei J. 2008 *Astrophysical Journal* **679** 86
[10] Popović L Č, Kovačević J 2011 *Astrophysical Journal* **738** 68
[11] Baldwin J A 1977 *Astrophysical Journal* **214** 679

Laboratory Studies of Charging Properties of Dust Grains in Astrophysical/Planetary Environments

D. Tankosic[1] **and M. M. Abbas**[2]
[1] NASA Postdocotral Program-NASA/MSFC
[2] NASA/MSFC

E-mail: Dragana.Tankosic-1@nasa.gov

Abstract. Dust grains immersed in ambient plasmas and radiation, are charged and coupled to the plasma through electric and magnetic fields. Dust grains in various astrophysical/planetary environments are generally charged by: (a) photoelectric emissions with incident radiation at photon energies higher than the work function of the material and (b) sticking of low energy electrons and ions of the surrounding plasma or by secondary electron emissions induced by incident electrons/ions at sufficiently high energies. Consequenly, the particle charge is an important parameter that influences physical and dynamical processes in the interplanetary and interstellar medium, planetary rings, interstellar dust clouds, comets and the outer atmospheres of planets. The charging properties of individual micron-size dust grains are expected to be substantially different from the bulk materials. However, no viable models for calculation of the charging properties of individual micron size dust grains are available at the present time. Currently, very limited experimental data are available for charging of individual micron-size dust grains. In this paper we give a review of the results of the measurements on charging of analogs of the interstellar as well as Apollo 11 and 17 lunar dust grains carried out on the Electrodynamic Balance Facility at the NASA-MSFC.

1. Introduction

Dust grains in various astrophysical environments are generally charged electrostatically by photoelectric emissions with UV/X-ray radiation, as well as by electron/ion impact. Knowledge of physical and optical properties of individual dust grains is required for understanding of the physical and dynamical processes in space environments and the role of dust in formation of stellar and planetary systems.

In this paper we focus on charging properties of individual micron/submicron dust grains by processes that include:

(a) UV photoelectric emissions involving incident photon energies higher than the work function of the material,

(b) Electron impact at low energies with the incident electrons sticking to the dust grains, charging the dust grains negatively; and

---

1 To whom any correspondence should be addressed.

(c) Electron impact at sufficiently high energies with the incident electrons penetrating the dust grain and leading to excitation and emission of electrons referred to as secondary electron emission (SEE).

It is well accepted that the charging properties of individual micron/submicron size dust grains are expected to be substantially different from the bulk materials (e.g. [1-2]). However, no viable models for calculation of the charging properties of individual micron size dust grains are available at the present time. Therefore, the photoelectric yields, and secondary electron emission yields of micron-size dust grains have to be obtained by experimental methods. Currently, very limited experimental data are available for charging of individual micron-size dust grains (e.g. [3]). Our experimental results, obtained on individual micron-size dust grain levitated in an EDB facility at NASA-MSFC, indicate the following significant features:

(1) The measured photoelectric yields are substantially higher than the bulk values given in the literature [e.g. 3-7], and indicate a particle size dependence with larger particles having order-of-magnitude higher values than for submicron-size grains [e.g. 4-5];

(2) Dust charging by low energy electron impact is a complex process. Also, our measurements indicate that the electron impact may lead to charging or discharging of dust grains depending upon the grain size, surface potential, electron energy, electron flux, grain composition, and configuration (e.g. [6-7]).

A review of the laboratory measurements on charging of analogs of the interstellar dust as well as Apollo 11 and 17 dust grains conducted at the NASA-MSFC Dusty Plasma Laboratory is presented in the following.

2. Experimental set-up for the dust charging experiments

The research work presented here is based on the experimental investigations conducted on a facility referred to as EDB that provides a unique capability to conduct a wide range of experiments on micron size dust grains in simulated space environments.

The experimental apparatus consists of: particle generator, EDB (with the top and bottom DC electrode and the AC ring electrode), DC and AC voltage power supplies, vacuum system, electron gun, the monitoring equipment, the UV source and monochomator and the electron gun . The experimental setup is shown in Fig. 1 (a).

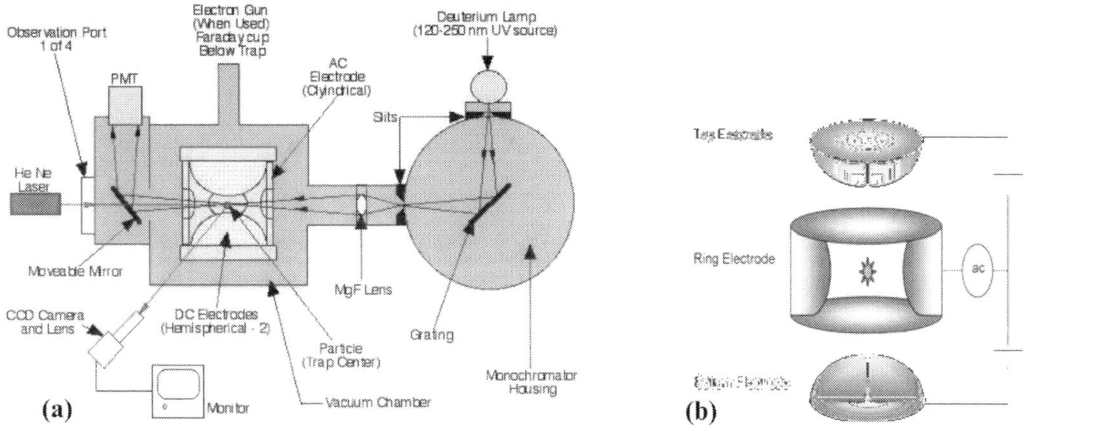

Fig.1. (a) Experimental Setup; **(b)** Electrodynamic Balance

The particle injector employing a pressure impulse technique is used to charge the particles (e.g. [8]). The balance itself consists of spherically shaped DC top and bottom electrode and a ring AC electrode with apertures made to allow optical access to the trapped particle as shown in Fig.1(b). Such an electric field configuration creates a null potential at the geometric centre of the electrodes. A charged micron-size dust grain can be levitated against the gravity by adjusting the dc potential (V_{DC}) applied between the two cap end electrodes. The trap is placed in a vacuum chamber. A 5 mW-HeNe laser and a CCTV camera with a zoom microscope lens are used to observe the particle by projecting the scattered light from the dust grain onto a TV monitor.

The above experimental facility has been used to conduct the measurements presented in the following sections.

3. Laboratory measurements on dust charging by photoelectric emissions

The photoelectric emission process is considered to be the dominant process in many astrophysical environments with radiation from nearby sources. Photoelectric emission occurs at the incident photon energies higher than the work function of the material. The photoelectric efficiencies and yields have to be determined experimentally. In Section 3.1 we describe the experimental technique employed in the Dusty Plasma Laboratory at NASA/MSFC, while in Section 3.2 we present and discuss some of the results of the measurements on dust charging by photoelectric emissions.

3.1 Experimental technique for measurements on dust charging by photoelectric emissions

Dust grains of desired composition and size are injected into the balance at atmospheric pressure. Once the particle is stably trapped and the particle generator is removed, the chamber is closed and the system is evacuated to pressures of ~ 1-5 torr at which the effective diameter is determined by measurements based on marginal stability conditions ("spring point" measurements) (e.g.[8]). The particle diameter has to be determined separately since the direct measurements on the EDB provide the charge to mass ratio only. This technique is based on slowly varying the electrical parameters of the EDB to a point near an unstable regime when the particle begins to oscillate. The system is then evacuated to pressures of ~ 10^{-5} torr at which the photoelectric emission measurements on the levitated particles are carried out. The particle is then exposed to UV radiation in the range of 120-160 nm. As the particle charge changes, its position in the trap is balanced against gravity by adjusting V_{DC}. The change in particle charge is then determined as a function of time in accordance with equation:

$$q(t) = \frac{g z_o m}{C_o} \frac{1}{V_{DC}(t)} \tag{1}$$

where g is gravitational constant, z_o is the distance between the DC electrodes and the center of the trap, and C_o is a geometrical constant of the trap.With the value of V_{DC} needed to balance the gravity and the mass m that is calculated by using the effective particle diameter determined by the "spring point" technique and the known particle mass density (ρ), the particle charge $q(t)$ is calculated from equation (1) as a function of time. With this measurement technique change in particle charge by one electron can be detected in certain regimes.

Journal of Physics: Conference Series **399** (2012) 012024 doi:10.1088/1742-6596/399/1/012024

3.2. Experimental results and discussion on dust charging by photoelectric emissions

The above measurements were conducted on a number of SiO_2 spherical particles in the size range of 0.2-6.62μm that represent the analogs of the interstellar dust grains as well as a number of selected lunar dust grains collected during Apollo 11 and 17 missions that were obtained from NASA-JSC. In the following figures we present some of the results from the measurements dust charging by UV photoelectric emissions that are published in [4-5]

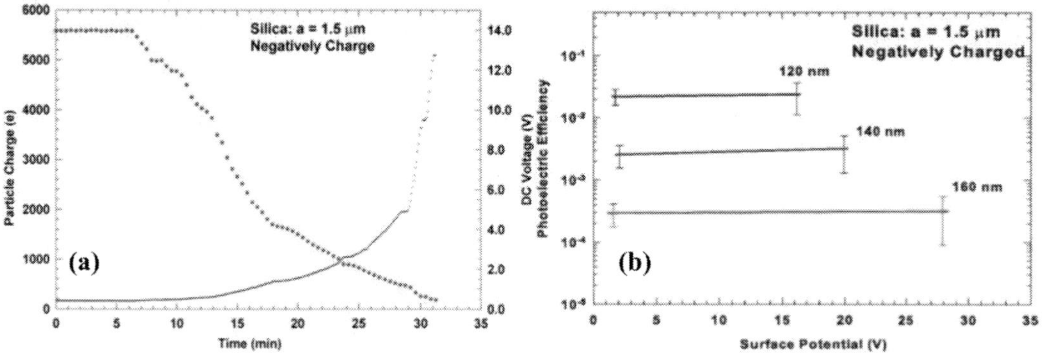

Fig.2(a) Discharging of a 1.5μm SiO_2 particle exposed to UV radiation of λ=120nm (red dots) and the corresponding change of Vdc during the experiment (blue curve); (b) The average photoelectric efficiencies as a function of the particle surface potential for silica particles illuminated with UV radiation at λ=120, 140, and 160 nm.

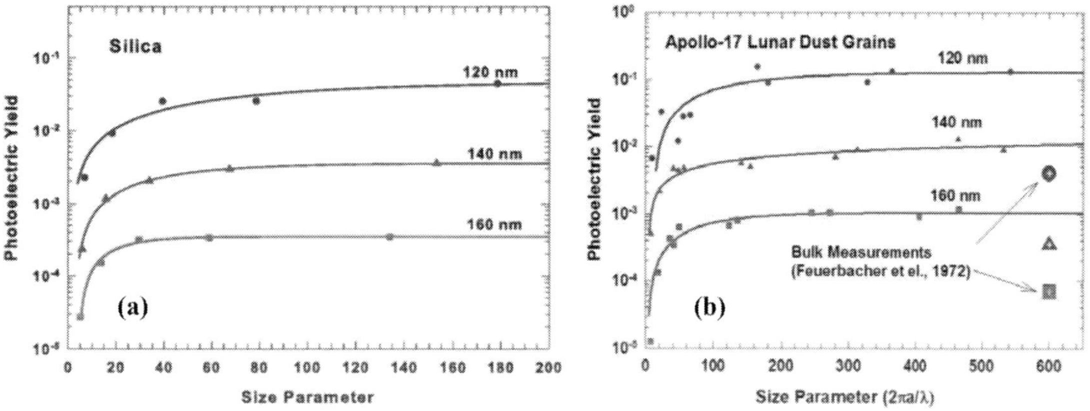

Fig.3 Composite plot of the photoelectric yields Y measured at UV wavelengths of 120, 140, and 160 nm as a function of the size-parameter $x = (2\pi a/\lambda)$ for **(a)** SiO2 particles and **(b)** Apollo 17 lunar dust grains.

Fig.2(a) shows discharging of a 1.5μm SiO_2 particle when exposed to UV radiation of λ=120nm (red dots) and the corresponding change of V_{dc} during the course of the experiment (blue curve). Fig.2(b) represents the average photoelectric efficiencies as a function of the particle surface potential based on measurements on silica particles illuminated with UV radiation at 120, 140, and 160 nm wavelength. The error bars indicate the 1-σ standard deviation due to variability of the particles size as well as the measurements errors. Fig. 3

shows the composite plot of the photoelectric yields Y measured at UV wavelengths of 120, 140, and 160 nm (10.3, 8.9, and 7.9 eV) as a function of the size-parameter $x = (2\pi a / \lambda)$ for (a) SiO_2 particles and (b) Apollo 17 lunar dust grains. In Fig.4(b) the corresponding data for the bulk materials [9] are shown for comparison. The details of the calculations of photoelectric efficiencies and yields are given in [4-5].

3.3. Conclusions from measurements of dust charging by photoelectric emissions

Some of the most important conclusions from the above presented experiments on dust charging by photoelectric emissions are:

(1) Measurements on individual 0.1 to 11.8 μm silica, carbonaceous, olivine, and lunar dust grains exposed to 120-160 nm UV radiation indicate size dependence of the photoelectric yields, with yields of submicron size particles being lower than for large size particles.
(2) Measured photoelectric yields for carbonaceous particles and lunar dust grains are found to be larger by ~an order of magnitude compared to bulk measurements.
(3) This is in contrast with the existing models that indicate higher values for smaller particles.
(4) However, surprisingly size dependence of the yields is found to be in qualitative agreement with the size dependence predicted by classical electrostatic models applied to the atomic clusters.

4. Laboratory measurements on dust charging by electron impact

The dust charging by electron impact is an important dust charging process in planetary and astrophysical environments. Low energy electrons are reflected or stick to the grains charging the dust grains negatively. At sufficiently high energies, electrons penetrate the grain leading to excitation and emission of electrons referred to as secondary electron emission (SEE). Available theoretical models for the calculation of the SEE yield applicable for neutral, planar, bulk surfaces, are generally based on Sternglass Equation [10]:

$$\delta_{\sec}(E) = 7.4\delta_M (E / E_M)\exp\left[-2(E / E_M)^2\right] \qquad (2)$$

where δ_M is the yield at maximum energy E_M.

However, laboratory experiments on micron-size dust grains carried out on our facility indicate that the SEE by electron impact is a very complex process. Incident electrons may lead to positive or negative charging of dust grains depending upon the grain size, surface potential, electron energy, electron flux, grain composition, and configuration. In Section 4.1 we give a brief description of the experimental technique employed at Dusty Plasma Laboratory at NASA/MSFC, and in Section 4.2 we present and discuss some of the results of the measurements on dust charging by low energy electron impact with the conclusions given in Section 4.3.

4.1 Experimental technique for measurements of dust charging by electron impact

The experimental setup for dust charging by the electron impact is essentially same as for the measurements discussed in the previous section. Once the particle is stably trapped and the particle generator is removed, the electron gun (Kimball Physics ELG-5/EGPS-5A) is installed on the top of the chamber. After the determination of the grain effective diameter, the system is evacuated to pressures of ~ 10^{-5} torr, and the levitated lunar dust grain is exposed to 25 or 100 eV electron beam. The electron beam current is measured by a Faraday cup located below the bottom electrode of the trap. The change in the particle charge is determined by manually adjusting V_{DC} to position the particle at the trap center by balancing gravity as described in Section 3.1.

4.2. Experimental results and discussion on dust charging by electron impact

The above measurements were conducted on a number of selected lunar dust grains collected during Apollo 11 and 17 missions that were obtained from NASA-JSC and some SiO_2 particles. The results of this research were presented in [6-7]

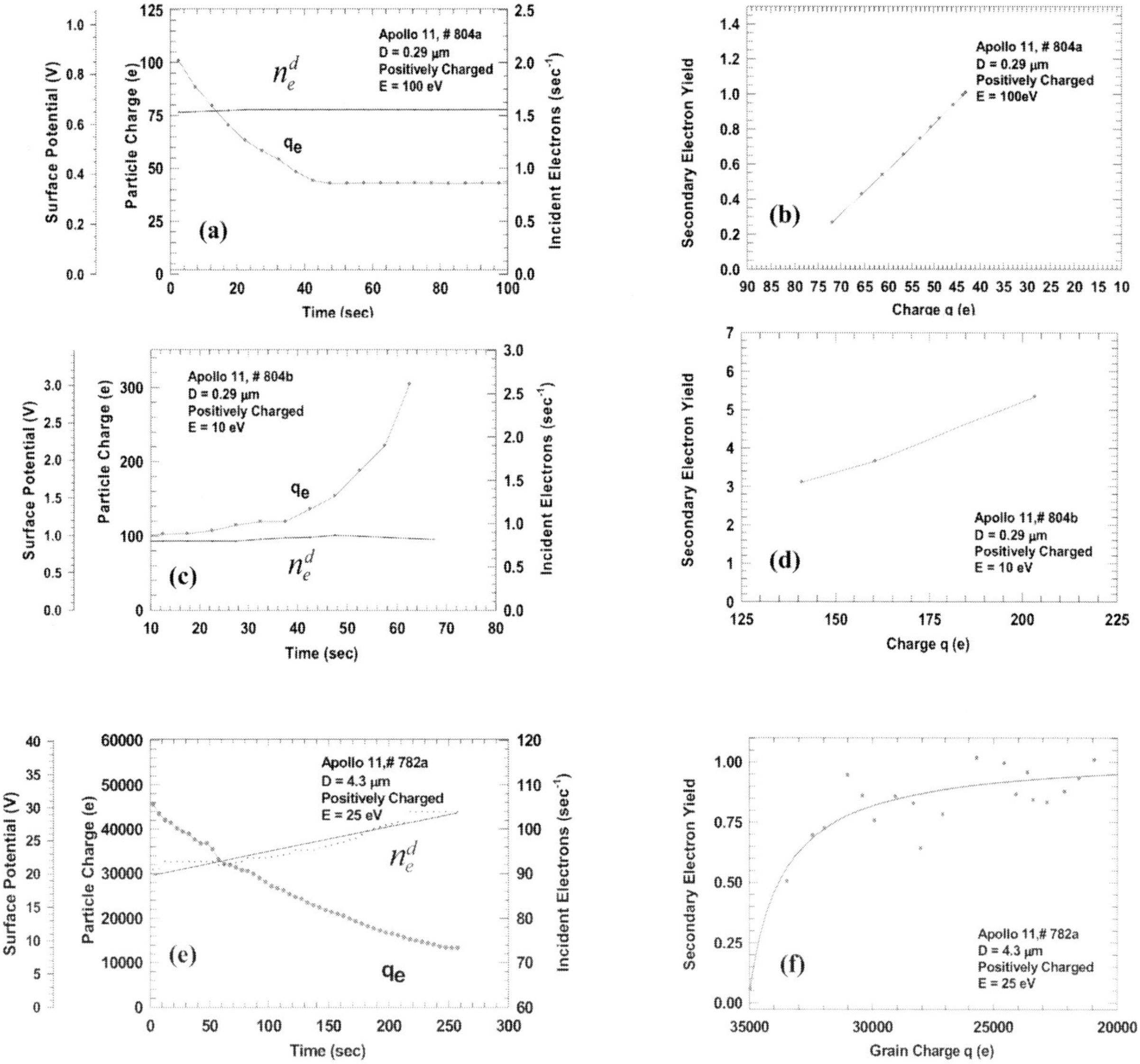

Fig.4 (a,b) Discharging of Apollo 11 positively charged dust grain of D = 0.29 μm, exposed to an electron beam of E = 100 eV with a corresponding calculated SEE yield; **(c,d)** Charging of the same particle when exposed to a beam of electrons of energy E = 10 eV and the corresponding calculated

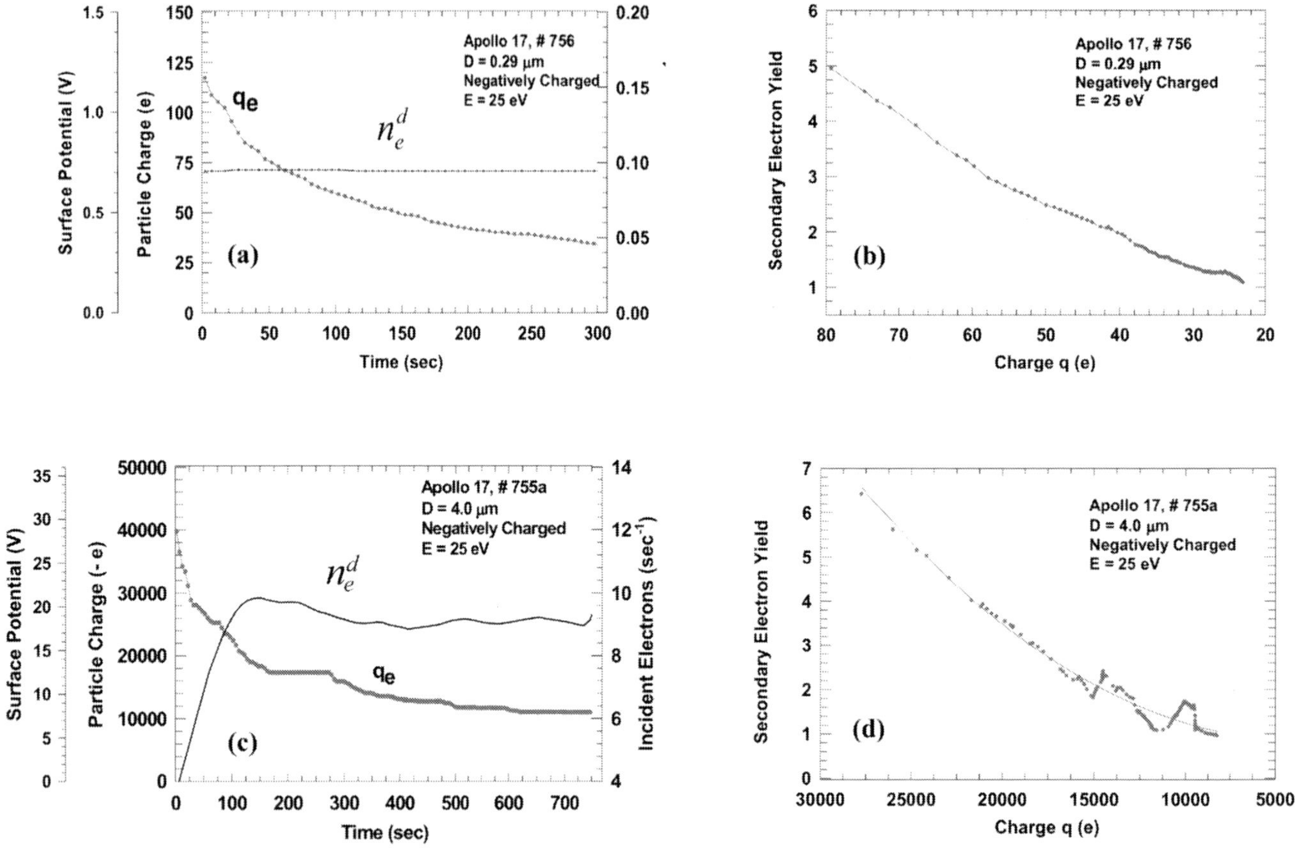

Fig.5 (a,b) Discharging of Apollo 17 negatively charged dust grain of D = 0.29 μm, exposed to a 25 eV electron beam with the corresponding calculated SEE yield; **(c,d)** Discharging of Apollo 17 negatively charged dust grain of D= 4.0 μm when exposed to a E = 25 eV electron beam with the corresponding calculated SEE yield.

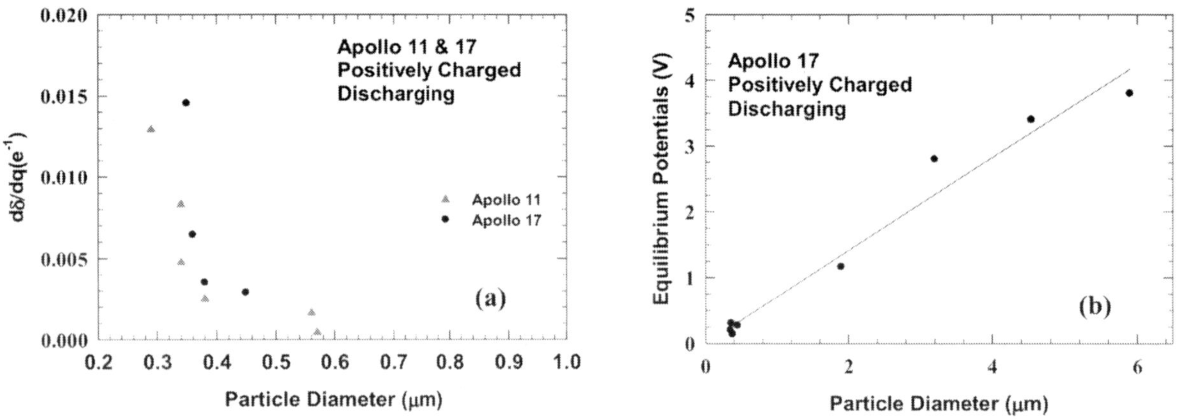

Fig.6 (a) Time averaged change in the SEE yield normalized to the change in the particle charge over the course of the experiment vs the grain diameter for discharging of positively charged lunar dust grains; **(b)** Equilibrium potentials for small size dust grains vs dust grain diameter.

Fig.4 (a,b) shows discharging of Apollo 11 positively charged dust grain of diameter D = 0.29 μm, exposed to a beam of electrons of energy E = 100 eV with the corresponding calculated secondary electron yield (SEE) while Fig.4 (c,d) represents charging of the same particle when exposed to a beam of electrons of energy E = 10 eV and the corresponding calculated SEE yields. Discharging of Apollo 11 positively charged particle of diameter D = 4.3μm, exposed to a beam of electrons of energy E = 25 eV with the corresponding calculated SEE is shown in Fig. 4 (e,f).

In Fig. 5 we present some of the results for Apollo 17 negatively charged particles. Fig.5 (a,b) shows discharging of Apollo 17 negatively charged dust grain of diameter D = 0.29 μm, exposed to a beam of electrons of energy E = 25 eV with the corresponding calculated secondary electron yield (SEE) while Fig.5 (c,d) represents discharging of Apollo 17 negatively charged dust grain of diameter D= 4.0 μm when exposed to a beam of electrons of energy E = 25 eV and the corresponding calculated SEE yield.

Fig.6(a) shows the size dependence of the time averaged change in the SEE yield normalized to the change in the particle charge over the course of the experiment for discharging of positively charged lunar dust grains. The equilibrium potentials for small size dust grains vs dust grain diameter presented in Fig. 6(b) indicate a general linear dependence. The details of the calculations of SEE yields are given in (e.g. [6]).

4.3. Conclusions from measurements of dust charging by the electron impact

Following are some important conclusions based on the measurements on dust charging by SEE presented here:

(1) The current models & measurements of SEE by low energy electron impact on planar/bulk/neutral materials are not valid for individual micron/submicron size dust grains. Rigorous theoretical models applicable for individual micron-size particles are not available yet.

(2) Charging/Discharging processes and equilibrium potentials of small size dust grains are complex and depend on the conditions of: Particle size, Polarity, surface potential, and electron flux density.

(3) Our results are validated with the model calculations of Chow et al [11], as well as the independent experimental measurements by Ziemann et al. (e.g. [3]), that indicate strong size dependence and higher SEE yields for submicron-size particles.

References:

[1] Gallo C F and Lama W L 1976a, *IEEE Trans. Ind. Appl.*1A-12
[2] Gallo C F and Lama W L 1976b, *J. Electrost,* 2 145
[3] Ziemann P J et al. 1996 *J. Aerosol. Sci. 27* 587
[4] Abbas, M M et al. 2006 *ApJ* 645 324
[5] Abbas, M M.et al. 2007 *Planet. Space Sci.* 55 953
[6] Abbas, M M et al 2010 *ApJ* 718 795
[7] Abbas, M M et al 2012 *ApJ* in press
[8] Spann J F et al. 2001 *Phys. Scr.* T89 147
[9] Feuerbacher B and Fitton B 1972 *J. Appl. Phys.* 43 1563
[10] Sternglass E J 1954 *Sci/Pap* 1772, Westinghouse Lab., Pittsburgh, PA
[11] Chow, V W et al. 1993 *J. Geophys. Res.* 98 19065

AUTHOR INDEX

Abbas, M. M. 190
Aerts, R. 82
Al Shorman, M. M. 11
Annaloro, Julien 112
Antonelli, L. 32
Antoniou, A. 171
Aparicio, Juan A. 94
Avaldi, L. 128
Axente, E. 40
Batani, D. 32
Belmonte, Maria T. 94
Bizau, J. M. 11
Blancard, C. 11
Bogaerts, A. 82
Bolognesi, P. 128
Bon, Edi 178
Bon, Natasa Gavrilovic 178
Brunger, Michael J. 143
Bultel, Arnaud 112
Burgdorfer, J. 75
Campbell, Laurence 143
Capes, H. 153
Carsky, P. 17
Cubaynes, D. 11
Curik, R. 17
Danezis, E. 171
Djurovic, Stevica 94
Dudzak, R. 32
Galijas, S. M. D. 61, 68
Giuffrida, L. 32
Giuliani, A. 11
Gizzi, L. A. 32
Godbert-Mouret, L. 153
Guilbaud, S. 11
Hammami, R. 153
Hermann, J. 40
Ikeda, T. 52
Ingolfsson, Oddur 143
Kanai, Y. 52
Kobayashi, T. 52
Koester, P. 32
Kojima, T. M. 52
Kotov, V. 153
Koubiti, M. 153
Kovacevic, Jelena 184
Krousky, E. 32
Krus, M. 32
Labate, L. 32
Lazovic, Sasa 121
Lebel, E. 32
Lemaire, J. 11
Lemell, C. 75
Lyratzi, E. 171
Mackel, V. 52
Majkic, M. D. 61, 68
Malka, G. 32
Mar, Santiago 94
Marandet, Y. 153
Margarone, D. 32
Meissl, W. 52
Mercadier, L. 40
Mihelic, A. 128
Milosavljevic, A. R. 11
Moise, A. 128
Morel, Vincent 112
Nakata, M. 161
Nedeljkovic, N. N. 61, 68
Nejdl, J. 32
Neyts, E. 82
Nicolas, C. 11
Noel, S. 40
Nunami, M. 161
O'Keeffe, P. 128
Patria, A. 32
Pelaez, Ramon J. 94
Popovic, Luka C. 178, 184
Reiter, D. 153
Ribeyre, X. 32
Richetta, M. 32
Richter, R. 128
Rosato, J. 153
Sawicka, M. 32
Schurtz, G. 32
Skala, J. 32
Skoro, N. 135
Snoeckx, R. 82
Somers, W. 82
Spasojevic, Dj 102
Stamm, R. 153
Stathopoulos, D. 171
Stockli, Martin P. 1
Sugama, H. 161
Tankosic, D. 190
Thissen, R. 11
Tikhonchuk, V. 32
Tosic, S. D. 24
Van Gaens, W. 82
Volpe, L. 32
Wachter, G. 75
Watanabe, T. H. 161
Yamazaki, Y. 52
Yusupov, M. 82
Zitnik, M. 128

Institute of Physics
Dirac House, Temple Back
Bristol BS1 6BE UK

CURRAN ASSOCIATES INC
proceedings.com